Nanofluid Flow in Porous Media

Edited by Mohsen Sheikholeslami Kandelousi, Sadia Ameen, M. Shaheer Akhtar and Hyung-Shik Shin

Published in London, United Kingdom

IntechOpen

Supporting open minds since 2005

Nanofluid Flow in Porous Media
http://dx.doi.org/10.5772/intechopen.77588
Edited by Mohsen Sheikholeslami Kandelousi, Sadia Ameen, M. Shaheer Akhtar and Hyung-Shik Shin

Contributors
Zhanbo Sun, Wenyu Shi, Haiyang Wang, Dong Duan, Yi-Lung Cheng, Chih-Yen Lee, Masikana Mdleleni, Ebrahim Mohiuddin, Yusuf Isa, David Key, Gamal M. Abdel-Rahman, Faiza El-Fayez, Mohsan Hassan, Ayesha Siddiqui, Bandari Shankar, Nadeem Abbas, Sohail Nadeem, Hamid Farrokhi, David O. Otuya, Anna Khimchenko, Jing Dong, Yangguang Xu, Vladimir S. Chekanov, Natalya V Kandaurova, Galina V. Shagrova, Victoria I. Drozdova, Veniamin V. Romantsev, Mikhail Yu. Shevchenko, Han He, Sadia Ameen, Mohammad Shaheer Akhtar, Rhushikesh Godbole, Hyung-Shik Shin, Guojiang Liao, Taixiang Liu

First published in London, United Kingdom, 2020 by IntechOpen
IntechOpen is the global imprint of INTECHOPEN LIMITED, registered in England and Wales, registration number: 11086078, 7th floor, 10 Lower Thames Street, London, EC3R 6AF, United Kingdom
Printed in Croatia

British Library Cataloguing-in-Publication Data
A catalogue record for this book is available from the British Library

Additional hard and PDF copies can be obtained from orders@intechopen.com

Nanofluid Flow in Porous Media
Edited by Mohsen Sheikholeslami Kandelousi, Sadia Ameen, M. Shaheer Akhtar and Hyung-Shik Shin
p. cm.
Print ISBN 978-1-78923-837-2
Online ISBN 978-1-78923-838-9
eBook (PDF) ISBN 978-1-83968-342-8

For EU product safety concerns:
IN TECH d.o.o., Prolaz Marije Krucifikse Kozulić 3, 51000 Rijeka, Croatia, info@intechopen.com or visit our website at intechopen.com.

Meet the editors

Dr. Mohsen Sheikholeslami works at the Babol Noshirvani University of Technology's Department of Mechanical Engineering in Iran. He is Head of the *Renewable energy systems and nanofluid applications in heat transfer* Laboratory at Babol Noshirvani University of Technology. His research interests are nanofluid, CFD, simulation, mesoscopic modeling, nonlinear science, magnetohydrodynamic, ferrohydrodynamic, electrohydrodynamic, and heat exchangers. He has written several papers and books in various fields of mechanical engineering. He is the first scientist to develop a new numerical method (CVFEM) and he published the reference book with title: "Application of Control Volume Based Finite Element Method (CVFEM) for Nanofluid Flow and Heat Transfer". He is also the first author of the following books: "Applications of Nanofluid for Heat Transfer Enhancement", "Application of semi analytical methods for nanofluid flow and heat transfer", "Hydrothermal Analysis in Engineering Using Control Volume Finite Element Method", and "External Magnetic Field Effects on Hydrothermal Treatment of Nanofluid", which are published in ELSEVIER. According to the reports of Thomson Reuters (Clarivate Analytics), he has been selected as a Web of Science *Highly Cited Researcher* (Top 0.01%) in 2016, 2017, and 2018.

Professor Sadia Ameen obtained her Ph.D. in Chemistry (2008) and then moved to Jeonbuk National University. Presently she is working as an Assistant Professor in the Department of Bio-Convergence Science, Jeongeup Campus, Jeonbuk National University. Her current research focuses on dye-sensitized solar cells, perovskite solar cells, organic solar cells, sensors, catalyst, and optoelectronic devices. She specializes in manufacturing advanced energy materials and nanocomposites. She has achieved a gold medal in academics and is the holder of a merit scholarship for the best academic performance. She is the recipient of the Best Researcher Award. She has published more than 120 peer-reviewed papers in the field of solar cells, catalysts and sensors, contributed to book chapters, edited books, and is an inventor/co-inventor of patents.

Professor M. Shaheer Akhtar completed his Ph.D. in Chemical Engineering (2008) from Jeonbuk National University, Republic of Korea. Presently, he is working as an Associate Professor at Jeonbuk National University, Republic of Korea. His research interests constitute the photo-electrochemical characterizations of thin film semiconductor nanomaterials, composite materials, polymer based solid-state films, solid polymer electrolytes and electrode materials for dye-sensitized solar cells (DSSCs), hybrid organic-inorganic solar cells, small molecules based organic solar cells, and photocatalytic reactions.

Professor Hyung-Shik Shin received his Ph.D. in the kinetics of initial oxidation Al (111) surface from the Cornell University, USA, in 1984. He is a Professor in the School of Chemical Engineering, Jeonbuk National University and the President of the Korea Basic Science Institute (KBSI), Gwahak-ro, Yuseong-gu, Daejon, Republic of Korea. He has been a promising researcher and visited several universities as a visiting professor/invited speaker worldwide. He is an active executive member of various renowned scientific committees such as KiChE, copyright protection, KAERI, etc. He has extensive experience in electrochemistry, renewable energy sources, solar cells, organic solar cells, charge transport properties of organic semiconductors, inorganic-organic solar cells, biosensors, chemical sensors, nano-patterning of thin film materials, and photocatalytic degradation.

Contents

Preface

Nanoporous materials are porous solids that possess pores in the range of 0.2 nm up to 50 nm, for example zeolites and zeolite-like materials, mesoporous oxides such as silica, metal organic frameworks, pillared clays, porous carbons, and related materials. These materials provide unique template platforms for the synthesis of other nanostructures with controlled geometric features. Nanoporous materials are outstanding due to their unique chemical and physical properties such as high specific surface areas, well-defined pore sizes, and functional sites, which make them suitable candidate for a plethora of applications, including catalysis, energy harvesting and storage, photonics and optoelectronics, nanomedicine, sensing etc.

This book is a collection of comprehensive chapters of lasting value to the field. The contributions cover all aspects of nanoporous materials, including their preparation, structure, post-synthetic modification, characterization, and potential applications. The book covers the new trends and challenges in the synthesis and surface engineering of a wide range of nanoporous materials. All the chapters are written by professionals that make this book an especially valuable and reliable source of information. To illustrate the scope of nanoporous materials, the chapters are presented with simple diagrams, charts, and graphs. We hope this book will benefit the scientists and engineers to enhance the understanding of nanoporous materials to a new level and its subsequent utilization to accelerate the advancement in technology. We are deeply thankful to the authors for their hard work and contributions. We wish to express our gratitude to the staff of IntechOpen publishers for their patience during the development of this project and for encouraging us during the various stages of preparation. Without all this support, it would not have been possible for us to publish this book.

Mohsen Sheikholeslami Kandelousi
Babol Noshirvani University of Technology,
Islamic Republic of Iran

Sadia Ameen
Advanced Materials and Devices Laboratory,
Department of Bio-Convergence Science,
Jeongeup Campus,
Jeonbuk National University,
Republic of Korea

M. Shaheer Akhtar
New & Renewable Energy Material Development Center (NewREC),
Jeonbuk National University,
Republic of Korea

Hyung-Shik Shin
School of Chemical Engineering,
Jeonbuk National University,
Republic of Korea

Korea Basic Science Institute (KBSI),
Daejon, Republic of Korea

Chapter 1

Introductory Chapter: An Introduction to Nanoporous Materials

Sadia Ameen, Mohammad Shaheer Akhtar, Rhushikesh Godbole and Hyung-Shik Shin

1. Introduction

In the last few decades, research interests and efforts on the synthesis, characterization, functionalization, molecular modeling, and designing of new and novel nanoporous materials have exceedingly grown. In general, the materials having sizes smaller than 100 nm in at least one dimension are considered to be nanomaterials. The materials possessing porous morphology with porous features comparable to 100 nm are termed as nanoporous materials [1]. The properties of nanoporous materials are not only governed by the arrangement of atoms within the crystal but also by the porosity and specific surface area. These materials contain several voids with the controllable dimensions in atomic, molecular, or nanometer scales which enable them to interact more effectively with their environment. According to the International Union of Pure and Applied Chemistry (IUPAC) classification, nanoporous materials can be categorized on the basis of pore sizes as microporous materials (pore size <2 nm), mesoporous materials (pore size between 2 and 50 nm), and macroporous materials (pore size >50 nm) [2–6].

The main challenge in this research pertains to the tailor designing of nanoporous materials to obtain uniform particle/pore size and shape to suit a particular application. For nanoporous materials, it is also essential to achieve a precise composition in their chemical buildup which eventually becomes responsible for carrying out any chemical interaction with its surroundings. Moreover, based on the permeability of the pores to any fluid, these could be classified as closed pores and open pores. Materials properties especially the macroscopic properties like mechanical strength, density, thermal conductivity, etc. are associated to closed pores. On the other hand, the pores which are connected with the surfaces of the material through channels are called open pores. These are active in the processes like absorption and flow of fluids as well as adsorption of gases.

Nanoporous materials have potential applicability to obtain robust, miniaturized, and portable devices. Structural and morphological properties of these materials can be reliably characterized using X-ray diffraction technique (XRD), electron crystallography, field emission scanning electron microscopy (FESEM), and transmission electron microscopy (TEM). The oxidation state, coordination, and optical properties can be studied using X-ray absorption spectroscopy, UV-Vis spectroscopy, solid-state nuclear magnetic resonance (NMR), etc. For elemental analysis of nanoporous materials, energy-dispersive analysis of X-rays (EDAX), inductively coupled plasma mass spectrometry (ICP-MS),

 IntechOpen

Auger electron spectroscopy (AES), and X-ray photoelectron spectroscopy (XPS) are generally adopted, and the surface area and pore size are evaluated by N_2 adsorption-desorption isotherm.

The structure, pore size and hence the porosity, specific surface area, water insolubility, density, pH environment, hydrophilicity/hydrophobicity, charge distribution, conductivity, and catalytic activity of nanoporous materials could be tailored using different fabrication techniques such as precipitation, solid-state reaction (usually performed at high temperature), sol-gel, hydrothermal, and solvothermal routes.

Precipitation synthesis of nanoporous materials involves the creation of a solid from the solution. The synthesis occurs in a liquid solution, and the solid formed is called the “precipitate,” whereas the precursor that causes solid formation is called the “precipitant.” More importantly, without sufficient settling force to bring the solid particles together, the precipitate remains suspended in the solution and can be separated by centrifugation.

Unlike precipitation, sol-gel process is a colloidal route which is effectively used to synthesize nanoporous materials with an intermediate stage known as sol and/or a gel state. In this technique, a colloidal solution known as “sol” is formed which normally advances into a diphasic system called as gel featuring both a liquid phase and solid phase whose morphology could vary from discrete particles to continuous polymeric networks. During the colloidal synthesis, the volume fraction or density of the particles might be very low such that initially a desirable amount of fluid might need to be washed out to recognize its gel-like nature.

The hydrothermal synthesis is a solution-based reaction approach carried out in an autoclave for the synthesis of nanoporous materials. In a broader sense, this method involves the synthesis from the room temperature to a high-temperature in which the morphology of the as-synthesized nanoporous materials can be controlled. The hydrothermal process at the elevated temperatures provides stability for the as-synthesized nanoporous materials. Another variant of the solution-based synthesis is solvothermal route which is very similar to the hydrothermal with the only difference that the precursor solution is usually nonaqueous. The solvothermal route allows precise control over the size, shape, and crystallinity of nanoporous materials. These characteristics could be altered by changing certain experimental parameters which include temperature, time, type, and the concentration of solvent and the surfactant [7].

Another simple and rapid method of preparation of highly porous materials, namely, metal oxides, is the auto-combustion route wherein some precursors act as oxidizers, while others as fuel. In a typical experiment, a homogeneous aqueous solution containing the necessary reaction mixture of oxidizers and fuel is kept under constant stirring on a hot plate in air until drying. Subsequently, the fuel initiates the combustion reaction at specific temperature which results in porous and spongy nanocrystalline material in powder form. The particle and pore size can be controlled by optimizing the fuel concentration in the reaction mixture. The reaction herein is exothermic, and the heat evolved in the combustion process is sufficient for the proper phase formation of the as-synthesized material. Hence, it is noteworthy that the as-synthesized porous material does not require any high-temperature sintering/annealing [8].

In modern technology, porous thin films are attracting tremendous attention due to their applicability in gas sensors, chemical and energy conversion devices, catalysis, and so on. In view of this, spray pyrolysis is a simple and the effective technique to synthesize thin films possessing porous morphology. In this technique aqueous or alcoholic solutions of inorganic precursors are sprayed onto a hot substrate which then pyrolytically decompose into a desired compound in thin-film form featuring excellent adhesion. The thickness and morphology of the porous films can be varied by maneuvering various process parameters such as molarity of the solution, substrate temperature, flow rate of the carrier gas, distance from spray nozzle to substrate, etc [9].

Research efforts in the field of nanoporous materials have been driven by the emerging and rapidly growing applications of nanoporous materials as adsorbents and ion exchangers. The nanomaterials with uniform pores and a high surface-area-to-volume ratio produce large amount of catalytic sites for efficient catalysis, guest-host interaction for immobilized homogeneous catalysts, enzymes, nonlinear optics, low dielectric constant mediate, etc. Nanoporous materials have shown excellent application in several bioanalyses such as in biomedical diagnosis and the monitoring of food, as decontaminants and environmental quality, antibacterial agents in drug release, enzyme mimetics, biosensors, and adjuvant in anticancer therapy [10]. The applicability of nanoporous material shows several exotic new and novel opportunities for the researchers and scientists to develop new methodologies, strategies, and techniques for synthesizing the nanoporous materials.

In this book, a series of systematic chapters based on the recent developments in nanoporous materials is presented. To facilitate the understanding and to highlight the essential concepts, the chapters are designed to explain the basics of nanoporous materials. The chapters are strongly interconnected and intermingled as well, with the fundamental principles forming a framework that permeates throughout the book. The important properties of the nanoporous materials are covered under the topics of nanoporous oxides and nanoporous composites, porous low dielectric constant material for semiconductor microelectronics, synthesis and application of kaolin-based ZSM-5 in petrochemical industry, nanoporous silicon materials fabricated by electrochemical method for bio-applications, etc. Our approach is to emphasize the essential concepts and to develop a basic textbook for the students and researchers of chemistry, physics, and materials science as well as engineers of chemical industry, environmental protection, automotive industry, refining technology, civil engineering, and electronics. The book covers a selected range of topics covering the most significant aspects involved in the understanding, properties, preparation methods, and emerging applications of nanoporous materials. Some of the topics have not been addressed in detail in a view to keep the length of the book manageable. Hopefully, this book fulfills its purpose of inspiring young brains and guiding them into the challenging realms of nanoporous materials.

Author details

Sadia Ameen[1]*, Mohammad Shaheer Akhtar[2], Rhushikesh Godbole[3] and Hyung-Shik Shin[3]

1 Advanced Materials and Devices Laboratory, Jeongeup Industry-Academic Cooperation Support Centre, Chonbuk National University, Republic of Korea

2 New and Renewable Energy Material Development Center (New REC), Chonbuk National University, Jeonbuk, Republic of Korea

3 Energy Materials and Surface Science Laboratory, Solar Energy Research Center, School of Chemical Engineering, Chonbuk National University, Jeonju, Republic of Korea

*Address all correspondence to: sadiaameen@jbnu.ac.kr

References

[1] Sammons R. Biological responses to hydroxyapatite In: Mucalo M, editor. Hydroxyapatite (HAp) for Biomedical Applications. Woodhead Publishing. 2015. pp. 53-83

[2] Sotomayor F, Cychosz K, Thommes M. Acc. Mater. Surf. Res. 2018;**3**:34-50

[3] Kim E-B, Ameen S, Akhtar MS, Shin HS. Sensors and Actuators B: Chemical. 2018;**275**:422-431

[4] Ameen S, Akhtar MS, Shin HS. Materials Letters. 2016;**183**:329-333

[5] Ameen S, Kim E-B, Akhtar MS, Shin HS. Materials Letters. 2017;**209**:571-575

[6] Jang G-S, Ameen S, Akhtar MS, Shin H-S. Ceramics International. 2018;**44**:588-595

[7] Nyamukamba P, Okoh O, Mungondori H, Taziwa R, Zinya S. Synthetic Methods for Titanium Dioxide Nanoparticles: A Review. In: Yang D, editor. Titanium Dioxide - Material for a Sustainable Environment. IntechOpen; 2018

[8] Godbole RV, Rao P, Alegaonkar PS, Bhagwat S. Materials Chemistry and Physics. 2015;**161**:135-141

[9] Godbole RV, Godbole VP, Alegaonkar PS, Bhagwat S. New Journal of Chemistry. 2017;**41**:11807

[10] Catalano E. Magnetic Nanomaterial-based Anticancer Therapy. In: Tiwari A, Iyer P, Kumar V, Swart H, editors. Advanced Magnetic and Optical Materials. USA: Wiley-Scrivener Publishing; 2016

Chapter 2

Nanoporous Oxides and Nanoporous Composites

Dong Duan, Haiyang Wang, Wenyu Shi and Zhanbo Sun

Abstract

Nanoporous oxides, such as cupric oxide (CuO), nickelous oxide (NiO), titanium dioxide (TiO_2), cobaltosic oxide (Co_3O_4), and cerium oxide (CeO_2), and noble-metal-based nanoporous composites, such as silver (Ag) ligaments loaded with CeO_2, TiO_2, zirconium dioxide (ZrO_2) or NiO and palladium (Pd) ligaments loaded with TiO_2 or ZrO_2, are described in the chapter. Oxide-based nanoporous composites, such as Au loaded on CuO and CeO_2 or platinum (Pt) loaded on TiO_2, are also summarized. The structures, microstructures, and microstructure parameters of these materials are reviewed. The performance of the noble-based nanoporous composites is presented, including the catalytic oxidation of methanol and ethanol. Environmental protection applications, such as catalytic oxidation of carbon monoxide (CO) for the oxide-based nanoporous composites, have also been developed. Applications of rare earth elements in nanoporous materials are also reviewed.

Keywords: precursor alloy, dealloying, nanoporous oxides, nanoporous composites, rare earth elements

1. Introduction

Research on nanoporous materials prepared by dealloying originated from work with nanoporous noble metals. During dealloying, the active metals in precursor alloys composed of corrosion-resistant noble metals, such as Ag and gold (Au), and cheaper metals, such as manganese (Mn), zinc (Zn), aluminum (Al), and magnesium (Mg), were selectively removed. Simultaneously, ligaments of corrosion-resistant noble metals and pores were formed [1]. At the nanoscale, both the preparation and the performance of noble metals have unparalleled advantages due to the excellent catalytic activity and chemical stability. However, the high price and the scarce resources of noble metals have prompted investigation of new paths to reduce their usage, resulting in the formation of alloy nanoporous materials containing less noble metals [2].

Certain metals, such as copper (Cu), nickel (Ni), and cobalt (Co), can remain in dilute acid or alkali solutions and maintain acceptable catalytic properties. These metals are potential substitutes for noble metals. Therefore, the preparation of cheaper nanoporous metals, such as Cu, was investigated [3]. However, it was found that the nanoscale ligaments inevitably oxidized in air or corrodents due to small scale and large surface effects after dealloying, which means that the materials must be kept in a special environment. These materials cannot be used in ideal environmental conditions, which will inevitably lead to their limited use. Certain nanooxides, such as CuO, NiO, cobaltic oxide (Co_2O_3), and CeO_2, have acceptable catalytic activities. In acid and alkali solutions, certain stable metals, such as Cu, Ni, and Co, can form nanoporous oxides after further oxidation [4–6]. These metals could be used to produce nanoporous composites with excellent catalytic performance after they are loaded with noble metals.

Research [7] has shown that the interface between the noble metal and loaded oxides and the interaction between the two phases can significantly enhance the catalytic activity at the nanoscale. Certain oxides, such as CeO_2, could prevent poisoning and failure of noble metals when loaded on the surface of noble metals. To improve the performance of noble metals, the proper amount of oxides should be loaded on the ligament surface of the noble metal. The direct solution mechanism is to infiltrate the target metal saline solution into the nanoporous metals. The metal ions are transformed into oxides after drying or dewatering. However, the added matter concentrates on the surface layer of the sample during drying of the aqueous solution, causing little oxide to be distributed in the inner region. In 2013, Li et al. added cerium (Ce) to the Al-Ag precursor alloy. With the removal of Al elements during corrosion, Ce compounds are formed during the formation of nanoporous Ag. After heat treatment in air or oxygen (O_2), CeO_2 is formed and loaded on the surface of Ag ligaments to create a nanoporous composite material of Ag ligament loaded with CeO_2 [8]. This material's performance was significantly better than that of pure nanoporous Ag. Nanoporous Ag loaded with TiO_2 [9] and ZrO_2 [10] and nanoporous Pd loaded with NiO [11] and TiO_2 [12] have also been prepared.

Certain metals, such as Ce and titanium (Ti), are active elements. However, they do not evolve into ions that are removed in alkaline solutions, such as sodium hydroxide (NaOH) and potassium hydroxide (KOH). Instead, they evolve into solid compounds containing water, and the nanoporous structure is formed after crystallization. Some of them, such as Ce and samarium (Sm), also form nanorods because the crystallization has a clear direction. After high-temperature dehydration, they evolved into rare earth oxide nanorods [5, 13]. The nanorods support each other and form a supporting pore structure or framework with a large number of pores. If noble metals, such as Au and Ag, are added to the precursor alloys in small quantities, the noble metal nanoparticles or atoms are loaded on the surface of the nanorods. These metals presented excellent gas catalytic activity [6, 14].

Focusing on the work of our group, this chapter introduces the recent development of nanoporous oxides and nanoporous composites prepared by dealloying. The design principle of the precursor alloys and preparation process and dealloying methods and effects of the preparation process on the morphologies and physical parameters of nanoporous oxides and nanoporous composites are reviewed.

2. Design and preparation of precursor alloys and the dealloying process

Generally, the more active elements, that is, the elements removed by corrosion in the precursor alloys, are the higher the porosity of the nanoporous material is and the thinner the ligaments are after dealloying. However, if the stable metal content is too low, that is, less than 5 at%, the nanoporous structure may not form. Therefore, a stable metal content of 8–15 at% is recommended [4–14]. In addition, it is important to consider whether the components can form alloys. If the solubility between the components is not good in the melt, it is likely that the alloy will have a serious tendency to macrosegregate, and the ideal nanoporous structure cannot form after dealloying. Some systems, such as the Al-Ti system, have a large temperature difference between the liquidus and the solidus on the Al-rich side (when

10% Ti is used, the distance between liquidus and the solidus is approximately 585°C), which would result in a serious macrosegregation tendency. Therefore, the Ti content should be minimized if possible. Thus, a eutectic system, such as Al-Cu, is an ideal precursor alloy. For some systems, such as Al-Ce and Al-Pt systems, the temperature difference between the liquidus and solidus is not very large. The melt quenching can be used to prepare precursor alloys with homogeneous composition. In addition, systems with a little or no intermetallic compounds in the precursor alloys are suitable for preparing nanoporous materials by dealloying. Some intermetallic compounds, such as Ag_2Al in the Al-Ag system, do not effectively decompose in alkaline solutions at room temperature, resulting in residual Ag_2Al and decreased Ag content in the nanoporous structures [8–10]. The properties of the nanoporous structures were less optimal. Certain intermetallic compounds, such as Al_2Cu, easily decompose in corrosion fluids, and its content has little effect on nanostructures. If a binary alloy cannot form an ideal nanoporous material, a third component could be added to the binary system. However, the third component should be completely removed in the dealloying. For example, Li et al. added an appropriate amount of rare earth element Ce to Cu-Ag alloys, resulting in significant microstructure refinement of the precursor alloys [15]. After electrochemical dealloy removal in aqueous cupric sulfate ($CuSO_4$), all of the Al and Ce were removed, and the microstructure of the nanoporous Ag was obviously refined.

For nanoporous oxides, binary precursor alloy systems, such as Al-Cu [3, 4], Al-Ni [5], and Al-Ce [5], can be used. For the noble metal-loaded oxide systems, Al-noble metal-oxide formation elements can be used to form ternary alloy systems, such as Al-Ag-Ce [8] and Al-Pd-Ni [11]. Their noble metal content is approximately 10–15%, and the oxide-forming element content is generally less than 2 at%. In the dealloying, elemental Al is dissolved or removed and the two other elements remain. Using components with higher noble metal content, noble metal would form ligaments. After heat treatment, oxide particles form and load on the surface of the ligaments. In oxide-loading noble metal systems, the precursor alloys are Al-oxide formation element-noble metal ternary systems, such as Al-Cu-Au [4], Al-Ce-Au [14], and Al-Ti-Pt [16]. Their oxide-formation element content is approximately 8–10%, and their noble metal content is less than 2%. In the dealloying, Cu, Ce, and Ti are oxidized into oxides and form ligaments. The noble metals are loaded on the surface of ligaments after forming nanoparticles.

Intermediate alloy ingots can be synthesized by melting and powder metallurgy. Melting is the most recommended method. However, the direct dealloying of alloy ingots results in a longer dealloying time due to their larger ingot size. The size of ligaments and holes might be too large. Some systems, such as Cu-Ag and Al-Cu systems, can be made into thin strips by cold rolling. However, other systems, such as the Al-Ti system, do not allow for creating a high-quality precursor by cold rolling due to poor ductility. Thus, it is necessary to adopt melt spinning. Melt spinning is a rapid solidification process. In this method, alloy melts can directly solidify or evolve into thin ribbons at cooling rates of 10^6–10^9 K/s. Based on the conditions adopted, the thickness of the thin ribbons can be controlled from 20 to 100 μm, and the width can be adjusted. For most alloys, alloy ribbons with a highly homogeneous distribution of elements can be obtained by melt spinning or melt quenching. Binary Al-Ti and Al-Ce and ternary Al-Ti-Pt and Al-Ce-Au systems could be prepared into high-quality precursor ribbons. If the required thin ribbons cannot be prepared in this manner, the system might not be an optimal precursor alloy for nanoporous materials.

Hydrochloric acid or aqueous KOH or NaOH is often used as dealloying solvents. The dealloying temperature should be room temperature. However, at higher temperatures such as 80°C, nanorod support pore structures can form [5, 6, 14] in the Al-Ce precursor system. Thus, the dealloying temperature is important for products with specific morphologies. In dealloying, the smaller components in the precursor alloys are distributed in the ligaments by mechanical mixing. During subsequent heat treatment, the ligament begins to crystallize or crystallizes further. The loaded elements cannot be fixed or dissolved in the ligaments, and they are forced to diffuse to the surface of the ligament. This promotes the formation of nanoparticles and atomic load on the ligament surface. As a result, composite materials from loading the ligament surface with other particles evolve well [4]. Regardless of whether the noble metal is loaded on oxides or the oxides are loaded on noble metals, designing the proper composition of precursor alloys is critical. Notably, the binding between the load and the carrier in the nanoporous composite prepared by dealloying is similar to that of metallurgical binding or a semi-embedding structure [4]. Compared with other methods, they have a stronger binding force, which is conducive to the stability of structures and stronger interface effects.

Heat treatment or calcination is an indispensable procedure in preparation of nanoporous oxides and nanoporous composites. During heat treating, some elements are oxidized into oxides and the desired structures can be formed [4, 5, 8, 9, 13, 14]. In certain composites, such as Co_3O_4 [17] and CeO_2 loaded with Au [14], interface and crystalline defects occurred. This can be useful to enhance the catalytic properties. The calcination technologies indicated that the choice of atmosphere should be optimized according to the materials. This approach provides a starting point for researchers. However, certain composites, such as nanoporous Pd loaded with oxides [11, 12, 18], are not suitable for calcination because the nano Pd easily oxidizes at high temperatures.

Rare earth elements are important metallurgical additives in the industry. In the past 5 years, rear earth elements, such as Ce, have been used in research examining nanoporous materials [5, 6, 8, 14, 15]. The Ce in melt-spun Cu-Ag-Ce alloys can observably refine the microstructures of precursors. The nanoporous structures were obviously fine after electrochemistry dealloying, and all Cu and Ce were removed [15]. The most compounds with rare earth elements could be decomposed during dealloying. Some rare earth elements, such as lanthanum (La), Ce, neodymium (Nd), and Sm, could not be removed by NaOH and KOH solutions at up to 100°C. When their contents are lower, that is, less than 2%, they can be loaded by ligaments of noble metals, resulting in improved catalytic properties [8]. In alkali liquors, these elements can form compounds with water and crystallize. The crystallization has specific directions, which results in the formation of nanorods at temperatures above 80°C. The compounds with water can transform into rare earth oxidations after dehydration, and a framework structure constructed by support nanorods forms [5, 8, 14]. Rare earth oxidation nanorods are important carriers loaded by nobles and other oxidations because they have special interactions with other matter. The composites present excellent catalytic properties. In 5% nitric acid (HNO_3) solution, the CeO_2 can be removed. Ultrafine nanoporous Pt with excellent properties was also achieved [19].

3. Nanoporous oxides

Transition metal oxides (TMOs) are important nanoporous materials exhibiting a wide variety of structures and electronic and magnetic properties due to the nature of the outer d states [20]. Nanoscale transition metal oxides exhibit favorable catalytic activity for catalytic oxidation of harmful gases due to the presence of variable valency ions [21–23]. Transition metal oxide semiconductor materials, such as TiO_2, have important applications in photocatalytic hydrogen production and photodegradation due to their nontoxicity, high chemical stability, low cost, and light stability [24–26]. Thus, research has been devoted to the creation of stable ordered nanoporous transition metal oxides.

3.1 Nanoporous CuO

In 2015, Zhang et al. reported that nanoporous CuO was prepared by a simple method of dealloying Al-Cu alloy ribbon in alkaline solution and calcining in air [4]. The precursor ingots were obtained by melting alloys composed of pure Al and pure Cu. The ribbons were prepared by melt spinning at 33 m s^{-1} and protected by 0.05 MPa argon (Ar). The as-quenched ribbons were immersed into aqueous 5 wt% NaOH at room temperature. The dealloyed ribbons were calcined at 600°C for 1 h in air. The results showed that the samples are mainly composed of Cu and copper (I) oxide (Cu_2O) after dealloying and only CuO could be detected after calcination at 600°C. The prepared CuO exhibits a three-dimensional interpenetrating nanoporous structure with mesoporous properties. The pore size is approximately 25 nm, and the ligament size is approximately 50 nm. The materials have good redox capacity and can be used for the catalytic oxidation of CO. The reaction temperature for catalyzing 50% of CO is approximately 170°C, and the temperature for completely converting CO is 240°C. The long-term stability did not degrade even after 60 h, indicating that its performance is favorable.

In 2017, Li et al. reported the fabrication of a 3D free-standing CuO nanowire array supported by a nanoporous CuO network by dealloying $Cu_{60}Zr_{35}Al_5$ glassy ribbons in a dilute HF aqueous solution at 0.05 mol L^{-1} for 12 h at room temperature [27]. The CuO nanocomposite exhibits a hierarchical nanostructure containing a well-aligned CuO nanowire array and nanoporous substrate with continuous nanoporosity. The nanoporous CuO exhibits superior degradation performance for methylene blue in the presence of hydrogen peroxide (H_2O_2) compared to that of commercial CuO nanoparticles. The remarkable catalytic activity and good reusability and stability during degradation make the as-prepared nanocomposite a promising candidate for purifying wastewater with organic dyes. The high degradation efficiency of the nanocomposite mainly results from the uniform nanowire array structure and its high internal surface area, which provide more effective contact between the MB and catalysts [27].

3.2 Nanoporous NiO

In 2013, Liang et al. reported that a three-dimensional (3D) nanoporous NiO film was fabricated via a two-step process using an electrochemical route [28]. The dealloying process included electrodeposition of the Ni/Zn alloy film and electrochemical dealloying using a direct-current power source. The NiO film had an irregular 3D interconnected nanosheet structure with open channels. The specific capacitance of the NiO reached 1670 F g^{-1} at a discharge current density of 1 A g^{-1} for the supercapacitor. In addition, the NiO exhibited high performance during a long-term cycling. The maximum specific energy and specific power at the 1.1 V potential window were 170 and 27.5 kW kg^{-1}, respectively.

In 2016, Zhang et al. successfully prepared nanoporous NiO by dealloying $Al_{85}Ni_{15}$ alloy and calcining in air [5]. The precursor alloys were prepared by similar methods with Al-Cu [4]. The surface and pores of the sample present a nanoporous structure. The pore diameter is approximately 30–50 nm, the pore walls are composed of island-like barriers, and the width is approximately 50 nm. The entire material presents a porous frame structure with a uniform pore size. It achieved good performance in the catalytic oxidation of CO, and the active temperature of fully catalyzed oxidation of CO was approximately 340°C.

3.3 Nanoporous Co_3O_4

In 2011, Xu et al. employed a simple fabrication method for Co_3O_4 nanosheets through dealloying Al-Co alloy in alkaline solutions [17]. The precursor alloy was $Al_{95}Co_5$, and ribbons were prepared by melt spinning. The microstructure was a hierarchical flower-like aggregate structure with the typical size at the micron scale, where each nanoflower is composed of many irregular interlaced nanoslices with thicknesses as small as 6 nm, which is a typical porous structure. The results indicated that Co_3O_4 nanosheets exhibited excellent catalytic activity for CO oxidation at ambient temperature, the reaction temperature for catalyzing 50% of the CO was approximately 80°C, and the temperature for completely converting the CO was 140°C. Calcination in an O_2 atmosphere was essential to achieve high CO oxidation activity for these nanostructures, which allowed for the generation of active species for surface reactions. In addition, the calcination temperature significantly affected the catalytic activity; 300°C was a more favorable calcination temperature than 200 or 450°C, considering the optimum balance between the active reaction species and the surface area upon calcination. In addition, Co_3O_4 nanosheets showed good time-onstream catalytic stability. It is expected that many other useful metal oxide materials can be fabricated similarly. Due to the evident advantages of simple processing, nearly perfect yield, and low fabrication cost, these functional nanomaterials may lead to applications in various catalytic processes.

3.4 Nanoporous TiO_2

In 2018, Shi et al. prepared nanoporous TiO_2 by dealloying an Al-Si-Ti precursor alloy and calcining [16]. The $Al_{61}Si_{30}Ti_9$ precursor alloy was prepared from pure Al, Ti, and silicon (Si) by arc-melting. The melt-spun ribbons were obtained by melt spinning at 33 m s^{-1} and protected by 0.05 MPa Ar. The as-quenched ribbons were dealloyed in aqueous 10% NaOH at 80°C. Subsequently, the ribbons were immersed in aqueous 3% (by weight) hydrogen chloride (HCl) at an ambient temperature for 7 h. The samples were then calcined at 400°C for 2 h. The results show that nanoporous TiO_2 exhibits a laminated sheet morphology, and the irregular sheets show a pore-ligament structure. The structure had large specific surface areas, small pore diameters, and large pore volumes. The measurements also revealed that nanoporous TiO_2 had favorable photocatalytic performance in degradation of methyl orange (MO).

3.5 Rare earth oxides

Due to the special electronic structure of rare earth elements, which have unique physical and chemical properties, and the continuous development of nanotechnology, the preparation of nanoporous rare earth oxides has attracted increasing attention [29, 30]. Rare earth oxides such as CeO_2 have heterogeneous catalytic ability and excellent storage/deoxygenation capacity, which is optimal for supporting noble metal nanoparticle carriers [31, 32]. Nanoporous rare earth oxides have a wide range of applications in energy, chemical, environmental, and magnetic materials.

In 2017, Zhang et al. reported a CeO_2 nanorod framework synthesized via dealloying Al-Ce alloys coupled with calcination treatment [5]. The precursor alloy was $Al_{90}Ce_{10}$, and ribbons were prepared by melt spinning at 33 m s^{-1} and protected by 0.05 MPa Ar. The precursor alloy ribbons were dealloyed in 20 wt% NaOH solution at 80°C. The dealloyed samples were heat treated at 400°C in O_2. After dealloying, the CeO_2 particles grew into nanorods with a diameter of approximately 20 nm. The nanorods piled up into frameworks containing pores. The nanorods and pores were homogeneously distributed and formed the framework structure. The material presented favorable CO catalytic oxidization properties: the reaction temperature for catalyzing 50% of CO was approximately 300°C, and the temperature for completely converting CO was 440°C. In 2018, Wang et al. reported that the CeO_2 nanorod framework also exhibits a high-specific capacitance and superior charge/discharge stability, which are mainly ascribed to its high-Brunauer-Emmett-Teller surface area [33].

In 2018, Duan et al. prepared nanorod samarium oxide (Sm_2O_3) by dealloying Al-Sm alloy and calcining in O_2 [13]. The preparation is the same as in the CeO_2 framework [5]. Similar to CeO_2, Sm_2O_3 also exhibits uniform nanorods, measuring approximately 20–30 nm in diameter, and a framework structure. The loose nanorod framework structure provides efficient transport of reactive gases and sufficient space for the activation of molecular oxygen, which improves catalytic performance. The Sm_2O_3 nanorod framework shows good performance for the catalytic oxidation of CO. The 50% CO conversion temperature was approximately 280°C, and 100% CO conversion occurred at approximately 360°C.

The above results show that a reasonable precursor alloy design and a combination of dealloying and calcination processes can be used to generate nanoporous oxide structures with ideal structural parameters. These materials have favorable catalytic properties and can act as a support for noble metals or oxides, providing good conditions for the development of higher performance nanoporous materials.

4. Noble ligaments loaded with metallic oxides

Nanoporous composites in which the noble ligaments are loaded with metallic oxides have recently been investigated. These composites include nanoporous $Ag\text{-}TiO_2$, $Ag\text{-}CeO_2$, Pd-NiO, and $Pd\text{-}TiO_2$ and exhibit improved catalytic properties compared with pure nanoporous nobles. In this section, their structures and performance are reviewed.

4.1 Nanoporous $Ag\text{-}CeO_2$

Nanoporous $Ag\text{-}CeO_2$ ribbons have been prepared through dealloying melt-spun $Al_{79.5}Ag_{20}Ce_{0.5}$ alloy in aqueous 5 wt% NaOH, followed by calcining in air [8]. Precursor ingots were obtained by melting the alloys composed of pure Al, Ag, and Ce. The ribbons were prepared by melt spinning at 33 m s^{-1} and protected by 0.05 MPa Ar. The melt-spun ribbons were dealloyed in aqueous 5 wt% NaOH at room temperature. The dealloyed ribbons were calcined for 2 h at 300–700°C. The SEM analyses indicated that the morphologies were not evaluated due to the addition of Ce in the precursor alloy. However, the CeO_2 particles were distributed on the surface of the Ag ligaments, and a composite with CeO_2 loaded on the nanoporous Ag was obtained. The electrochemical tests revealed that the nanoporous $Ag\text{-}CeO_2$ catalyst exhibited enhanced activity for the direct oxidation of sodium borohydride. The Raman analysis indicated that the intensity of the peak in the 2LO band (CeO_2) is stronger than that in F_{2g} and D bands (CeO_2) [8]. Therefore, the superior catalytic activity is attributed to the enhancement of the interfacial interaction between Ag and CeO_2. This research provided a novel method for preparing noble metal-oxide nanocomposites with high catalytic performance. The results also indicated that the structural stability of $Ag\text{-}CeO_2$ ribbons with a homogeneous pore/grain structure is improved due to the formation of the $Ag\text{-}CeO_2$ composite. This may aid in enhancing the longevity performance.

4.2 Nanoporous Ag-ZrO_2

After the Ag-ZrO_2 was prepared, nanoporous Ag-ZrO_2 composite catalysts were created by chemical dealloying of the melt-spun Al-Ag-Zr precursor alloys [10]. The composition of the precursor alloys was $Al_{80-X}Ag_{20}Zr_X$ (x = 1, 2, 3). The preparation of alloy ribbons and the dealloying method was the same as in the Ag-CeO_2 system. During dealloying, the zirconium (Zr) atoms released from the precursory alloys are oxidized into ZrO_2 and are loaded on the inner surface of the nanoporous Ag. The dealloyed ribbons exhibit an interpenetrating ligament-channel structure with nanometer-length scales. X-ray photoelectron spectroscopy (XPS) analysis indicated that the binding energy of the Zr 3d core levels decreases slightly with increasing calcination temperature because the Zr^{4+} in the ZrO_2 particles is partially reduced to Zr^{3+}, suggesting electron transfer occurring from the ZrO_2 to the metallic centers and the formation of oxygen vacancies in the thermal treatment. The electrochemical tests demonstrated that the current density peak increased with calcination temperature in a certain range, that nanoporous composites with the optimized ZrO_2 content exhibited higher catalytic activity, and that the optimized precursor alloy composition was $Al_{78}Ag_{20}Zr_2$. The oxidation current density increased by 91.3% compared with that of nanoporous Ag. The excellent catalytic activity can be attributed to the interfacial interaction and electron charge transfer between Ag and ZrO_2. In addition, nanoporous composites with high specific surface areas and pore volume can enhance their electrocatalytic activities, which can provide more transport channels and activity sites for conductive ions and reactant molecules.

4.3 Nanoporous Ag-TiO_2

Nanoporous Ag-TiO_2 composites were prepared by dealloying the melt-spun Al-Ag-Ti ribbons in aqueous NaOH [9] in 2014. The composition of the precursor alloys was $Al_{80-X}Ag_{20}Ti_X$ (X = 0.5, 1, 2, 3). The preparation of alloy ribbons and the dealloying method was the same as in the Ag-CeO_2 and Ag-ZrO_2 systems. The results revealed that TiO_2 formed in situ on the Ag ligaments. Ti^{3+} and Ag^+ species co-existed after the dealloyed samples were calcined at 600°C, which significantly influenced the catalytic oxidation of sodium borohydride. The electrochemical results showed that the nanoporous Ag-TiO_2 composites significantly promoted the direct oxidation of BH_4-superior to pure Ag. The current density of the oxidation peak for the nanoporous Ag-TiO_2 electrode prepared from the $Al_{79.5}Ag_{20}Ti_{0.5}$ alloy increased from 10.91 to 18.13 mA cm^{-2}. With increasing the Ti content from 0.5 to 1.0% in the precursor alloys, the current density of the oxidation peak increased to 28.86 mA cm^{-2}. With further increases in the Ti content, however, the current density of the nanoporous Ag-TiO_2 electrode obviously decreased, and the position of the oxidation peak became more positive. The enhanced catalytic activity could be attributed to the strong interfacial effects between the Ag ligaments and TiO_2 [9].

Among these Ag-based nanocomposites, the Ag-ZrO_2 nanocomposite catalyst exhibited the greatest improvement in the catalytic oxidation of sodium borohydride. The optimized precursor alloy composition was $Al_{78}Ag_{20}Zr_2$. However, Ag-CeO_2 may have the best anti-poisoning ability of CO.

4.4 Nanoporous Pd-TiO_2

Nanoporous Pd-TiO_2 composite catalysts were prepared by chemical dealloying of melt-spun Al-Pd-Ti precursor alloys [12] in 2014. The precursor alloy was $Al_{85-x}Pd_{15}Ti_x$ (x = 0, 0.3, 0.5, 0.7, 1.0, at%), and the preparation of alloy ribbons was the same as in the Ag-CeO_2 and Ag-ZrO_2 systems. However, calcination was not adopted to prevent oxidation of the Pd. The structure analysis revealed that the nanoporous Pd-TiO_2 composites exhibited a bicontinuous interpenetrating ligament-pore structure. The XPS analysis revealed that the Pd 3d spectrum of the nanoporous Pd-TiO_2 composites presented a slightly negative shift to lower binding energy compared to that of nanoporous Pd. The electrochemical measurements showed that the composites observably enhanced the electrocatalytic performance toward methanol/ethanol oxidation when the Ti content in the precursor alloys was 0.3–0.7 at%. Among these composites, the nanoporous Pd-TiO_2 composite dealloyed from $Al_{84.5}Pd_{15}Ti_{0.5}$ exhibited the largest catalytic activity, which was triple and double that of methanol and ethanol oxidation, respectively, compared with that of nanoporous Pd. This enhancement can be attributed to the synergistic effect between Pd and TiO_2. However, when the Ti content of the precursor alloys is 1 at%, the catalytic activity will obviously decline. These results imply that an electronic interaction between Pd and TiO_2 occurs in the nanoporous Pd-TiO_2 composites. The reduction of binding energy decreases the chemisorption ability of the adsorbate during electrochemical reactions. It will reduce the adsorption of CO-like intermediate species and result in superior poisoning tolerance.

4.5 Nanoporous Pd-NiO

Nanoporous Pd-NiO composites were prepared for methanol oxidation in alkaline media by one-step dealloying from melt-spun Al-Pd-Ni precursor alloys [11]. The precursor alloy was $Al_{85-x}Pd_{15}Ni_x$ (x = 0, 0.1, 0.3, 0.5, 0.7, 1.0 at%). The preparation of alloy ribbons was the same as in the Pd-TiO_2 systems. The structure, morphology, composition, and electrocatalytic activities of the composites were characterized. The results demonstrated that the composites exhibited a uniform bicontinuous and interpenetrating three-dimensional nanoporous structure. Pore channels that were less than 10 nm ran throughout the ribbons. The XPS analysis revealed that the nanoporous Pd-NiO composites significantly increased compared to that of nanoporous Pd due to the reduction of the PdO content in the Pd-NiO composites. The nanoporous Pd-NiO composites possessed better electrocatalytic performance compared to that of nanoporous Pd, and the composite dealloyed from $Al_{84.7}Pd_{15}Ni_{0.3}$ showed the highest catalytic activity of the catalytic oxidation of methanol; the mass activity of the mesoporous Pd-NiO composite from $Al_{84.7}Pd_{15}Ni_{0.3}$ was 289.0 mA mg^{-1}, which was approximately 4 times that of nanoporous Pd. The improvement in the electrocatalytic properties was attributed to the increased electrochemically active specific surface areas and the synergistic effect between Pd and NiO. The reduction of the PdO content in the nanoporous Pd-NiO composites means that there are more active sites (Pd atoms) in the composites involved in the electrocatalytic process. There is also a synergistic effect between Pd and NiO. Oxidative states of Ni in the nanoporous Pd-NiO composites serve as oxophilic sites for the oxygen source required for the surface removal of CO and promoting methanol oxidation via a surface redox process.

4.6 Nanoporous Pd-ZrO_2

Composition-controllable three-dimensional mesoporous Pd-ZrO_2 composites were synthesized through simple one-step dealloying of melt-spun Al-Pd-Zr ribbons [18]. The precursor alloy was $Al_{85-x}Pd_{15}Zr_x$ (x = 0, 0.5, 1, 2, 3 at%). The preparation of alloy ribbons was the same as in the Pd-TiO_2 systems. The Zr atoms in the precursor alloy transformed into approximately 3 nm ZrO_2 nanoparticles after dealloying without being calcined. The ZrO_2 particles embedded in the Pd ligaments. The XPS analysis revealed that the Pd 3d spectrum of the dealloyed

$Al_{83}Pd_{15}Zr_2$ sample was negatively shifted to a lower binding energy after the addition of Zr in comparison with the Pd 3d spectrum of the dealloyed $Al_{85}Pd_{15}$. This finding suggests the existence of electronic interactions between Pd and ZrO_2. The composites exhibited remarkable catalytic activity and stability for methanol oxidation in an alkaline electrolyte due to the synergistic effect and electronic interactions between Pd and ZrO_2 [18]. Among the composites, the Pd-ZrO_2 sample dealloyed from the $Al_{85}Pd_{13}Zr_2$ precursor alloy had the highest peak current density, and the mass activity of the mesoporous Pd-ZrO_2 composite from $Al_{85}Pd_{13}Zr_2$ was 254.24 mA mg^{-1}, which was approximately 3.6 times that of the pure mesoporous Pd sample. In one study [18], Zr in the form of ZrO_2 was embedded in the surface of the Pd ligaments in mesoporous Pd-ZrO_2 composites. This arrangement endowed ZrO_2 with more convenient and effective absorption toward OH^- species to more rapidly transform CO-like poisoning species on the surface of the active Pd sites into carbon dioxide or other dissolvable cleansing products. The synergistic effect between Pd and ZrO_2 results in more active Pd atoms being released to the surface of the samples for the use in additional methanol oxidation processes.

Among these Pd-based nanocomposites, the Pd-NiO nanocomposite catalyst showed the greatest improvement in the catalytic oxidation of methanol. The optimized precursor alloy composition was $Al_{84.7}Pd_{15}Ni_{0.3}$.

5. Nanoporous oxides loaded by noble metals

Nanoporous metal oxides have been widely applied in catalysis due to their unique porous support structure and large specific surface area, including in carbon monoxide catalytic oxidation, electrochemical catalytic oxidation of methanol, supercapacitors, and photocatalytic fields. As an intermediary, noble metals are irreplaceable in the field of catalysis. They can effectively reduce catalytic activation energy or promote electron transfer, improving the catalytic performance. Loading noble metals on oxide substrates can improve the utilization rate of noble metals and enhance the catalytic performance of oxides. Nanoporous oxides loaded or modified by noble metals can be prepared by simply dealloying melt-spun alloys and subsequent calcination in air. Nanoporous oxide materials obtained via simple and inexpensive preparation methods have homogeneously distributed pores, structural stability, large specific surface area, and excellent dispersion of supported noble metals. The large specific surface area provides more active sites for catalysis, which can effectively improve the catalytic performance of the materials. The good dispersion of loaded noble metal nanoparticles maximizes the utilization of noble metals. Typical oxide-based porous catalytic materials and catalytic applications are as follows.

5.1 Nanoporous CuO loaded with Au nanoparticles

Ultrafine nanoporous CuO ribbons loaded by Au nanoparticles were prepared by dealloying melt-spun Al-Cu-Au alloys and calcinating in air [4]. The precursor ingots were obtained by melting alloys composed of pure Al, Cu, and Au. The ribbons were prepared by melt spinning at 33 m s^{-1} and protected by 0.05 MPa Ar. The as-quenched ribbons were immersed in aqueous 5 wt% NaOH at room temperature. The dealloyed samples were rinsed with reverse osmosis water and dehydrated alcohol. The dealloyed ribbons were calcined in a muffle furnace at 300–900°C for 1 h.

The formation mechanism of the nanoporous Au-CuO composites is as follows: after the Al-Cu-Au alloys are immersed in NaOH solutions, Al is preferentially etched, and the released Cu atoms in the dealloyed layer are rearranged to form a nanoporous structure. The Cu and Au atoms are released simultaneously because the Au atoms are mainly dissolved in the fcc Al. A Cu (Au) solid solution forms due to the existence of unlimited solid solution between Au and Cu according to the Cu-Au binary phase diagram. After the samples are calcined in air, Cu ligaments are oxidized to CuO, and the phase segregation occurs because the Au cannot be dissolved by CuO. During calcination, Au atoms diffuse to the surface of the CuO layers and form Au clusters. With the absorbed O atoms diffusing to the inner Cu and Cu_2O, the Au atoms dissolved in the Cu solution diffuse to the Au clusters and form Au particles, leading to a slight growth of Au particles. As a result, a portion of Au particles emboss from the surface of CuO ligaments, where the as-generated Au nanoparticles are partially embedded in CuO ligaments. Due to the unique porous structure, Au nanoparticles are immobilized on the surface of the ligaments, and the aggregation of particles can be effectively prevented.

The experimental [4] results indicate that the composites have large surface areas and high activity and stability. The CO conversion as a function of reaction temperature for the dealloyed $Al_{79.7}Cu_{20}Au_{0.3}$ ribbons calcined at different temperatures indicated that the CO conversion rate can reach 100.0% at 180°C for samples treated with the optimized process. The X-ray photoelectron spectroscopy (XPS) results indicate that the superior performance at low temperature is ascribed to the presence of Au^{+1} species and the interfacial interaction between Au nanoparticles and CuO ligaments. The binding energy of Cu 2p in the catalyst prepared from $Al_{79.7}Cu_{20}Au_{0.3}$ alloy shifts lower due to an interaction between Au and CuO_x. Notably, Au^{+1} is present on the surface of the CuO, which acts as an active site for CO oxidation and can lead to more activated CO than Au^0. The catalytic measurement showed that the valence of Au changes whereas the valence of Cu does not change, indicating that the Au state is vital to maintaining a high CO conversion rate and long-term stability. It is probable that Au and Au^{+1} sites located at the interface enhance the absorption of CO and O and that the presence of Cu^{2+} would result in high mobility and reactivity of surface lattice oxygen to participate in the reaction, which are favorable to the catalytic reaction of CO oxidation.

5.2 CeO_2 nanorod framework loaded with Au

CeO_2 nanorod frameworks (NFs) with the porous structure loaded with Au were prepared by dealloying melt-spun $Al_{89.7}Ce_{10}Au_{0.3}$ ribbons and calcination [14]. The preparation of alloy ribbons was the same as that in the Au-CuO system. The as-quenched ribbons were immersed in aqueous 20 wt% NaOH at a room temperature for 2 h and treated at 80°C for 10 h. The dealloyed ribbons were pretreated at 200–800°C for 2 h in pure O_2. After the dealloyed sample was calcined at 400°C in O_2, the XPS results demonstrated that the Au peaks slightly shifted lower due to the interaction of gold with oxygen vacancies. This phenomenon indicates that Au species interacted with the CeO_2 nanorods via charge transfer between Au and CeO_2. The $Au^{\delta+}/Au^0$ ratio of calcined Au-CeO_2 (0.39) was much higher than that of the dealloyed sample (0.21). Therefore, Au^0 and $Au^{\delta+}$ species coexisted in the calcined Au-CeO_2 due to strong interactions, which can effectively enhance the catalytic activity. The Au-CeO_2 nanorod catalyst calcined at 400°C exhibited much higher catalytic activity for CO oxidation than the dealloyed sample or pure CeO_2 nanorods [14]. Moreover, its complete reaction temperature was as low as 91°C. The designed Au-CeO_2 catalyst possessed extreme sintering resistance and exhibited high performance due to the enhanced interaction between the Au clusters/NPs and CeO_2 nanorods during calcination. The CeO_2 nanorod framework structure can be retained, and some Au nanoparticles supported on the nanorod CeO_2 surfaces did not evolve after catalytic testing.

The formation process of the Au-CeO_2 nanorod catalyst can be deduced as: during the formation of the porous CeO_x framework, the Au atoms may mix with the nanorods. After the nanorods are calcined in O_2, CeO_x is oxidized to CeO_2 NF, whereas the Au atoms diffuse to the surface of the nanorods and form NPs/clusters or nanoparticles, which are in situ supported and immobilized on the surface of the CeO_2 nanorods. The dispersion of Au NPs or clusters on CeO_2 NF could create many nanoscale contact interfaces and prevent NP/cluster sintering and migration of active Au species during calcination and catalytic reaction. This result was significantly different from those of other preparation methods, in which the NP aggregation typically leads to unexpected activity loss due to the collapse of the nanostructure during the annealing and catalytic processes.

5.3 Nanoporous Pt-TiO_2 nanocomposites

Nanoporous TiO_2 and Pt-TiO_2 composites were prepared by dealloying melt-spun Al-Si-Ti and Al-Si-Ti-Pt alloys, respectively [16]. The preparation of alloy ribbons was the same as in the Au-CuO system. The as-quenched ribbons were dealloyed in aqueous 10 wt% NaOH at 80°C. After the samples were dried in air, the ribbons were immersed in aqueous 3 wt% HCl at an ambient temperature for 7 h. The samples were then calcined at 400°C for 2 h. The mechanism of formation of the structure is as follows: Al and Si are etched away in the NaOH solution, and Ti and Pt are retained. During dissolution, the remaining Ti and Pt move freely along the interface between the alloy and the dissolution medium. The Ti reacts with NaOH to form insoluble Na-titanate (Ti + NaOH → Na-titanate). The Pt-Na-titanate reorganizes into a three-dimensional network exhibiting an open porosity. After acid treatment, there is no significant evolution in the microstructure morphology in the H-titanate (pt-Na-titanate + H^+ → Pt-H-titanate + Na^+). Nanoporous Pt-TiO_2 can be obtained through the calcination of Pt-H-titanate. Measurement of the structural parameters indicated that the nanoporous Pt-TiO_2 nanocomposites exhibit large specific surface areas, small pore diameters, and large pore volumes.

The performance measurements revealed that nanoporous TiO_2 showed favorable photocatalytic performance in the degradation of methyl orange (MO). Using Pt-TiO_2 nanocomposites, approximately 98% of the MO solution degraded in 50 minutes. The significant enhancement of the photocatalytic performance of Pt-TiO_2 was a result of the large specific surface area of approximately 98.06 $m^2\ g^{-1}$ and synergistic effect between TiO_2 and metallic Pt, which reduced the band gap of TiO_2 and expanded the absorption range. With a high work function, Pt enhances the Schottky barrier effect; the Schottky barrier formation at the interfaces of Pt-TiO_2 composites reduced the rate of electron-hole (e-h^+) pair recombination, which contributed to the improvement of the photocatalytic performance.

Noble metal-modified nanoporous oxides prepared by the dealloying method and their supported noble metals have the following advantages: (1) unique stable structure, which results in stability of the catalyst, (2) refined and homogeneous distribution of nanopores provides a larger specific surface area for the catalyst, exposing more active sites and effectively promoting catalytic performance, (3) loaded noble metals avoid agglomeration and evenly disperse on oxide surfaces or ligaments, which improves the availability of noble metals and enhances catalytic performance, and (4) the tight interaction between the supported noble metal and the oxide enhances the synergy between the noble metal and the oxide.

6. Conclusions

By the reasonable design of precursors and adoption of dealloying process, the nanoporous oxides, such as CuO, NiO, TiO_2, Co_3O_4, and CeO_2, noble-metal-based nanoporous composites, such as Ag ligaments loaded with CeO_2, TiO_2, ZrO_2, or NiO, and Pd ligaments loaded with TiO_2 or ZrO_2 could be obtained. Adding the noble into the melt-spun Al-M (M = Ce, Cu, and Ti) ribbons and dealloying in NaOH or KOH solutions, oxide-based nanoporous composites, such as Au loaded on CuO and CeO_2 or Pt loaded on TiO_2, have been achieved. The catalytic performances of the noble-based nanoporous composites, including the catalytic oxidation of methanol and ethanol, could be improved obviously comparing with single nanoporous nobles. The catalytic oxidation activities of CO for the oxide-based nanoporous composites have also been increased much. The rare earth elements could play an important role in the nanoporous materials. Sometimes, they could be the key metals for the nanoporous composites with high properties.

Acknowledgements

This work was supported by the National Natural Science Foundation of China (Grant no. 51371135, 51771141). The authors of this chapter thank G.J. Li, Y.Y. Song, and X.L. Zhang for their contributions.

Author details

Dong Duan, Haiyang Wang, Wenyu Shi and Zhanbo Sun*
School of Science, Xi'an Jiaotong University, Xi'an, P.R. China

*Address all correspondence to: szb@mail.xjtu.edu.cn

References

[1] Wittstock A, Zielasek V, Biener J, et al. Nanoporous gold catalysts for selective gas-phase oxidative coupling of methanol at low temperature. Science. 2010;**327**(5963):319-322. DOI: 10.1126/science.1183591

[2] Chen A, Holt-Hindle P. Platinum-based nanostructured materials: Synthesis, properties, and applications. Chemical Reviews. 2010;**110**(6):3767-3804. DOI: 10.1021/cr9003902

[3] Kou T, Jin C, Zhang C, et al. Nanoporous core–shell Cu@Cu_2O nanocomposites with superior photocatalytic properties towards the degradation of methyl orange. RSC Advances. 2012;**2**(33):12636-12643. DOI: 10.1039/c2ra21821f

[4] Zhang X, Li G, Yang S, Song X, Sun Z. Nanoporous CuO ribbons modified by Au nanoparticles through chemical dealloying and calcination for CO oxidation. Microporous and Mesoporous Materials. 2016;**226**:61-70. DOI: 10.1016/j.micromeso.2015.12.028

[5] Zhang X, Li K, Shi W, Wei C, Song X, Yang S, et al. Baize-like CeO_2 and NiO/CeO_2 nanorod catalysts prepared by dealloying for CO oxidation. Nanotechnology. 2017;**28**(4):045602. DOI: 10.1088/1361-6528/28/4/045602

[6] Zhang X, Li G, Song X, Yang S, Sun Z. Three-dimensional architecture of Ag/CeO_2 nanorod composites prepared by dealloying and their electrocatalytic performance. RSC Advances. 2017;**7**(52):32442-32451. DOI: 10.1039/c7ra04651k

[7] Tauster SJ, Fung SC, Garten RL. Strong metal-support interactions. Group 8 noble metals supported on titanium dioxide. Journal of the American Chemical Society. 1978;**100**(1):170-175. DOI: 10.1021/ja00469a029

[8] Guijing L, Feifei L, Xin W, Xiaoping S, Zhanbo S, Zhimao Y, et al. Nanoporous Ag-CeO_2 ribbons prepared by chemical dealloying and their electrocatalytic properties. Journal of Materials Chemistry A. 2013;**1**(16):4974-4981. DOI: 10.1039/c3ta01506h

[9] Li G, Sun Z, Zhang X, Song X, Sun Z, Feng W. Preparation of nanoporous Ag@TiO_2 ribbons through dealloying and their electrocatalytic properties. Journal of Solid State Electrochemistry. 2015;**19**(4):967-974. DOI: 10.1007/s10008-014-2702-x

[10] Zhang X, Wei C, Song Y, Song X, Sun Z. Nanoporous Ag-ZrO_2 composites prepared by chemical dealloying for borohydride electro-oxidation. International Journal of Hydrogen Energy. 2014;**39**(28):15646-15655. DOI: 10.1016/j.ijhydene.2014.07.102

[11] Song Y, Zhang X, Yang S, Wei X, Sun Z. Electrocatalytic performance for methanol oxidation on nanoporous Pd/NiO composites prepared by one-step dealloying. Fuel. 2016;**181**:269-276. DOI: 10.1016/j.fuel.2016.04.086

[12] Yanyan S, Caihua W, Xiaolong Z, Xin W, Xiaoping S, Zhanbo S. Nanoporous Pd/TiO_2 composites prepared by one-step dealloying and their electrocatalytic performance for methanol/ethanol oxidation. Materials Chemistry and Physics. 2015;**161**:153-161. DOI: 10.1016/j.matchemphys.2015.05.030

[13] Duan D, Hao C, Shi W, Wang H, Sun Z. Sm_2O_3/Co_3O_4 catalysts prepared by dealloying for low-temperature CO oxidation. RSC Advances. 2018;**8**(21):11289-11295. DOI: 10.1039/C8RA01219A

[14] Zhang X, Duan D, Li G, Feng W, Yang S, Sun Z. Monolithic Au/CeO_2

nanorod framework catalyst prepared by dealloying for low-temperature CO oxidation. Nanotechnology. 2018;**29**(9):095606. DOI: 10.1088/1361-6528/aaa726

[15] Li G, Song X, Lu F, Sun Z, Yang Z, Yang S, et al. Formation and control of nanoporous Ag through electrochemical dealloying of the melt-spun Cu-Ag-Ce alloys. Journal of Materials Research. 2012;**27**(12):1612-1620. DOI: 10.1557/jmr.2012.121

[16] Shi W, Song Y, Zhang X, Duan D, Wang H, Sun Z. Nanoporous Pt/TiO_2 nanocomposites with greatly enhanced photocatalytic performance. Journal of the Chinese Chemical Society. Nov. 5, 2018. DOI: 10.1002/jccs.201700251

[17] Xu C, Liu Y, Zhou C, Wang L, Geng PH, Ding PY. An in situ dealloying and oxidation route to Co_3O_4 nanosheets and their ambient-temperature CO oxidation activity. ChemCatChem. 2011;**3**:399-407. DOI: 10.1002/cctc.201000275

[18] Song Y, Duan D, Shi W, Wang H, Yang S, Sun Z. Promotion effects of ZrO_2 on mesoporous Pd prepared by a one-step dealloying method for methanol oxidation in an alkaline electrolyte. Journal of the Electrochemical Society. 2017;**164**:F1495-FF505. DOI: 10.1149/2.1821713jes

[19] Zhang X, Li G, Duan D, Wang H, Sun Z. Formation and control of nanoporous Pt ribbons by two-step dealloying for methanol electro-oxidation. Corrosion Science. 2018;**135**:57-66. DOI: 10.1016/j.corsci.2018.02.030

[20] Ganduglia-Pirovano MV, Hofmann A, Sauer J. Oxygen vacancies in transition metal and rare earth oxides: Current state of understanding and remaining challenges. Surface Science Reports. 2007;**62**:219-270. DOI: 10.1016/j.surfrep.2007.03.002

[21] Guo X, Zhou R. A new insight into morphology effect of ceria on CuO/CeO_2 catalysts for CO selective oxidation in hydrogen-rich gas. Catalysis Science & Technology. 2016;**6**:3862-3871. DOI: 10.1039/C5CY01816A

[22] Maitarad P, Han J, Zhang D, Shi L, Namuangruk S, Rungrotmongkol T. Structure–activity relationships of NiO on CeO_2 nanorods for the selective catalytic reduction of NO with NH_3: Experimental and DFT studies. Journal of Physical Chemistry C. 2014;**118**:9612-9620. DOI: 10.1021/jp5024845

[23] Lin Z, Li X, Yao Z, Chen Z, Mei H, Zhu R, et al. transition-metal doped ceria microspheres with nanoporous structures for CO oxidation. Scientific Reports. 2016;**6**:23900. DOI: 10.1038/srep23900

[24] Liu S, Yu J, Jaroniec M. Anatase TiO_2 with dominant high-energy {001} facets: Synthesis, properties, and applications. Chemistry of Materials. 2011;**23**(18):4085-4093. DOI: 10.1021/cm200597m

[25] Ong WJ, Tan LL, Chai SP, Yong ST, Mohamed AR. Highly reactive {001} facets of TiO_2-based composites: Synthesis, formation mechanism and characterization. Nanoscale. 2014;**6**:1946-2008. DOI: 10.1039/c3nr04655a

[26] Zhao W, Sun Y, Castellano FN. Visible-light induced water detoxification catalyzed by PtII dye sensitized titania. Journal of the American Chemical Society. 2008;**130**:12566-12567. DOI: 10.1021/ja803522v

[27] Li R, Chan KC, Liu XJ, Zhang XH, Liu L, Li T, et al. Synthesis of well-aligned CuO nanowire array integrated with nanoporous CuO network for oxidative degradation of methylene blue. Corrosion Science. 2017;**126**:37-43. DOI: 10.1016/j.corsci.2017.06.001

[28] Liang K, Tang X, Wei B, Hu W. Fabrication and characterization of a nanoporous NiO film with high specific energy and power via an electrochemical dealloying approach. Materials Research Bulletin. 2013;**48**:3829-3833. DOI: 10.1016/j.materresbull.2013.05.086

[29] Li S, Zhu H, Qin Z, Wang G, Zhang Y, Wu Z, et al. Morphologic effects of nano CeO_2—TiO_2 on the performance of Au/CeO_2—TiO_2 catalysts in low-temperature CO oxidation. Applied Catalysis B: Environmental. 2014;**144**:498-506. DOI: 10.1016/j.apcatb.2013.07.049

[30] Kim HY, Lee HM, Henkelman G. CO oxidation mechanism on CeO_2-supported Au nanoparticles. Journal of the American Chemical Society. 2012;**134**:1560-1570. DOI: 10.1021/ja207510v

[31] Yang S, Zhu W, Wang J, Chen Z. Catalytic wet air oxidation of phenol over CeO_2—TiO_2 catalyst in the batch reactor and the packed-bed reactor. Journal of Hazardous Materials. 2008;**153**:1248-1253. DOI: 10.1016/j.jhazmat.2007.09.084

[32] Huang PX, Wu F, Zhu BL, Gao XP, Zhu HY, Yan TY, et al. CeO_2 nanorods and gold nanocrystals supported on CeO_2 nanorods as catalyst. Journal of Physical Chemistry B. 2005;**109**:19169-19174. DOI: 10.1021/jp052978u

[33] Wang H, Liang M, Zhang X, Duan D, Shi W, Song Y, et al. Novel CeO_2 nanorod framework prepared by dealloying for supercapacitors applications. Ionics. 2018;**24**:2063-2072. DOI: 10.1007/s11581-018-2443-4

Chapter 3

Measurement of Heat and Mass Flow Characteristics of Nanofluid in a Porous Parallel-Plate Channel by Darcy-LTNE/LTE, Brinkman-LTNE/LTE Models

Mohsan Hassan

Abstract

In current study, fully developed flow of Al_2O_3-water nanofluid with forced convection heat transfer in channel is investigated with different models of heat and mass distribution. The channel is filled with porous media of open-celled Cu metal foams. The Darcy and Brinkman models are used for the mass flow; however, the heat transfer distribution is examined through the local thermal equilibrium (LTE) and the local thermal non-equilibrium (LTNE) models. Exact solutions for Darcy-LTE, Brinkman-LTE, Darcy-LTNE, and Brinkman-LTNE models are obtained. Temperature profiles by these different models are discussed under effect of nano-particle concentration and compare the profiles with each other.

Keywords: nanofluid, porous media, Darcy-LTE, Brinkman-LTE, Darcy-LTNE, Brinkman-LTNE models

1. Introduction

Studies on heat transfer in porous media have been augmented rapidly. The porous media likely filled with metallic foams with open cells, metallic sintered fiber felts, or metallic lattice frame structures have high thermal conductivity and have much importance in different engineering areas such as heat transport enhancement, thermal storage, solar thermal utilization, modeling of biological tissues, etc. There are two main models, the local thermal equilibrium (LTE) and the local thermal non-equilibrium (LTNE), which can be used to represent heat transfer phenomena in a porous medium. The LTE model can be used effectively to examine the heat transfer in porous media when the temperature difference between the fluid phase and that of the solid phase is small. But not in all situations the differences between temperatures are ignored. In these cases, the influence of the interfacial surface and interstitial convective heat transfer coefficient became major factors of heat exchange among the fluid and solid phases. In such cases, LTNE model needs to be utilized.

A number of researchers investigated heat transfer through LTE model in porous media that are made of low-conductive materials. Poulikakos and Kazmierczak [1] present a fully developed convective flow in a channel with partially filled porous matrix. They modeled the problem of heat and mass flow in the porous medium by using the Brinkman-LTE model to see the effect of parameters on the flow field and on the heat. Vafai and Kim [2] investigated the forced convection flow in a porous medium channel through Forchheimer-LTE model and found a significant increase in the rate of heat transfer as the inertia parameter increases especially for high- to medium-permeability porous media. Gong et al. [3] studied the influence of the thermal conductivity, permeability, and the porous material's thickness on heat transfer by using a Brinkman-LTE model in annular duct. Their obtained results display that heat transfer is improved by increasing either the thermal conductivity or the permeability. Cheng and Hsu [4] studied the heat and mass flow in an annulus enclosed with porous media by using Brinkman-LTE model under the influence of permeability and porosity. Mitrovic and Maletic [5] got the results for heat and mass flow in a parallel-plate porous channel by using the LTE model with respect to asymmetrical conditions. Sheikholeslami et al. [6] investigated the free convection flow in permeable enclosure by using Darcy-LTE model. In another study, he used the Darcy-LTE model for free convection flow in porous cavity [7].

In highly thermal conductive porous media, solid has usually higher thermal conductivity three to five times more than the fluid. In this case, the LTE model no longer satisfies the requirements of modeling. Then two-energy equation LTNE model is used to measure accurate temperature distribution in fluid and solid phases. Kuznetsov [8] got the results of heat transport phenomena in narrow annulus by using Darcy-LTNE model. Xu et al. [9] studied the Forchheimer-LTE/Forchheimer-LTNE models for heat transfer in pore channel having different thermal conductivities of fluid and solid phases and found the maximum heat distribution in the case of LTE model as compared to LTNE model. Lu et al. [10] described the force convection flow in cylinder fill with pore spume by using Brinkman-LTNE models. Zhao et al. [11] discussed the heat and mass flow through porous media in annulus by using Brinkman-LTNE. Ouyang et al. [12] briefly studied the heat transport phenomena in equidistant-plate channel fill with porous matrix. Xu et al. [13] investigated the mass flow by Brinkman model and LTNE model for heat transfer in equidistant plates filled with mini-spume. Shaikh and Memon [14] provided the numerical results for heat transport in round pipe with or without pore medium by using Darcy-Brinkman-Forchheimer models along LTNE model. Sheikholeslami and Houman [15] reported the transportation of fluid inside a porous cavity through LTNE model.

An important heat transfer fluid is nanofluid that is used in industries because of high rates of heat transfer. The main purpose of nanofluids is to achieve great enhancement in thermal or rheological properties. In a continuation of achieving better thermal conductivity and thermal performance of nanomaterial, many studies have been done [16–20]. Recently, Sheikholeslami [21] demonstrated the nanofluid flow in a porous enclosure by Darcy law model. In another study, he investigated the nanofluid flow in a porous media through non-Darcy law model [22]. In current study, keeping in mind these thermal properties of nanofluid, its fully developed convective flow in a parallel-plate channel filled with highly porous media is investigated. For thermal transport in porous media, LTE and LTNE models are utilized with Darcy and Brinkman models. In addition, the temperature profiles under nanoparticle concentration for both different approaches of heat transfer with constant heat flux at wall are calculated and compared.

2. Modeling and formulation

Consider the fully developed forced convection flow of nanofluid through a parallel-plate porous channel filled with open-celled metallic foams. The schematic diagram of the channel is shown in **Figure 1**.

Two infinite plates with height $2H$ are exposed by constant heat flux q_w. In this problem, Darcy-LTE, Brinkman-LTE, Darcy-LTNE, and Brinkman-LTNE models are employed for mass flow and heat transfer process in porous media with hydraulically and thermally fully developed conditions.

For fully developed fluid flow, momentum equation with Brinkman term is

$$0 = -\frac{dp}{dx} + \frac{\mu_e}{\varepsilon}\frac{d^2u}{dy^2} - \frac{\mu_{nf}}{K}u. \tag{1}$$

Here, u is the velocity, μ_e is the effective viscosity, μ_{nf} is the viscosity of nanofluid, K is the permeability, and ε is the porosity.

Temperature distribution in porous media owns two basic models LTE and LTNE. The LTE model containing one energy equation that treats the same value of temperature for fluid and solid phases is given in Eq. (2), while the LTNE model having two-energy equations that treats the different values of the temperatures for solid and fluid phases is shown in Eqs. (3) and (4):

$$\left(\rho C_p\right)_e u\frac{\partial T_f}{\partial x} = k_e\frac{\partial^2 T_f}{\partial y^2}, \tag{2}$$

$$0 = k_{se}\frac{\partial^2 T_s}{\partial y^2} - hA\left(T_s - T_f\right), \tag{3}$$

$$\left(\rho C_p\right)_e u\frac{\partial T_f}{\partial x} = k_{fe}\frac{\partial^2 T_f}{\partial y^2} + hA\left(T_s - T_f\right). \tag{4}$$

In the above equation, T_f is the temperature of fluid phase, T_s is the temperature of solid phase, h is the heat transfer coefficient, A is the specific surface area, ρC_p is

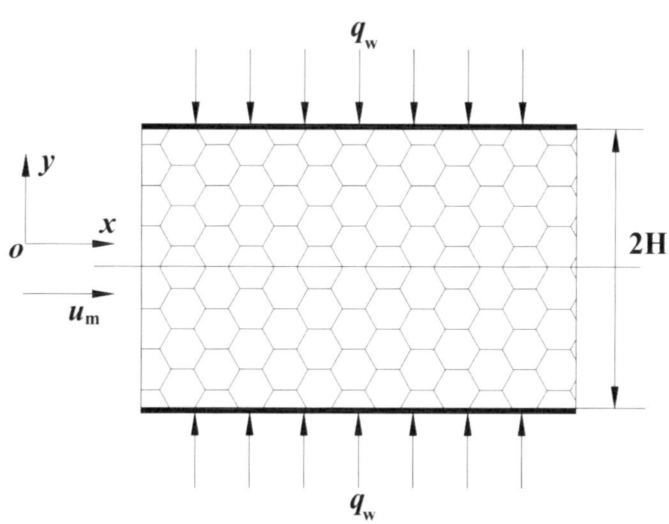

Figure 1.
Geometry of problem.

the heat capacity, k_e is the effective thermal conductivity, k_{fe} is the fluid phase's thermal conductivity, and k_{se} is the solid phase's thermal conductivity.

For corresponding boundary conditions, the temperature of the solid and the fluid at the wall interface will be the same, and velocity is considered to be zero:

$$\begin{aligned} y = H: \quad u = 0, T_f = T_s = T_w, \\ y = -H: \quad u = 0, T_f = T_s = T_w, \end{aligned} \tag{5}$$

where T_w is the temperature at interface.

The total heat flux q_w is shared among the solid and fluid phases' subject to their temperature gradients and effective conductivities at the wall:

$$\begin{aligned} y = H: \quad k_{fe}\frac{\partial T_f}{\partial y} + k_{se}\frac{\partial T_s}{\partial y} = q_w, \\ y = -H: \quad k_{fe}\frac{\partial T_f}{\partial y} + k_{se}\frac{\partial T_s}{\partial y} = q_w. \end{aligned} \tag{6}$$

The boundary condition cross-ponding to symmetry condition can be used as

$$y = 0: \quad \frac{\partial u}{\partial y} = 0, \frac{\partial T_f}{\partial y} = \frac{\partial T_f}{\partial y} = 0 \tag{7}$$

In obtaining the analytical solutions of the above governing equations, the following dimensionless parameters are employed:

$$\begin{aligned} Y = \frac{y}{H}, U = \frac{u}{u_m}, P = \frac{K}{\mu_f u_m}\frac{dp}{dx}, \theta_s = \frac{T_s - T_w}{q_w H/k_s}, \\ \theta_f = \frac{T_f - T_w}{q_w H/k_s}, s = \frac{K}{H^2}, C = \frac{k_f}{k_s}, D = \frac{hAH^2}{k_s}. \end{aligned} \tag{8}$$

2.1 Darcy-LTNE model

In this case, Darcy model for fluid flow and LTNE model for temperature distribution in solid and fluid phases are utilized. In the Darcy model, the velocity distribution is taken to be uniformed. So, the energy equations for solid and fluid phases are dimensionless as

$$\frac{k_{se}}{k_s}\frac{d^2\theta_s}{dY^2} - D(\theta_s - \theta_f) = 0, \tag{9}$$

$$C\frac{k_{fe}}{k_f}\frac{d^2\theta_f}{dY^2} + D(\theta_s - \theta_f) = 1. \tag{10}$$

2.2 Darcy-LTE model

In this case, the energy equation for the LTE model is normalized as

$$C\frac{k_e}{k_f}\frac{d^2\theta_f}{dY^2} = 1 \tag{11}$$

2.3 Brinkman-LTNE model

In this part, the Brinkman and LTNE models for heat and mass transfer are utilized. The dimensionless governing equations are obtained as

$$\frac{1}{\varepsilon}\frac{\mu_e}{\mu_f}\frac{d^2U}{dY^2} - s^2\left(\frac{\mu_{nf}}{\mu_f}U + P\right) = 1, \tag{12}$$

$$\frac{k_{se}}{k_s}\frac{d^2\theta_s}{dY^2} - D(\theta_s - \theta_f) = 0, \tag{13}$$

$$C\frac{k_{fe}}{k_f}\frac{d^2\theta_f}{dY^2} + D(\theta_s - \theta_f) = U. \tag{14}$$

2.4 Brinkman-LTE model

For the Brinkman and LTE models, governing equations in dimensionless form are

$$\frac{1}{\varepsilon}\frac{\mu_e}{\mu_f}\frac{d^2U}{dY^2} - s^2\left(\frac{\mu_{nf}}{\mu_f}U + P\right) = 1, \tag{15}$$

$$C\frac{k_e}{k_f}\frac{d^2\theta_f}{dY^2} = U. \tag{16}$$

The dimensionless boundary conditions are

$$\begin{aligned} Y = 1: &\quad \mathrm{U} = 0, \theta_f = \theta_s = 1, \\ y = 0: &\quad \frac{dU}{dY} = 0, \frac{d\theta_f}{dY} = \frac{d\theta_s}{dY} = 0, \\ y = -1: &\quad \mathrm{U} = 0, \theta_f = \theta_s = 1. \end{aligned} \tag{17}$$

2.5 Physical properties

In the above equations, the effective viscosity μ_e is defined as

$$\mu_e = (1 + 2.5\varepsilon)\mu_{nf}, \tag{18}$$

where

$$\mu_{nf} = \frac{\mu_f}{(1-\phi)^{2.5}}. \tag{19}$$

In the above equation, ϕ is the nanoparticle volume fraction and μ_f is the viscosity of base fluid.

Since the heat is transferred via nanofluid in porous media, the effective thermal conductivity is as follows:

$$k_e = \frac{k_m k_{nf}}{\varepsilon k_m + (1-\varepsilon)k_{nf}}, \tag{20}$$

$$k_{fe} = \varepsilon k_{nf}, \tag{21}$$

$$k_{se} = (1-\varepsilon)k_{nf}. \tag{22}$$

In the above equation, k_{nf} is given as

$$k_{nf} = \frac{k_p + 2k_f + 2(k_p - k_f)\phi}{k_p + 2k_f - (k_p - k_f)\phi} k_f, \tag{23}$$

where k_f is the thermal conductivity of fluid, k_p is the thermal conductivity of nanoparticle, and k_m is the thermal conductivity of medium.

3. Results and discussion

The behavior of nanoparticle concentration on temperature distributions in solid and fluid phases is displayed in section. For fluid phase, consider the nanofluid which is repaired by water and alumina oxide nanoparticles. The porous medium is taken as solid phase that is made by open-celled copper metallic foams. The governing equations for physical problem are demonstrated by taking Brinkman-LTNE/Brinkman-LTE and Darcy-LTNE/Darcy-LTE models and converted into non-dimensionless form to find its exact solution. The exact solutions of these equations are obtained by using computational software Mathematica 9. To see the effects of nanoparticle concentration, the values of other embedding parameters are taken to be fixed as $\varepsilon = 0.7$, $s = 0.64$, $C = 0.0015$, and $D = 1$.

The influence of nanoparticle concentration on the temperature profiles of fluid and solid phases for Brinkman-LTNE and Darcy-LTNE models is shown in **Figure 2**. It shows that the temperature profiles of fluid and solid are increased due to improvement in thermal physical properties of fluids especially thermal conductivity through increasing the concentrations of nanoparticles. Here negative sign shows that fluid transfers the heat to the wall. In this regard, temperature profiles of solid phase in both models are increased because of increasing the temperature of fluid phase by nanoparticle concentrations. The temperature variation for Brinkman-LTNE and Darcy-LTNE models has similar trend, but the temperature profile of Brinkman-LTNE models is found maximum as compared to the Darcy-LTNE models.

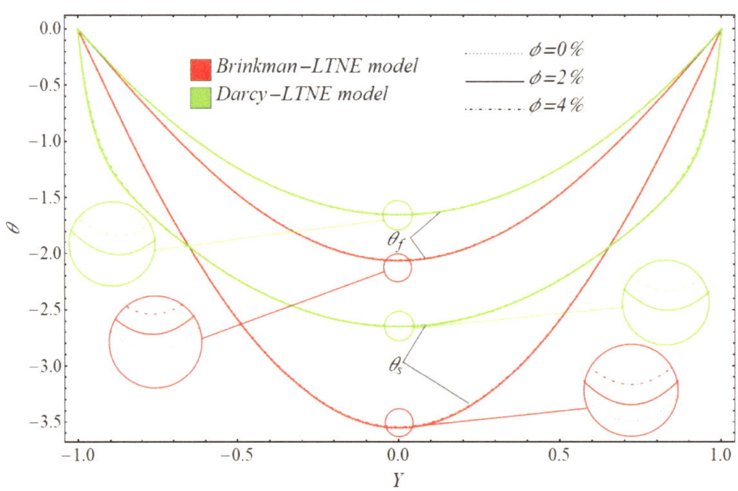

Figure 2.
Effect of nanoparticle concentrations on the temperature profiles of Brinkman-LTNE and Darcy-LTNE models.

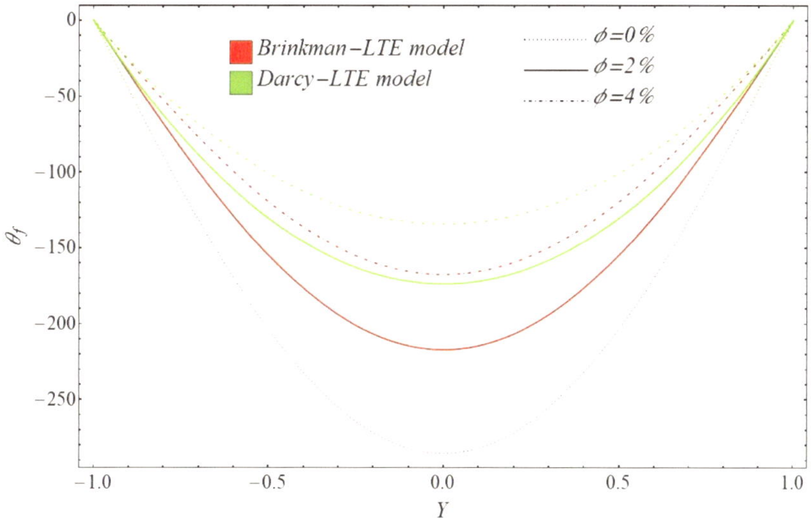

Figure 3.
Effect of nanoparticle concentrations on the temperature profile of Brinkman-LTE and Darcy-LTE models.

The temperature profiles for Brinkman-LTE and Darcy-LTE models with effects of nanoparticle concentration are displayed in **Figure 3**. In **Figure 3**, it is seen that the temperature profiles for both models are amplified by increasing the nanoparticle concentrations. In comparison of models, it is noted that the effects of nanoparticle concentrations are dominant in the case of LTE as compared to LTNE. Moreover, the temperature at wall for LTE and LTNE models is the same but with maximum boost in the case of LTNE models at center. Moreover, the heat transfer with LTNE approach is smaller as compared to LTE approach because of thermal resistance due to solid phase.

4. Conclusions

A fully developed heat and mass flow of Al_2O_3-water nanofluid in a parallel-plate channel filled with porous media of Cu material is investigated by using Darcy-LTE/Darcy-LTNE and Brinkman-LTE/Brinkman-LTNE models. It is found that the distribution of temperature is improved in both approaches of heat transfer by using nanofluid. But temperature distribution is overestimated in LTE approach as compared to LTNE approach. This overestimate results are due to neglecting the difference between thermal conductivities of fluid and solid phase.

Nomenclature

u	velocity
μ_e	effective viscosity
μ_{nf}	viscosity of nanofluid
μ_f	viscosity of base fluid
ε	porosity
T_f	temperature of fluid
T_s	temperature of solid phase
T_w	temperature at interface

ρC_p	heat capacity
ϕ	nanoparticle volume fraction
K	permeability
k_e	effective thermal conductivity
k_m	thermal conductivity of medium
k_{fe}	thermal conductivity of fluid phase
k_{se}	thermal conductivity of solid phase
k_p	thermal conductivity of nanoparticle
k_f	thermal conductivity of base fluid
q_w	heat flux
h	heat transfer coefficient
H	height
A	specific surface area

Author details

Mohsan Hassan
Department of Mathematics, CUI, Lahore, Pakistan

*Address all correspondence to: mohsan.hassan@cuilahore.edu.pk

References

[1] Poulikakos D, Kazmierczak M. Forced convection in a duct partially filled with a porous material. Journal of Heat Transfer. 1987;**109**:653-662

[2] Vafai K, Kim SJ. Forced convection in a channel filled with a porous medium: An exact solution. Journal of Heat Transfer. 1989;**111**:1103-1106

[3] Gong W, Han J, Cheng F. Theoretical study of heat transfer enhancement in pipe with porous body. Journal of Hydrodynamics, Series B. 2001;**13**: 111-116

[4] Cheng P, Hsu CT. Fully-developed, forced convective flow through an annular packed-sphere bed with wall effects. International Journal of Heat and Mass Transfer. 1986;**29**:1843-1853

[5] Mitrovic J, Maletic B. Effect of thermal asymmetry on laminar forced convection heat transfer in a porous annular channel. Chemical Engineering and Technology. 2006;**29**:750-760

[6] Sheikholeslami M, Shahzad SA, Li Z, Ahmad S. Numerical modeling for alumina nanofluid magnetohydrodynamic convective heat transfer in a permeable medium using Darcy law. International Journal of Heat and Mass Transfer. 2018; **127**:614-622

[7] Sheikholeslami M. CuO-water nanofluid free convection in a porous cavity considering Darcy law. European Physical Journal Plus. 2017;**132**:55-66

[8] Kuznetsov AV. Analysis of a non-thermal equilibrium fluid flow in a concentric tube annulus filled with a porous medium. International Communications in Heat and Mass Transfer. 1996;**23**:929-938

[9] Xu HJ, Gong L, Zhao CY, Yang YH, Xu ZG. Analytical considerations of local thermal non-equilibrium conditions for thermal transport in metal foams. International Journal of Thermal Sciences. 2015;**95**:73-87

[10] Lu W, Zhao CY, Tassou SA. Thermal analysis on metal-foam filled heat exchangers. Part I: Metal-foam filled pipes. International Journal of Heat and Mass Transfer. 2006;**49**: 2751-2761

[11] Zhao CY, Lu W, Tassou SA. Thermal analysis on metal-foam filled heat exchangers. Part II: Tube heat exchangers. International Journal of Heat and Mass Transfer. 2006;**49**: 2762-2770

[12] Ouyang XL, Vafai K, Jiang PX. Analysis of thermally developing flow in porous media under local thermal non-equilibrium conditions. International Journal of Heat and Mass Transfer. 2013;**67**:768-775

[13] Xu HJ, Zhao CY, Xu ZG. Analytical considerations of slip flow and heat transfer through microfoams in mini/microchannels with asymmetric wall heat fluxes. Applied Thermal Engineering. 2016;**93**:15-26

[14] Shaikh AW, Memon GQ. Analytical and numerical solutions of fluid flow filled with and without porous media in circular pipes. Applied Mathematics and Computation. 2014;**232**:983-999

[15] Sheikholeslami M, Houman BR. CVFEM for effect of Lorentz forces on nanofluid flow in a porous complex shaped enclosure by means of non-equilibrium model. Journal of Molecular Liquids. 2018;**254**:446-462

[16] Hassan M. Impact of iron oxide particles concentration under a highly oscillating magnetic field on ferrofluid

flow. European Physical Journal Plus. 2018;**133**:230-244

[17] Hassan M, Marin M, Ellahi R, Alamri SZ. Exploration of convective heat transfer and flow characteristics synthesis by Cu-Ag/Water hybrid-nanofluids. Heat Transfer Research. 2018;**49**:1837-1848

[18] Hassan M, Marin M, Alsharif A, Ellahi R. Convective heat transfer flow of nanofluid in a porous medium over wavy surface. Physics Letters A. 2018; **382**:2749-2753

[19] Sheikholeslami M, Ellahi R, Hassan M, Soleimani A. A study of natural convection heat transfer in a nanofluid filled enclosure with elliptic inner cylinder. International Journal of Numerical Methods for Heat and Fluid Flow. 2014;**24**:1906-1927

[20] Sheikholeslami M, Ganji DD. Application of Nanofluid for Heat Transfer Enhancement. Elsevier Science; 2017. DOI: org/10.1016/B978-0-08-102172-9.00016-2

[21] Sheikholeslami M. Numerical approach for MHD Al_2O_3-water nanofluid transportation inside a permeable medium using innovative computer method. Computer Methods in Applied Mechanics and Engineering. 2019;**344**:306-318

[22] Sheikholeslami M. New computational approach for exergy and entropy analysis of nanofluid under the impact of Lorentz force through a porous media. Computer Methods in Applied Mechanics and Engineering. 2019;**344**:319-333

Chapter 4

Magneto-Sensitive Smart Materials and Magnetorheological Mechanism

Yangguang Xu, Guojiang Liao and Taixiang Liu

Abstract

Magneto-sensitive smart materials, also named as magnetorheological (MR) materials, are a class of smart composites prepared by dispersing nanometer- or micrometer-sized ferromagnetic fillers into the different carrier matrix. As the rheological properties can be controlled by an external magnetic field rapidly, reversibly, and continuously, magneto-sensitive smart materials have great application potential in construction, automotive industry, artificial intelligence, etc. In this chapter, a brief history and classification of magneto-sensitive smart materials are firstly summarized. Next, we discuss the state-of-the-art of the magnetorheological mechanism through experimental and theoretical studies, respectively. Finally, the prospect for this material in the future is presented.

Keywords: smart soft material, magnetorheological material, magnetorheology, magnetic dipole theory, viscoelasticity

1. Introduction

Most smart materials imitate natural biological materials, which can respond to the stimuli (like mechanical, thermal, electrical, photic, acoustic, magnetic, chemical, etc.) by changing one or multiple properties to adapt to the changing environment [1]. So far, bioinspired smart materials have become an important research direction in material science. It is difficult for a homogenous material to possess multifunction, so it generally combines the materials with the functions of perception, actuation, control, etc., together in a specific way to design a novel composite with multiple characteristics. The smart material is multilevel with different components; each component has different characteristics and microstructures, and the coupling effect exists between different components, and these components make the smart material show complicated responses to external stimuli. Generally, the adapting ability of smart material to external environment, which is similar to the activated function of biological material, can be dynamically adjusted through the transportation of substance and energy.

Magnetorheological (MR) materials can be regarded as a kind of bioinspired smart materials because their viscoelastic properties can be easily adjusted by an external magnetic field. Magnetorheological materials can be generally classified into MR fluids, MR elastomers, and MR gels according to the type of the carrier matrix and the physical state in the absence of magnetic field [2–4]. The MR fluid-based practical devices mainly include damper, buffer, clutch, artificial muscle, and so on [5]. Some application examples of MR fluid working under different modes can be found in various dampers, which take MR fluid as working medium. Besides, MR fluid has also found wide application potential in the fields of thermal conduction [6], sound transmission [7], precision machining [8], and biomedicine [9]. The applications of MR elastomer have been widely reported in the fields of adaptive tuned vibration absorber [10], impact absorber [11], active noise abatement barrier system [12], vibration isolator [13], sensor [14–16], and so on. MR elastomer mainly works by changing its modulus through magnetic field during pre-yield stage, which shows different working mechanism with MR fluid (MR fluid mainly works at post-yield stage). The examples for the application of MR gel are relatively less in comparison with MR fluid or MR elastomer, but the unique magneto-induced phenomena have attracted more and more attentions; some work concentrating on the potential applications of MR gel have been reported [17–20].

Here, we will give a brief introduction to different kinds of MR materials, discuss the related experimental and theoretical work when investigating the MR mechanism, and finally propose some future prospects for these magneto-sensitive smart materials aiming at practical applications.

2. The development of magnetorheological materials

2.1 Magnetorheological fluid

Magnetorheological (MR) fluid is the earliest developed magneto-sensitive smart material, which is a particulate suspension by mixing micrometer-sized ferromagnetic fillers, non-magnetic fluid, and some additives together. After applying an external magnetic field, the MR fluid will change from Newtonian-like fluid to semi-solid material quickly (within several microseconds) [21], as shown in **Figure 1a** [22]. The randomly dispersed magnetic fillers are rearranged to form chain-like ordered microstructure through the magnetic interaction. Moreover, the ordered degree of the microstructure is relevant with the magnetic field strength. That is, a stronger magnetic field will induce a more ordered chain-like microstructure parallel with the direction of external magnetic field (**Figure 1b**) [23].

The rheological measurements indicate that MR fluid shows Bingham fluidic behavior under magnetic field, an obvious yield stress exists, and the maximum of yield stress has exceeded 100 kPa [24]. Moreover, the apparent viscosity of MR fluid changes typically 3–4 orders of magnitude by changing the magnetic field strength, presenting a typical MR effect. However, the sedimentation problem due to the density mismatch between carrier matrix and ferromagnetic fillers become one of the bottlenecks to hinder the development of MR fluid. To solve the problems of particle sedimentation and the re-dispersion after the particle aggregation in the carrier matrix, many effective methods have been proposed. Generally, these methods can be classified into two main categories from the aspects of filler and carrier matrix.

Carbonyl iron particle is an ideal candidate to prepare MR fluid due to its low coercive force and high saturation magnetization ($\mu_0 M_s$ = 2.1 T). However, a serious sedimentation problem exists in the MR fluid with carbonyl iron particle. Sometimes, it has to abandon carbonyl iron particle with excellent magnetic property and choose some other magnetic material which could improve the sedimentation of MR fluid for the balance between performance and stability.

Decreasing the size of magnetic fillers is an effective method to improve the stability of MR fluids. Submicron- or nanometer-sized particle shows better

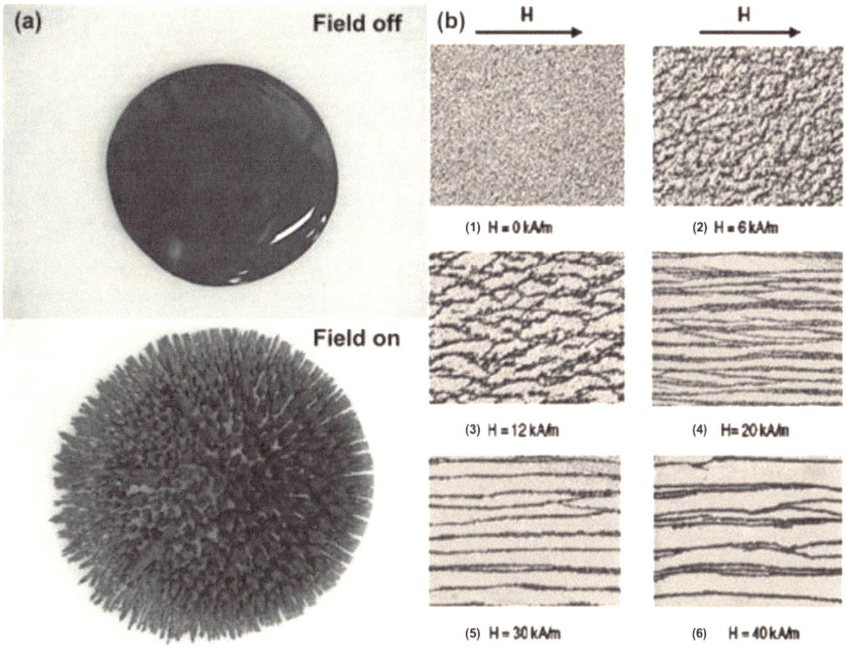

Figure 1.
The images of MR fluid without and with magnetic field [22] (a). The microstructures of MR fluid with different magnetic fields [23] (b).

stability in the non-magnetic carrier matrix in comparison with the micrometer-sized particle because of the Brownian movement and Van der Waals force. Especially, the particulate suspension totally using nanometer-sized ferromagnetic fillers as dispersed phase is named as magnetic fluids or ferrofluids [25–32]. Each ferromagnetic filler in ferrofluids includes single magnetic domain and could disperse in the carrier matrix randomly due to the Brownian movement, so the sedimentation is greatly improved. However, the ferrofluids keep their liquid-like state even under a strong magnetic field, which indicates that it is impossible for ferrofluids to have a high magneto-induced yield stress. In addition, the particle aggregation caused by the particle's nanometer size (or the re-dispersion of particles after removing the magnetic field) is another challenge to promote the further development of ferrofluids [33]. Using the magnetic material with the shape of rod-like or fibroid is regarded as another effective method to improve the stability and MR performance of MR fluids [34].

Except for changing the size and shape of ferromagnetic fillers, the core-shell structured ferromagnetic particle, whose surface is chemically modified by various polymers, is also a good candidate. The core-shell structure not only decreases the density of the particle but also increases the static electrical repulsion between adjacent particles, so the stability and the redispersibility of MR fluid are evidently improved. Moreover, the antioxidation is improved because the surface of the particle is wrapped by polymer [35]. **Figure 2** shows the SEM images of carbonyl irons wrapped by various polymers, which were mainly made by Choi's research group at Inha University.

It is a new approach to balancing the stability and performance of MR fluid by dispersing the submicron- or nanometer-sized ferromagnetic fillers (such as nanotube, submicron- or nanometer-sized carbonyl iron particle, wrought monox, organic clay, clavate ferromagnetic Co-γ-Fe_2O_3, and CrO_2) into conventional MR fluid to change the property of continuous phase (i.e., the matrix) [36–39]. Both of

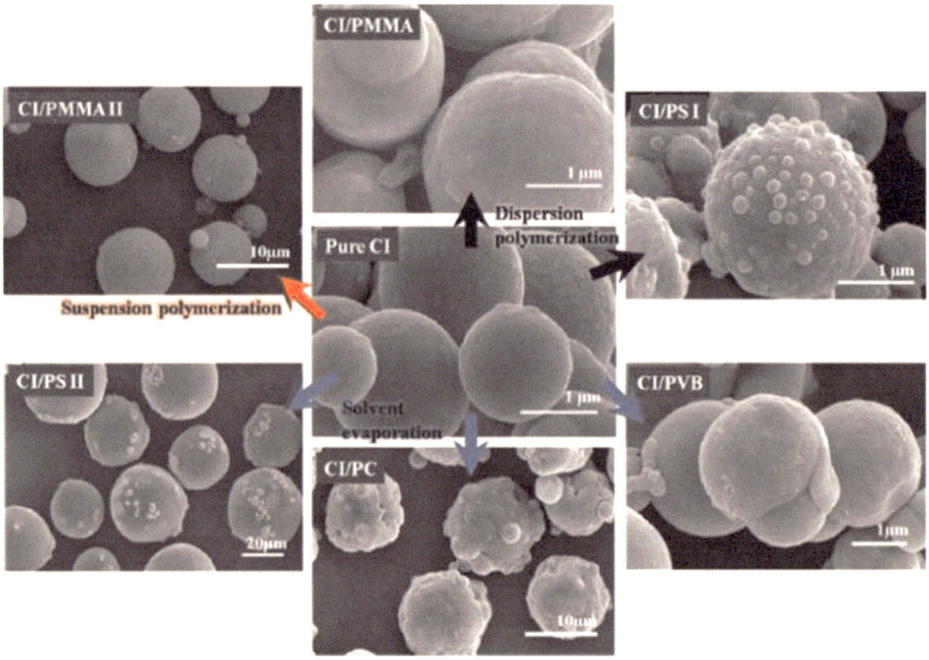

Figure 2.
Pure carbonyl iron and the carbonyl iron wrapped by various polymers [35].

the sedimentation rates of the dispersed phase and the MR effect are improved significantly due to the existence of ion. However, the modification to the carrier matrix, together with the modification to the dispersed particles, could only improve the sedimentation of MR fluid to some extent. To solve the particle sedimentation completely, the non-magnetic liquid matrix can be totally substituted by rubber-like polymer matrix. Thereout, a new magneto-sensitive soft material, that is, MR elastomer, appears later.

2.2 Magnetorheological elastomer

MR elastomer presents solid state even without magnetic field; it inherits the magneto-sensitive feature of MR fluids, but its working principle and application field are quite different from MR fluid. The magnetic fillers are fixed in the polymer matrix after the MR elastomer is prepared, the particles cannot move freely even exposed in a magnetic field, and no "phase transition" appears like happened in MR fluid. MR elastomer carries out the intelligent control mainly through changing its damping and modulus by magnetic field before yield [40]. The ordered chain-like (or column-like) microstructure aligned parallel to the direction of magnetic field generates if an external magnetic field is applied during the vulcanizing process of the polymer matrix. After vulcanization, the ordered microstructure can be solidified in the matrix, and the anisotropic MR elastomer is obtained.

Figure 3 shows the SEM images of MR elastomers pretreated by different magnetic fields. It is clear that chain-like structured anisotropic MR elastomer can be prepared if it is exposed under an external magnetic field during the pre-configuration process. The structure of particle chains can be further adjusted by the magnetic field. The ferromagnetic fillers aggregate more easily under a strong magnetic field, resulting in a higher degree of anisotropy of MR elastomer. Further magnetorheological characterization indicates that the MR elastomer with higher

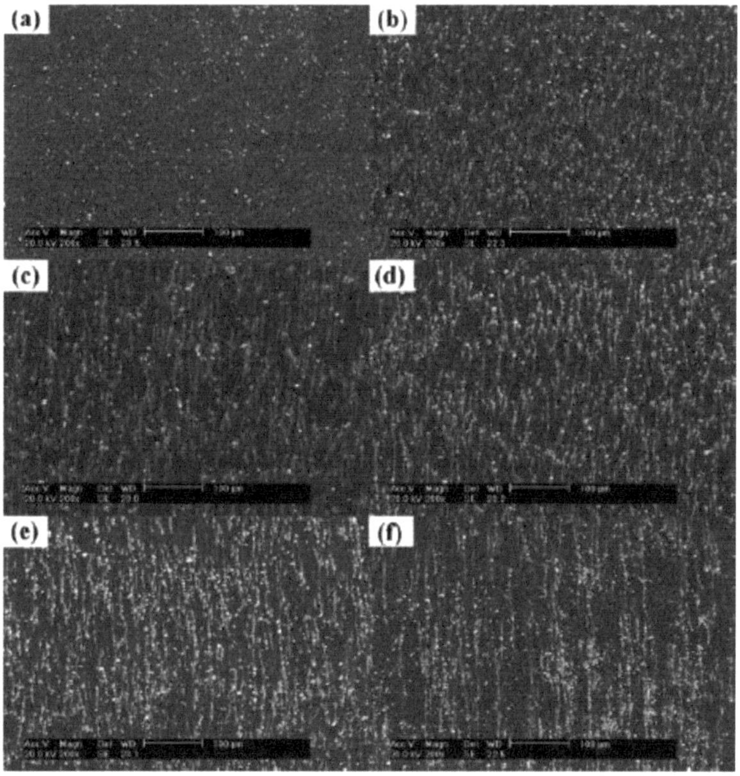

Figure 3.
The SEM images of MR elastomer pretreated by different magnetic fields [41]: (a) 0 mT; (b) 200 mT; (c) 400 mT; (d) 600 mT; (e) 800 mT; and (f) 1000 mT.

degree of anisotropy also possesses high MR effect [41]. So far, many kinds of rubber, such as natural rubber [41], silicon rubber [42], polyurethane rubber [43], cis-polybutadiene rubber [44], nitrile butadiene rubber [45], and thermoplastic elastomer blended by polypropylene and SEBS [46], have been chosen as the candidates for fabricating the MR elastomer.

Although the particle sedimentation is completely solved, the magneto-induced effect and MR effect constrain each other, making the requirement of engineering application not easy to be achieved. Moreover, MR elastomer loses the feature that the particulate microstructure can be easily controlled by magnetic field as like in MR fluid because the ferromagnetic fillers are fixed in the rubber matrix, which indicates that it loses the magneto-controllable flexibility. To pursue a stable MR material with a high MR effect and a strong magneto-induced effect aiming to the engineering application, it needs to abandon the conventional fabricating solutions and redesign a new magneto-sensitive smart material.

2.3 Magnetorheological gel

MR gel is another magneto-sensitive smart material whose continuous phase is viscoelastic. The inelastic matrix makes MR gel present typical viscoelastic characteristics even without external magnetic field. Shiga et al. firstly proposed the concept of MR gel in 1995 [47], and they prepared a series of MR gels with different particle contents by dispersing ferromagnetic fillers into silicone gel and investigated the magneto-dependent viscoelastic behaviors and microstructures of MR gel.

The MR gels can be further classified into liquid-like and solid-like MR gels according to their physical state without magnetic field.

The liquid-like MR gels can be regarded as a kind of special MR fluid with a little mount of polymer solution as additive. The polymer additive can generate a network structure in the matrix to modify the interfacial characteristic of ferromagnetic fillers and their interaction, greatly slowing down the sedimentation velocity of ferromagnetic fillers [48–50]. In comparison with MR fluids, except for improving the sedimentation stability, the off-state viscosity and yield stress of liquid-like MR gels can also be adjusted by adding a certain amount of polymer additives. However, the polymer network increases the moving resistance of ferromagnetic fillers in the matrix, and the response of MR gels to external magnetic field is then decreased accordingly. In addition, the sedimentation problem in the liquid-like MR gels is not completely solved.

Strictly speaking, the magnetic gel reported by Shiga et al. is a kind of solid-like MR gels [47]. The most distinct characteristic of solid-like MR gels is that there is no particle sedimentation problem exists, like MR elastomer. Yet the matrix is not the rubber-like elastic material, the solid-like MR gel cannot be classified into MR elastomer. Recently, these solid-like MR gels have been paid more and more attention [51–57]. A novel solid-like MR gel by mixing micrometer-sized magnetic particles and plasticine-like polyurethane was reported by Xu et al. [58]. As **Figure 4a** shows, this material presents like plasticine without magnetic field, can be changed into any shapes, and remains the status of plastic deformation, so it is named as MR plastomer. MR plastomer deforms along with the direction of the applied external magnetic field (**Figure 4b**). Further microstructure characterization indicates that the randomly dispersed magnetic particles can rearrange to generate chain-like (or column-like) orientated microstructure driven by magnetic force

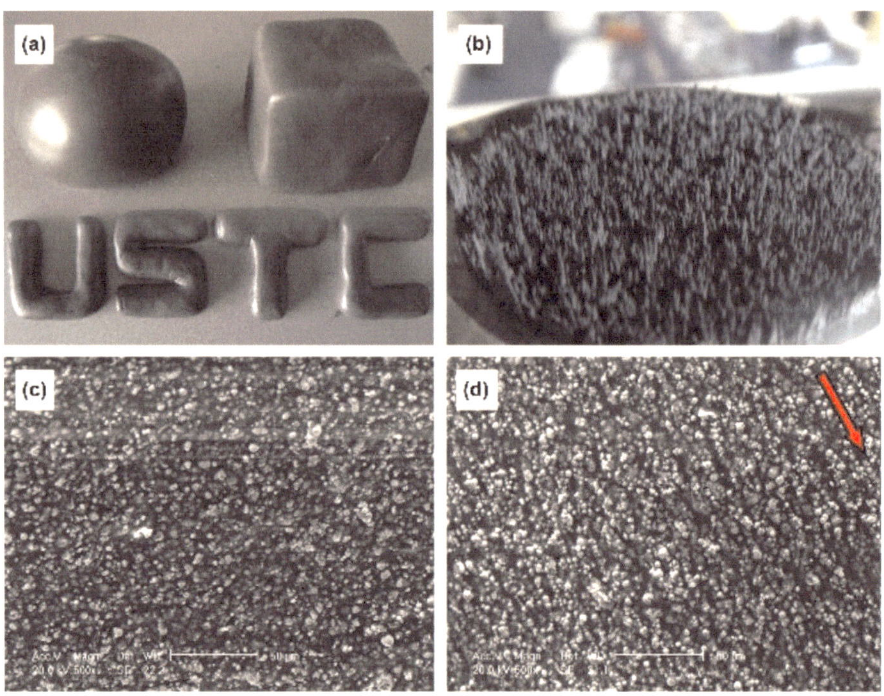

Figure 4.
The images of MR plastomer without (a) and with (b) magnetic field [58]. The SEM images of MR plastomer without (c) and with (d) magnetic field (the direction of magnetic field is marked by the red arrow) [59].

(**Figure 4c** and **d**). In addition, the anisotropic particulate microstructure is kept even the external magnetic field is removed [59].

In one word, an orientated particulate microstructure in the solid-like MR gel can be adjusted through an external magnetic field. Meanwhile, the chain-like (or column-like) microstructure can be fixed in the soft matrix after removing the magnetic field. This unique feature makes solid-like MR gel process the merits existed on MR fluid and MR elastomer (i.e., movability of particles and the "frozen" property of orientated microstructure) at the same time, which is significant to investigate the magneto-mechanical coupling mechanism.

2.4 Other magnetorheological materials

Some other special MR materials, which cannot be simply classified into the most well-known MR materials as mentioned above, were reported in the literature. MR foam is a kind of solid-like polymer composite by pouring MR fluid into porous polymer foam (**Figure 5**); the modulus of the MR foam can be controlled by changing the rheological property of MR fluid through a magnetic field [60–62]. Due to the special porous microstructure, MR foam presents the merits of light-weight, controllable modulus, excellent sound-absorbing property, and so on [63].

To solve the problem of particle sedimentation for MR fluid, Park et al. prepared a kind of novel MR material with excellent sedimentation resistance ability by substituting the fluidic matrix of MR fluid with commercial grease, and they named this MR material as MR grease [64]. MR grease shows typical Bingham fluid behavior; so strictly speaking, MR grease can be regarded as a special MR fluid. Byrom and Biswal reported a colloidal material system by adding micrometer-sized paramagnetic and diamagnetic particles into ferrofluid [65]. Different from the conventional MR fluid, the particles do not generate a chain-like orientated structure parallel with the direction of magnetic field but a fractal net-like microstructure in 2D direction. Further analysis indicated that the fractal net-like microstructure is induced by the magnetic dipole interaction between paramagnetic and diamagnetic particles, and the fractal dimension of the particle aggregates can be controlled by adjusting the concentration of ferrofluid and the ratio of paramagnetic and diamagnetic particles.

A multifunctional magnetic plasticine™ was developed by Xuan et al. [66], and they chose paraffin wax petroleum jelly as the matrix. Except for the high magneto-induced G' (4.23 MPa) and MR effect (305%), magnetic plasticine™ can be switched between liquid-like state and solid-like state by changing the temperature, which greatly enhances the regulation ability. Shahrivar and de Vicente also reported a thermo-responsive polymer-based magneto-sensitive material [67],

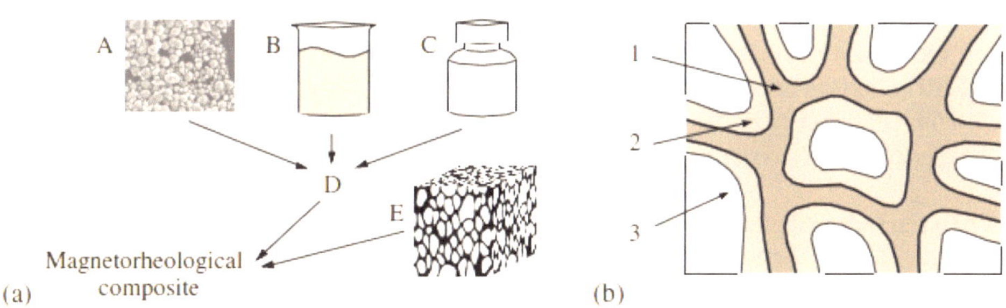

Figure 5.
The fabrication procedure of MR foam (a): pouring the MR fluid D into porous polymer material E. MR fluid D consists of carbonyl iron particles A, carrier fluid B, and additives C; (b) the microstructure inside the foam: microtubule wall 2 is surrounded by MR fluid 1, and the other space is full of air 3 [61].

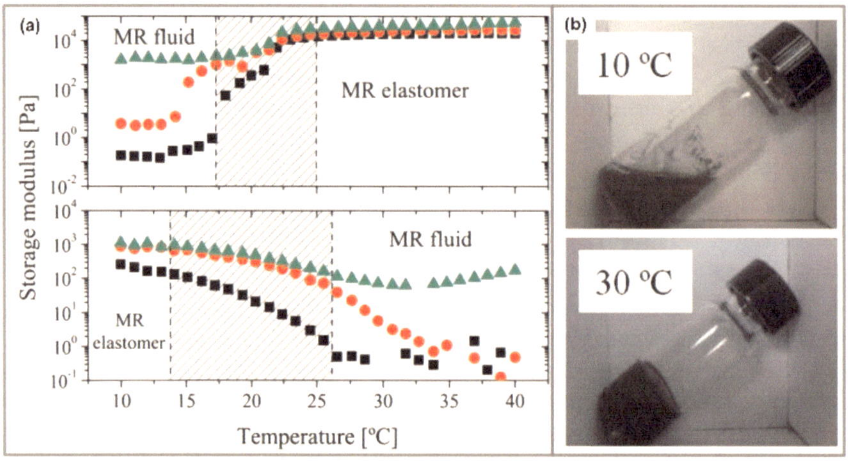

Figure 6.
Temperature dependent G′ of MR composites with PEO-PPO-PEO copolymer solution (upper subfigure) and P-NIPA microgel dispersion (lower subfigure) (a). Photos of the copolymer-based MR material at different temperatures after 1 h at rest (b) [67].

which can easily achieve phase transition by changing temperature. **Figure 6a** depicts that "liquid-to-solid" transition with increasing temperature appears in the MR composite with PEO-PPO-PEO solutions because a repulsive colloidal glass generates, while an inverse temperature-driven phase transition can be achieved for MR composite with P-NIPA microgel dispersions (**Figure 6b**). Besides, the critical temperature of phase transition can be changed by tuning polymer concentration. The multi-responsive MR material is a good effort in the frontier between conventional MR fluid and MR elastomer.

3. Magneto-sensitive properties and MR mechanism

The magnetic field usually leads to a structural rearrangement in soft MR material, and this process has significant influence on the physical properties of soft MR material. Due to the weak restriction of polymer matrix to the ferromagnetic fillers and complexity of polymer matrix, the responses of MR gel to external stimuli are more complicated than those of MR fluid and MR elastomer. For this reason, MR gel presents some unique magneto-electro-thermo-mechanical coupling phenomena. However, the realization on the coupling mechanism of MR gel is far from enough in comparison with those of MR fluid and MR elastomer, and more efforts need to be made through experimental and theoretical studies. Tight correlations exist between these three MR materials; so, there are some similarities in MR mechanism, and we can use the characterization techniques and theoretical models of MR fluid and MR elastomer for reference when studying the MR mechanism of MR gel. Next, we will briefly discuss the characterization methods and theoretical studies for different MR materials.

3.1 Experimental characterization of MR materials

Experimental characterization of MR materials can not only quantitatively evaluate their performance but also provide the necessary parameters for theoretical research or certify the accuracy of the theoretical results, which is the foundation for investigating the MR mechanism. The magneto-induced rheological properties of MR materials under different loading conditions (quasi-static shear, tensile, compressive loading, dynamic shear, and squeeze loading) are the most important properties, and this is the primary reason for naming this magneto-sensitive smart material as MR material. The influences of relevant factors (i.e., magnetic field, temperature, PH value, particle concentration, shape, size, and so on) on rheological properties were also widely investigated. Besides, the magnetization, electrical conduction, thermal conduction, and magnetostriction of MR materials are studied as well.

The rheological property of MR materials under shear loading is the most used characterization parameters at present. Many famous commercial rheometers, like the Physica MCR rheometer from Anton Paar Company and Discovery hybrid rheometer from TA Company, are designed for the rheological characterization under shear loading.

The magneto-dependent rheological behavior of MR fluid is usually described by Bingham model:

$$\begin{aligned} \tau &= \tau_y + \eta\dot{\gamma} \qquad |\tau| \geq \tau_y \\ \tau &= G_0\gamma \qquad |\tau| < \tau_y, \end{aligned} \tag{1}$$

where τ is shear stress, $\dot{\gamma}$ is shear strain rate, τ_y is magneto-dependent shear yield stress, G_0 is the shear modulus before yield, and η is the plastic viscosity. τ_y is defined as the minimum stress of MR fluid to resist the thixotropic effect and start to deform or flow, which can be calculated by fitting the shear stress-strain rate curves of MR fluid by using Bingham model (Eq. (1)). **Figure 7** depicts the typical magneto-sensitive relationship between shear stress and strain rate of MR fluid [68]. Usually, it can be approximately considered the fitting value at the strain rate of zero as τ_y, so it is easily found from **Figure 7** that τ_y significantly increases when the magnetic field strength increases. Therefore, the magneto-dependence of τ_y can be regarded as a characterization parameter to evaluate the magnetorheological effect of MR fluid [69]. This method can also be directly utilized to characterize the magnetorheological property of liquid-like MR gels [70].

Both of the solid-like and liquid-like MR material can be considered as the viscoelastic material, and the dynamic mechanical analysis under oscillatory shear is one of the most important characterization methods for viscoelastic materials. Normally, applying a sinusoidal shear strain to the viscoelastic material, if the amplitude of the strain is small enough, a sinusoidal response stress at the same frequency

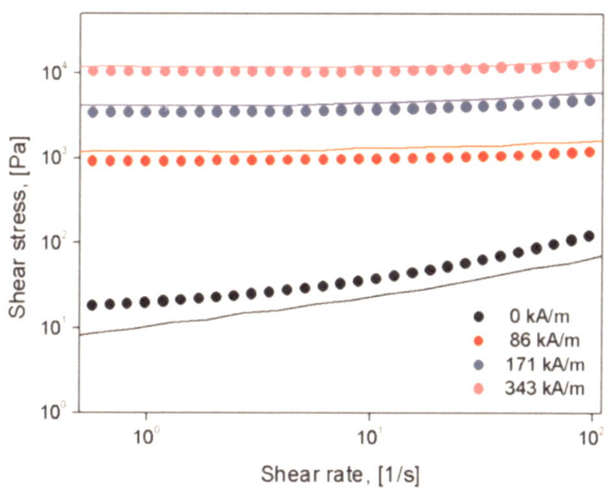

Figure 7.
Shear stress of MR fluid as a function of shear rate under different magnetic fields [68].

but a specified phase shift can be obtained; then, we define the dynamic mechanical properties (i.e., storage modulus G′, loss modulus G″, and loss factor tanδ) of material within the linear viscoelastic (LVE) range as the ratio of response stress to actuation strain in the complex plane. Dynamic mechanical properties are frequently used characterization parameters for investigating the magneto-induced microstructure evolution mechanism of MR materials. For the practical application, most MR materials in the devices are working under the oscillatory shearing mode. More importantly, both solid-like MR elastomer and liquid-like MR fluid can be characterized by dynamic mechanical analyzer (DMA) [71, 72], which indicates that dynamic mechanical analysis can be considered as a universal method to characterize MR materials. Therefore, as the intermediate material system between MR elastomer and MR fluid, the magnetorheological properties of MR gel are mostly studied by DMA [73].

The oscillatory shear mode can be further classified into simple shear and rotating shear according to different measurement principles of commercial devices. **Figure 8a** shows a typical DMA (Tritec 2000, provided by the Triton Technology Co. Ltd., UK). If an external magnetic field generator is added (**Figure 8b**), the magneto-mechanical coupling behaviors of MR elastomer under oscillatory simple shear mode can be investigated [74]. The deformation of the sample under simple shear mode is uniform, which is valuable for theoretical analysis. The rheometer can also be used for dynamic mechanical analysis. **Figure 8c** shows a parallel-plate rheometer (Physica MCR 301, Anton Paar Co., Austria) equipped with a MR accessory (MRD 180), which provides a controllable magnetic field when carrying out a rotating shear experiment (**Figure 8d**). The deformation of the sample under rotating shear mode is inhomogeneous (the shear strain increases linearly in the radial direction of disc-like sample; the shear strain at the center of the sample is zero). Although with different measurement principles, the measurement results obtained from these two kinds of devices show little difference if the amplitude of the actuating strain is small enough. Besides, the magneto-dependent creep and

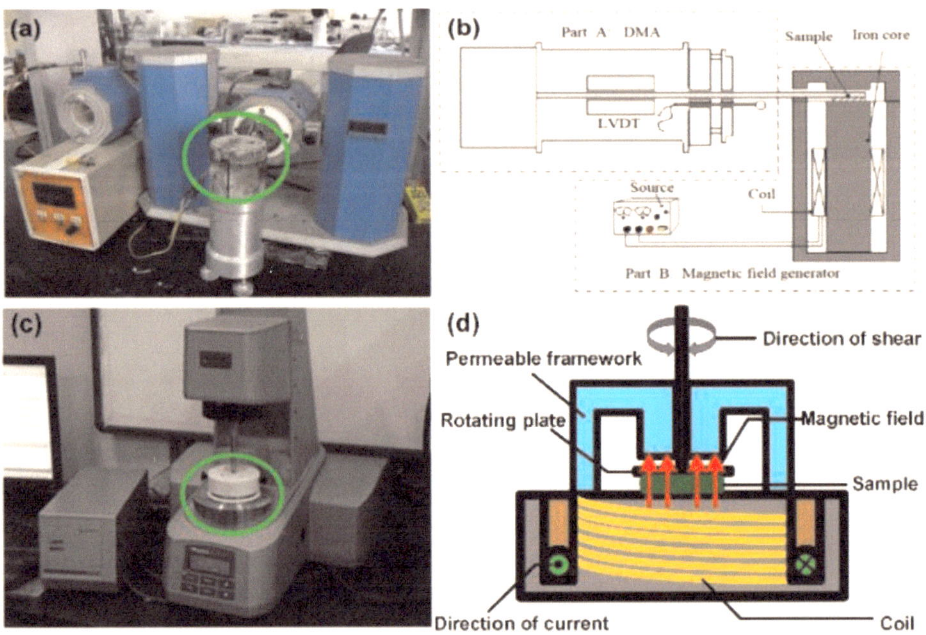

Figure 8.
A modified magneto-mechanical coupled DMA (a) and its measurement schematic diagram (b); a parallel-plate rheometer (c) with a magnetic field generation accessory (d).

recovery [75–77] and stress relaxation behaviors [78] of MR materials under shear loading are very helpful for investigating the magnetorheological mechanism and can be tested with the modified DMA and the rheometer as mentioned above.

The modulus variation of ferromagnetic filler-doped polymer composite can be reflected from the compressive and tensile properties of MR elastomer, which are also frequently used characterization parameters. Bellan and Bossis investigated the influences of magnetic field, particle concentration, and particle distribution on the tensile property of silicon rubber-based MR elastomer [79]. The compressive properties of MR elastomer with different particle distributions (randomly dispersed isotropic and chain-like orientated anisotropic structures) and its magnetic field dependency were investigated by Varga et al., as shown in **Figure 9**. It is found that the compressive modulus of MR elastomer is not only affected by the particle distribution (i.e., the compressive modulus of anisotropic MR elastomer is larger than that of isotropic one under the same loading condition) but also by the magnetic field and loading direction. When the directions of magnetic field, particle chain, and compressive loading are parallel (i.e., the middle test condition in the second row of **Figure 9**), the largest compressive modulus and magneto-induced effect of MR elastomer were observed [80]. In addition, the mechanical properties of MR elastomer under oscillatory squeezing mode were reported by Kallio et al. and Koo et al., respectively [81, 82], which provide valuable experimental data for the application of MR elastomer.

In recent years, many people pay attention to the rheological behaviors of MR fluid under tensile or squeezing loading, and the results suggest that the yield stress of MR fluids is significantly enhanced due to the squeeze-strengthen effect [83–85]. However, the tensile and squeezing behaviors of MR gel are rarely studied to date. Therefore, Xu et al. systematically investigated the squeeze flow behaviors (including quasi-static compressive and tensile behaviors and oscillatory squeeze behaviors) of MR gel [86]. It was concluded that the squeeze flow curve of the solid-like MR gel can be classified into three different deformation regions: elastic deformation, stress relaxation, and plastic flow regions. Yield stresses under both tension and compression are sensitive to the particle distributions, the filler concentration, and the magnetic field. In addition, the magneto-sensitive properties of MR

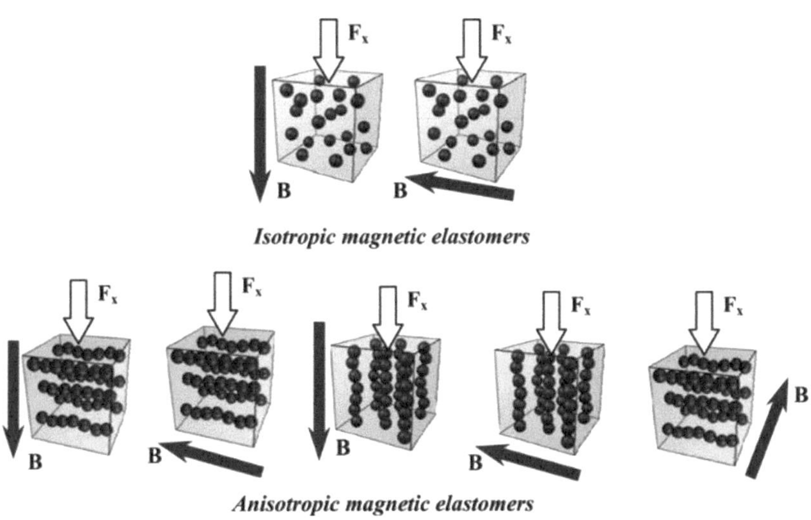

Figure 9.
The magneto-dependent compressive experiments of MR elastomers with different particle distribution states [80]. The white hollow arrow represents the direction of compressive loading, and the solid black arrow represents the direction of magnetic field.

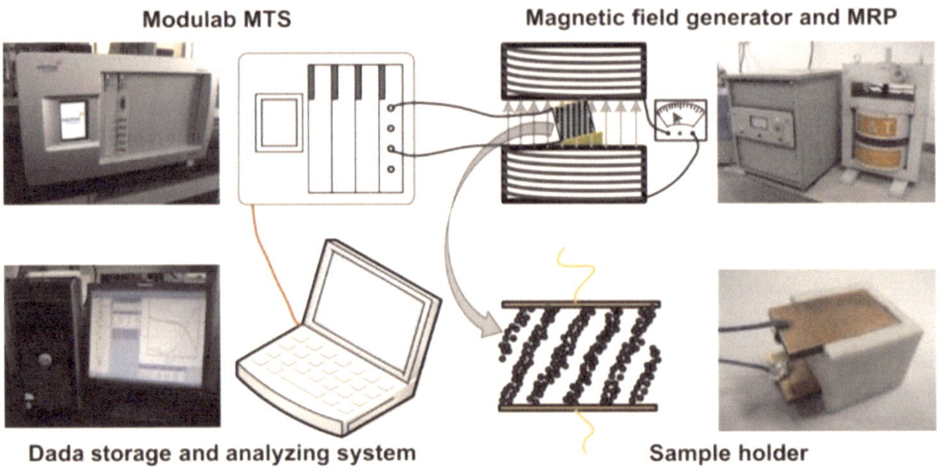

Figure 10.
The magneto-electrical coupling measurement system.

elastomer and MR gel under impact compression were also studied [11, 87]. The compressive modulus can still be strengthened by magnetic field even at high strain rate.

Most magnetic fillers (such as carbonyl iron particles [88], nickel particles [89], and Fe_3O_4 particles covered by silver [90]) are conductive at the same time, so most MR materials also belong to conductive polymer composite. The conductivity of this kind of magneto-sensitive conductive polymer composite is adjusted through external magnetic field except for the particle distribution and particle concentration, presenting a typical magnetoresistance effect [91–93]. Impedance spectroscopy testing is a nondestructive method to quantitatively detect the evolution of the microstructure, which is suitable for analyzing the microstructure evolution mechanism and the interfacial feature of material. **Figure 10** shows a typical experimental setup to test the magneto-sensitive impedance spectroscopy of conductive polymer composite. With this magneto-electrical coupling measurement system, the structure-dependent (**Figure 11a**) and the magneto-induced (**Figure 11b** and **c**) impedance spectroscopy are obtained, and the microstructure-dependent conduction mechanism (**Figure 11d**) can be further analyzed based on the related experimental results [94].

Besides, the investigations on the antioxidation [95], durability [96], and thermal conductivity [6] are helpful to the deep understanding on the magneto-induced mechanism of MR materials as well as some specific practical applications.

3.2 Magnetorheological mechanism

Magnetorheological effect essentially originates from the discrepancy of magnetic permeability between the continuous phase (the carrier matrix) and the dispersed phase (ferromagnetic fillers). Particular magnetization model (i.e., magnetic dipole model) is the most popular microstructural model to explain the magneto-induced effect of MR fluid [21]. If we ignore the multi-body magnetic interaction between particles (i.e., only the magnetic interaction between adjacent particles in a single particle chain is considered) and the multidirectional magnetization in a single magnetic particle (i.e., simplify the micrometer sized ferromagnetic particles as magnetic dipole), the magnetic moment of spherical ferromagnetic particle within the linear magnetization range is:

$$m = 4\pi\mu_0\mu_{cr}\beta a^3 H_0, \quad (2)$$

where a is the radius of magnetic particle, μ_0 is the permeability of vacuum, μ_{cr} is the relative permeability of carrier matrix, H_0 is magnetic field strength, and $\beta = (\mu_{pr} - \mu_{cr})/(\mu_{pr} + 2\mu_{cr})$ is the dimensionless coupling parameter of permeability (μ_{pr} is relative permeability of ferromagnetic particles). As the magnetic field increased, the magnetization of magnetic particle tends to saturation, and the magnetic moment of particle is independent on magnetic field strength, that is

$$m = \frac{4}{3}\pi\mu_0\mu_{cr}a^3 M_s, \quad (3)$$

where M_s represents the saturation magnetization. The parameter λ is usually introduced to represent the ratio of magnetic interaction energy to thermal energy between adjacent ferromagnetic particles [97],

$$\lambda = \frac{1}{4\pi\mu_0\mu_{cr}}\frac{m^2}{r^3}\frac{1}{\kappa T} = \frac{\pi\mu_0\mu_{cr}\beta^2 a^3 H_0^2}{2\kappa T}. \quad (4)$$

If λ is far larger than 1, the magnetic interaction between adjacent particles is far larger than the force induced by Brownian movement, and the particle will generate chain-like (or column-like) oriented microstructure parallel to the direction of magnetic field.

When it flows, the rheological properties of MR fluid are related to λ, the volume fraction of particles ϕ, and Mason number (a dimensionless parameter, Mn).

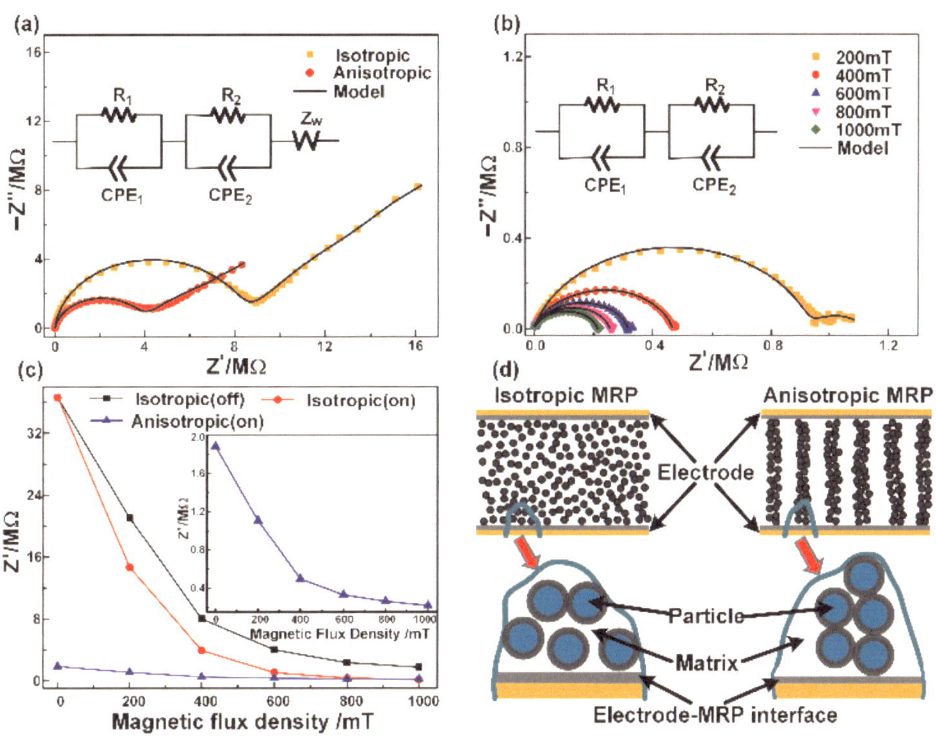

Figure 11.
Nyquist plots of magneto-sensitive conductive polymer composite with different particle distribution states (a) and under different magnetic fields (b); the conductivity of magneto-sensitive conductive polymer composite with different preprocessing methods (c); schematic of magneto-sensitive conductive polymer composites with different particle distributions and corresponding interfaces (d) [94].

In a stable shear flow, *Mn* is defined as the ratio of hydrodynamic drag applied on the ferromagnetic particle to the magneto-static force [98]:

$$Mn = \frac{8\eta_c \dot{\gamma}}{\mu_0 \mu_{cr} \beta^2 H^2}, \tag{5}$$

where η_c is the viscosity of carrier matrix and $\dot{\gamma}$ represents the shear strain rate.

As the most important rheological parameter of MR fluid, the magneto-induced yield stress can generally be explained from macroscopic and microscopic aspects. The macroscopic theoretical model is usually derived according to the minimum principle of energy. It assumes that the ferromagnetic fillers are spherical, cylindrical, or layered and are dispersed evenly in macroscopic theoretical model [99]. These models based on the sub-microstructure only consider the anisotropy of particle aggregation under small strain while the microscopic models consider the interaction between magnetic particles [100]. Most microscopic models ignore the interaction between the structured particle chains and believe that the magnetic interaction between adjacent particles is the main source of yield stress. Therefore, the yield stress of MR fluid can be well predicted when particle content is relatively low, but the assumption is untenable at high particle concentration, which makes a big deviation between theoretical and experimental results.

Figure 12 demonstrates the schematic of classic single-chain magnetic dipole model. When the material deforms by shearing, an affine deformation happens on the particle chain accordingly, which means that the particle moves horizontally along the direction of arrows as shown in **Figure 12a**. For this affine deformation, the distance between adjacent particles is identical before and after deformation. This assumption simplifies the deduction of magneto-induced yield stress. The magneto-induced shear yield stress of MR fluid within the linear magnetization range can be obtained based on the affine deformation assumption:

$$\tau_y = 2.31\phi\mu_0 M_s^{1/2} H_0^{3/2}. \tag{6}$$

In the saturated magnetization range, yield stress can be expressed as:

$$\tau_y^s = 0.086\phi\mu_0 M_s^2 \tag{7}$$

The details for the deduction process of magneto-induced yield stress and the explanation of relevant parameters can be found in the review article about the magnetorheological mechanism written by Bossis et al. [101].

The abovementioned single-chain magnetic dipole model was introduced directly into MR elastomer by Jolly et al. [102]. They predicted the

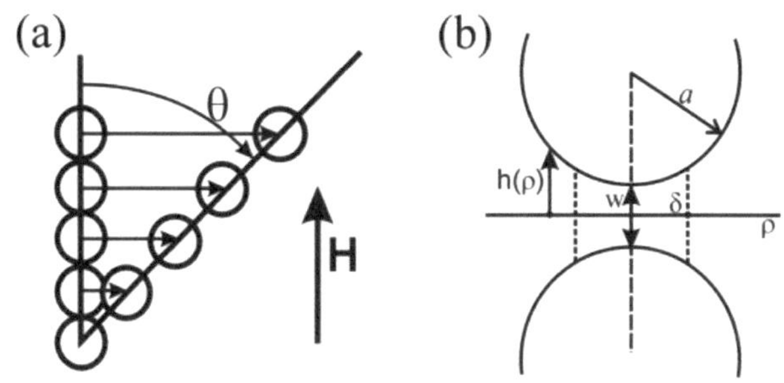

Figure 12.
The schematic of single-chain magnetic dipole model [101]. The affine deformation of particles within a single chain (a) and the geometrical relationship between adjacent particles (b).

magneto-induced elastic modulus by this model, and the theoretical results fit well with the experiment. Davis also calculated the magneto-induced shear modulus by this model, and he found that the magneto-induced shear modulus reaches a maximum if the volume ratio of the ferromagnetic fillers is 27% [103]. However, the influence of magnetized particle on the surrounding particles is not considered by the magnetic dipole model, and the bias of theoretical results from the fact is growing larger with the increasing particle concentration. Shen et al. realized this problem and modified the single-chain magnetic dipole model by considering all of the interaction of the particles within a single chain, which is more in line with the real situation [104]. A finite-column model based on the experimental results was proposed by Chen et al. [41]. However, this model can only predict the magneto-induced modulus of MR elastomer with low particle concentration.

The abovementioned models concentrate on the magnetic interactions between particles, but different from the MR fluid, a strong constraint effect exists between rubber matrix and magnetic particles for MR elastomer. This constraint will influence the magnetic interactions between particles to some extent. Therefore, the theoretical model of MR elastomer considering the coupling effect between rubber matrix and particles is more reasonable. Chen and Jerrams developed a more general magneto-mechanical coupling model of MR elastomer, which includes the magneto-induced mechanical property of particles, the interfacial slipping effect between ferromagnetic particles and rubber matrix, and the viscoelastic properties of rubber matrix [105]. This model could predict the dynamic mechanical properties of MR elastomer with different particle content or different kinds of rubber matrix, revealing the response mechanism of material to external magnetic field. It is a tendency in recent years that constructing the magneto-mechanical coupling model of MR elastomer bases on the theory of continuum medium mechanics [106], more details about the research progress on the theoretical modeling of MR elastomer can be found in the recent review article [107]. Although the numerical results can be obtained after complex mathematical derivation, these theoretical models can deeply reveal the complicated magneto-mechanism coupling mechanism of MR elastomer, which has guiding significance for the optimal design and the practical application of this smart material.

A lot of work has been done on the magnetorheological mechanism of MR fluid and MR elastomer, yet there are little reports on the magnetorheological mechanism of MR gel. On the one hand, MR gel (especially for the solid-like MR gel) has not attracted wide attentions as a new MR material, and the research on it is not enough. On the other hand, the investigation on the magnetorheological mechanism of MR gel is more difficult than those of MR fluid or MR elastomer due to its intrinsic complex MR characteristics. MR gel possesses both the characteristics of mobility of magnetic particles in MR fluid and the stability of oriented microstructure in MR elastomer. These two features are "contradictory" to some extent, but they indeed exist in solid-like MR gel at the same time. Although the magnetic particles are moveable in the carrier matrix of MR gel, the clustering phenomena of particles in MR gel cannot be interpreted by the theory used in MR fluid because the viscous resistance of polymer matrix is far larger than the resistance from the carrier fluid in MR fluid. It is mean that the Mason number of MR gel is far larger than that of MR fluid, the assumption in fluid is invalid in MR gel. In the meantime, the viscous resistance of polymer matrix to the particle is much less than the constraining force of the rubber matrix to the particle in MR elastomer. After applying a magnetic field, the "solidified" ferromagnetic fillers in MR elastomer can only move slightly from the original position, while the ferromagnetic fillers in MR gel can greatly move under a strong magnetic field and a large applied loading; if the direction of magnetic field is changed, the ferromagnetic fillers can even rearrange to generate chain-like or column-like structure along the new direction of magnetic field. These interesting characteristics that reflect the complexity of MR gel, the rheological behavior of polymer matrix, the magnetic interaction between adjacent fillers, and the interfacial problem due to the relative movement of particle and matrix have to be considered when studying the magnetorheological mechanism of MR gel. In addition, the "huge" change of microstructure after the rearrangement of particles will also make the modeling of the magneto-mechanical coupling behavior of MR gel more difficult.

It is not easy to fully describe the complicated magneto-mechanical coupling behaviors of MR gels. A field theory was developed by Han et al. to describe the magneto-sensitive viscoelasticity of ferrogel based on the principles of non-equilibrium thermodynamics [108]. The responses of ferrogel to different magnetic fields were analyzed by numerical calculation, and the theoretical results consistent with the experimental results under the cyclic magnetic field, which indicates that this theory is reasonable to some extent for the realization to magneto-mechanical coupling mechanism of MR gel. Zubarev evaluated the free energy of ferrogel after tension or compression along with the magnetic field direction by standard methods of statistical physics [109]. The analysis demonstrates that the magnetic field strength, the initial shape of the sample, as well as the particle concentration and the magnetic properties of particles determine the type of magneto-induced deformation (i.e., extension or shrunken). A particle-level molecular dynamics model was employed by Liu et al. to investigate the particle evolution in MR gel under a stable uniform magnetic field [110]. A modified magnetic dipole model is introduced to describe the magnetic interaction between adjacent particles, and this model presents higher precision than classic magnetic dipole model when processing the magnetic interaction of particles close to each other. The rheological behavior of the carrier matrix is described by the Bingham plastic model (Eq. (1)). Some complicated loading methods of magnetic field (such as the rotating magnetic field, as shown in **Figure 13**), which are difficult to be achieved by experiment, can be easily applied through simulation. With this, the 3D evolution of particular microstructure under complicated magnetic field loading conditions can be obtained, which is very important to understand the microstructure evolution mechanism of MR gel.

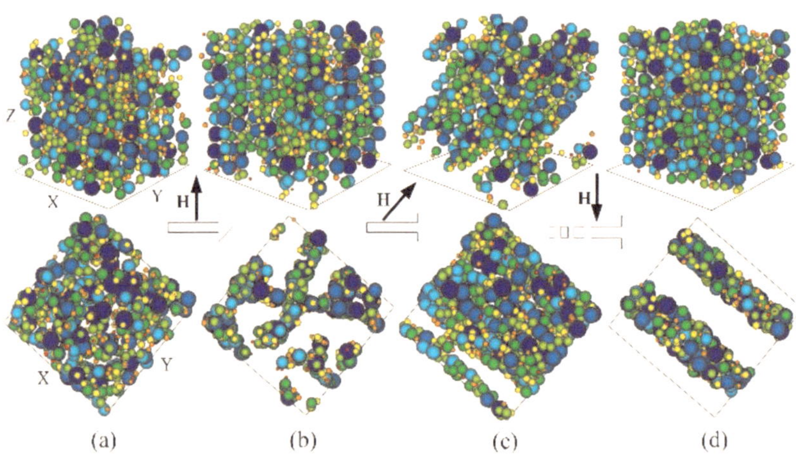

Figure 13.
The evolution process of 3D particular microstructure in MR gel: the initial state before applying magnetic field (a); the magnetic field is parallel with Z axis (b); the magnetic field rotates 45° in the clockwise direction (c); and the magnetic field further rotates 180° in the clockwise direction (d) [110].

The microstructure obtained by the particle-level molecular dynamics model matches well with the results by experimental observation under the same loading condition. However, the coupling effect between ferromagnetic fillers and carrier matrix is not considered in this model, and the Bingham plastic model is also too simple to describe the complicated rheological behavior of the carrier matrix; so, the governing equations which describe the physical behaviors of different components within the MR gel need to be further developed.

4. Conclusions and prospects

As a smart material whose physical properties can be easily controlled, magneto-sensitive composite (i.e., MR material) attracts more and more attentions in recent years. Many works concentrating on the magnetorheological mechanism and application have been reported. Various MR materials aiming at different practical applications were developed, which shows great application potential.

However, the inherent defects existed in conventional MR materials (e.g., the particle sedimentation in MR fluid, the microstructure control of MR elastomer, the bearing capacity of MR gel as structure unit) preclude their wide application. To this end, from the perspective of material preparation, on the one hand, we need to further improve the conventional MR material aiming at the inherent defects; on the other hand, the novel magneto-sensitive material system which meets the requirements of engineering applications should be developed, which means that we could develop multifunctional smart composite that has the magneto-controllable feature (e.g., magneto-sensitive impact-resistant composite, magneto-sensitive conductive composite, magneto-sensitive heat-conducting composite, and so on). From the perspective of mechanism, magneto-sensitive soft material refers to magneto-electro-thermo-mechanical coupling behavior, and it is difficult to describe the response to external stimuli. The difficulties can be summarized as follows: the description of the exact distribution state of dispersed phase before and after exposed under a magnetic field; the description of the discrepancy of size and shape of the dispersed ferromagnetic fillers; the interaction model between the dispersed phase and the continuous phase; the construction of constitutive model of the polymer matrix in the MR elastomer or solid-like MR gel; and the unification of multiscale model from microscale to macroscale. Considering the complexity of true situation, some necessary simplifications have to be made aiming at specific problem; then, the simplified model which could generally reflects the specific mechanism can be developed after ignoring the secondary factors. The numerical simulation is another effective method to investigate the microstructure evolution mechanism of MR material. It is an important research direction to construct the constitutive model of MR material, which could accurately describe the complicated coupling responses to different stimuli (magnetic field, temperature, strain rate, and so on). It is believed that the engineering applications of magneto-sensitive multifunctional material will be more widely concerned with the further realization on the magnetorheological mechanism and the enhancement of the performance of the material.

Acknowledgements

The research reported in this publication was supported by National Natural Science Foundation of China (Grant nos. 11502256, 11502255, 11602242), and the key project “Computational Solid Mechanics” of the China Academy of Engineering

Physics. The authors gratefully acknowledge Professor Xinglong Gong and Professor Shouhu Xuan at University of Science and Technology of China for their help on beneficial discussion.

Author details

Yangguang Xu[1,2]*, Guojiang Liao[2] and Taixiang Liu[3]*

1 College of Aerospace Engineering, Chongqing University, Chongqing, China

2 Institute of Systems Engineering, China Academy of Engineering Physics (CAEP), Mianyang, China

3 Research Center of Laser Fusion, China Academy of Engineering Physics (CAEP), Mianyang, China

*Address all correspondence to: xyg@mail.ustc.edu.cn and txliu@ mail.ustc.edu.cn

References

[1] Wegst U, Bai H, Saiz E, Tomsia A, Ritchie R. Bioinspired structural materials. Nature Materials. 2015;**14**: 23-36

[2] Tong Y, Dong X, Qi M. High performance magnetorheological fluids with flower-like cobalt particles. Smart Materials and Structures. 2017;**26**: 025023

[3] Wan Y, Xiong Y, Zhang S. Temperature effect on viscoelastic properties of anisotropic magnetorheological elastomers under compression. Smart Materials and Structures. 2019;**28**:015005

[4] Xu Y, Gong X, Xuan S. Soft magnetorheological polymer gels with controllable rheological properties. Smart Materials and Structures. 2013;**22**: 075029

[5] Klingenberg D. Magnetorheology: Applications and challenges. Aiche Journal. 2001;**47**:246-249

[6] Heine M, De Vicente J, Klingenberg D. Thermal transport in sheared electro- and magnetorheological fluids. Physics of Fluids. 2006;**18**:023301

[7] Donado F, Carrillo J, Mendoza M. Sound propagation in magneto-rheological suspensions. Journal of Physics-Condensed Matter. 2002;**14**: 2153-2157

[8] Kordonski W, Golini D. Multiple application of magnetorheological effect in high precision finishing. Journal of Intelligent Material Systems and Structures. 2002;**13**:401-404

[9] Liu J, Flores G, Sheng R. In-vitro investigation of blood embolization in cancer treatment using magnetorheological fluids. Journal of Magnetism and Magnetic Materials. 2001;**225**:209-217

[10] Deng H, Gong X, Wang L. Development of an adaptive tuned vibration absorber with magnetorheological elastomer. Smart Materials and Structures. 2006;**15**: N111-N116

[11] Liao G, Gong X, Xuan S. Magnetic field-induced compressive property of magnetorheological elastomer under high strain rate. Industrial & Engineering Chemisty Research. 2013; **52**:8445-8453

[12] Farshad M, Le Roux M. A new active noise abatement barrier system. Polymer Testing. 2004;**23**:855-860

[13] Blom P, Kari L. Smart audio frequency energy flow control by magneto-sensitive rubber isolators. Smart Materials and Structures. 2008; **17**:015043

[14] Tian T, Li W, Deng Y. Sensing capabilities of graphite based MR elastomers. Smart Materials and Structures. 2011;**20**:025022

[15] Leng J, Lan X, Liu Y, Du S, Huang W, Liu N, et al. Electrical conductivity of thermoresponsive shape-memory polymer with embedded micron sized Ni powder chains. Applied Physics Letters. 2008;**92**:014104

[16] Bica I. Magnetoresistor sensor with magnetorheological elastomers. Journal of Industrial and Engineering Chemistry. 2011;**17**:83-89

[17] Li Y, Huang G, Zhang X, Li B, Chen Y, Lu T, et al. Magnetic hydrogels and their potential biomedical applications. Advanced Functional Materials. 2013; **23**:660-672

[18] Pang H, Xuan S, Sun C, Gong X. A novel energy absorber based on magnetorheological gel. Smart Materials and Structures. 2017;**26**:105017

[19] Fuhrer R, Athanassiou E, Luechinger N, Stark W. Crosslinking metal nanoparticles into the polymer backbone of hydrogels enables preparation of soft, magnetic field-driven actuators with muscle-like flexibility. Small. 2009;**5**:383-388

[20] Nguyen V, Ahmed A, Ramanujan R. Morphing soft magnetic composites. Advanced Materials. 2012;**24**: 4041-4054

[21] De Vicente J, Klingenberg D, Hidalgo-Alvarez R. Magnetorheological fluids: A review. Soft Matter. 2011;**7**: 3701-3710

[22] Lee C, Jang M. Virtual surface characteristics of a tactile display using magneto-rheological fluids. Sensors. 2011;**11**:2845-2856

[23] Lloyd J, Hayesmichel M, Radcliffe C. Internal organizational measurement for control of magnetorheological fluid properties. Journal of Fluids Engineering-Transactions of the ASME. 2007;**129**:423-428

[24] Phule P. Synthesis of level magnetorheological fluids. MRS Bulletin. 1998;**23**:23-25

[25] Sheikholeslami M. Influence of magnetic field on Al_2O_3-H_2O nanofluid forced convection heat transfer in a porous lid driven cavity with hot sphere obstacle by means of LBM. Journal of Molecular Liquids. 2018;**263**:472-488

[26] Sheikholeslami M, Jafaryar M, Li Z. Second law analysis for nanofluid turbulent flow inside a circular duct in presence of twisted tape turbulators. Journal of Molecular Liquids. 2018;**263**: 472-488

[27] Sheikholeslami M. Application of Darcy law for nanofluid flow in a porous cavity under the impact of Lorentz forces. Journal of Molecular Liquids. 2018;**266**:495-503

[28] Sheikholeslami M, Haq R, Shafee A, Li Z. Heat transfer behavior of nanoparticle enhanced PCM solidification through an enclosure with V shaped fins. International Journal of Heat and Mass Transfer. 2019;**130**: 1322-1342

[29] Sheikholeslami M, Gerdroodbary M, Moradi R, Shafee A, Li Z. Application of neural network for estimation of heat transfer treatment of Al_2O_3-H_2O nanofluid through a channel. Computer Methods in Applied Mechanics and Engineering. 2019;**344**:1-12

[30] Sheikholeslami M. Numerical approach for MHD Al_2O_3-water nanofluid transportation inside a permeable medium using innovative computer method. Computer Methods in Applied Mechanics and Engineering. 2019;**344**:306-318

[31] Sheikholeslami M. New computational approach for exergy and entropy analysis of nanofluid under the impact of Lorentz force through a porous media. Computer Methods in Applied Mechanics and Engineering. 2019;**344**:319-333

[32] Sheikholeslami M, Mahian O. Enhancement of PCM solidification using inorganic nanoparticles and an external magnetic field with application in energy storage systems. Journal of Cleaner Production. 2019;**215**: 963-977

[33] Rinaldi C, Chaves A, Elborai S, He XW, Zahn M. Magnetic fluid rheology and flows. Current Opinion in Colloid & Interface Science. 2005;**10**:141-157

[34] Lopez-Lopez M, Vertelov G, Bossis G, Kuzhir P, Duran J. New magnetorheological fluids based on magnetic fibers. Journal of Materials Chemistry. 2007;**17**:3839-3844

[35] Park B, Fang F, Choi H. Magnetorheology: Materials and

application. Soft Matter. 2010;**6**: 5246-5253

[36] Iglesias G, Lopez-Lopez M, Duran J, González-Caballero F, Delgado A. Dynamic characterization of extremely bidisperse magnetorheological fluids. Journal of Colloid & Interface Science. 2012;**377**:153-159

[37] Lim S, Cho M, Jang I, Choi H. Magnetorheological characterization of carbonyl iron based suspension stabilized by fumed silica. Journal of Magnetism and Magnetic Materials. 2004;**282**:170-173

[38] Lim S, Choi H, Jhon M. Magnetorheological characterization of carbonyl iron-organoclay suspensions. IEEE Transactions on Magnetics. 2005; **41**:3745-3747

[39] Wereley N, Chaudhuri YJ, John S, Kotha S, Suggs A, Radhakrishnan R, et al. Bidisperse magnetorheological fluids using Fe particles at nanometer and micron scale. Journal of Intelligent Material Systems and Structures. 2006; **17**:393-401

[40] Chen L. The Development and Mechanical Characterization of Magnetorheological Elastomers [PhD thesis]. Hefei: University of Science and Technology of China; 2009

[41] Chen L, Gong X, Li W. Microstructures and viscoelastic properties of anisotropic magnetorheological elastomers. Smart Materials and Structures. 2007;**16**: 2645-2650

[42] Gong X, Zhang X, Zhang P. Fabrication and characterization of isotropic magnetorheological elastomers. Polymer Testing. 2005;**24**: 669-676

[43] Boczkowska A, Awietjan S, Wroblewski R. Microstructure-property relationships of urethane magnetorheological elastomers. Smart Materials and Structures. 2007;**16**: 1924-1930

[44] Sun T, Gong X, Jiang W, Li J, Xu Z, Li W. Study on the damping properties of magnetorheological elastomers based on cis-polybutadiene rubber. Polymer Testing. 2008;**27**:520-526

[45] Lokander M, Stenberg B. Performance of isotropic magnetorheological rubber materials. Polymer Testing. 2003;**22**:245-251

[46] Zajac P, Kaleta J, Lewandowski D, Gasperowicz A. Isotropic magnetorheological elastomers with thermoplastic matrices: Structure, damping properties and testing. Smart Materials and Structures. 2010;**19**: 045014

[47] Shiga T, Okada A, Kurauchi T. Magnetroviscoelastic behavior of composite gels. Journal of Applied Polymer Science. 1995;**58**:787-792

[48] Tiraferri A, Chen K, Sethi R, Elimelech M. Reduced aggregation and sedimentation of zero-valent iron nanoparticles in the presence of guar gum. Journal of Colloid and Interface Science. 2008;**324**:71-79

[49] Fuchs A, Hu B, Gordaninejad F, Evrensel C. Synthesis and characterization of magnetorheological polyimide gels. Journal of Applied Polymer Science. 2005;**98**:2402-2413

[50] Wei B, Gong X, Jiang W, Qin L, Fan Y. Study on the properties of magnetorheological gel based on polyurethane. Journal of Applied Polymer Science. 2010;**118**:2765-2771

[51] Zrinyi M. Intelligent polymer gels controlled by magnetic fields. Colloid and Polymer Science. 2000;**278**:98-103

[52] Wu J, Gong X, Fan Y, Xia H. Physically crosslinked poly(vinyl

alcohol) hydrogels with magnetic field controlled modulus. Soft Matter. 2011;7: 6205-6212

[53] Mitsumata T, Abe N. Magnetic-field sensitive gels with wide modulation of dynamic modulus. Chemistry Letters. 2009;**38**:922-923

[54] Mitsumata T, Honda A, Kanazawa H, Kawai M. Magnetically tunable elasticity for magnetic hydrogels consisting of carrageenan and carbonyl iron particles. Journal of Physical Chemistry B. 2012;**116**:12341-12348

[55] Mitsumata T, Kosugi Y, Ouchi S. Effect of particles alignment on giant reduction in dynamic modulus of hydrogels containing needle-shaped magnetic particles. Progress on Colloid and Polymer Science. 2009;**136**:163-170

[56] An H, Picken S, Mendes E. Enhanced hardening of soft self-assembled copolymer gels under homogeneous magnetic fields. Soft Matter. 2010;**6**:4497-4503

[57] An H, Picken S, Mendes E. Nonlinear rheological study of magneto responsive soft gels. Polymer. 2012;**53**: 4164-4170

[58] Xu Y, Gong X, Xuan S, Zhang W, Fan Y. A high-performance magnetorheological material: Preparation, characterization and magnetic-mechanic coupling properties. Soft Matter. 2011;7:5246-5254

[59] Gong X, Xu Y, Xuan S, Guo C, Zong L, Jiang W. The investigation on the nonlinearity of plasticine-like magnetorheological material under oscillatory shear rheometry. Journal of Rheology. 2012;**56**:1375-1391

[60] Carlson J, Jolly M. MR fluid, foam and elastomer devices. Mechatronics. 2000;**10**:555-569

[61] Kaleta J, Lewandowski D. Inelastic properties of magnetorheological composites: I. Fabrication, experimental tests, cyclic shear properties. Smart Materials and Structures. 2007;**16**: 1948-1953

[62] Zielinski T, Rak M. Acoustic absorption of foams coated with MR fluid under the influence of magnetic field. Journal of Intelligent Material Systems and Structures. 2010;**21**:125-131

[63] Gong Q, Wu J, Gong X, Fan Y, Xia H. Smart polyurethane foam with magnetic field controlled modulus and anisotropic compression property. RSC Advances. 2013;**3**:3241-3248

[64] Park B, Park B, Hato M, Choi H. Soft magnetic carbonyl iron microsphere dispersed in grease and its rheological characteristics under magnetic field. Colloid and Polymer Science. 2011;**289**:381-386

[65] Byrom J, Biswal S. Magnetic field directed assembly of two-dimensional fractal colloidal aggregates. Soft Matter. 2013;**9**:9167-9173

[66] Xuan S, Zhang Y, Zhou Y, Jiang W, Gong X. Magnetic Plasticine (TM): A versatile magnetorheological material. Journal of Materials Chemistry. 2012;**22**: 13395-13400

[67] Shahrivar K, De Vicente J. Thermo-responsive polymer-based magneto-rheological (MR) composites as a bridge between MR fluids and MR elastomers. Soft Matter. 2013;**9**:11451-11456

[68] Fang F, Choi H, Seo Y. Sequential coating of magnetic carbonyliron particles with polystyrene and multiwalled carbon nanotubes and its effect on their magnetorheology. ACS-Applied Materials & Interfaces. 2010;**2**: 54-60

[69] Kuzhir P, Magnet C, Bossis G, Meunier A, Bashtovoi V. Rotational diffusion may govern the rheology of magnetic suspensions. Journal of Rheology. 2011;**55**:1297-1318

[70] Fuchs A, Xin M, Gordaninejad F, Wang X, Hitchcock G, Gecol H, et al. Development and characterization of hydrocarbon polyol polyurethane and silicone magnetorheological polymeric gels. Journal of Applied Polymer Science. 2004;**92**:1176-1182

[71] Claracq J, Sarrazin J, Montfort J. Viscoelastic properties of magnetorheological fluids. Rheologica Acta. 2004;**43**:38-49

[72] Li W, Du H, Chen G, Yeo S, Guo N. Viscoelastic properties of MR fluids under oscillatory shear. Smart Structures and Materials 2001: Damping and Isolation. 2001;**4331**:333-342

[73] Rao P, Maniprakash S, Srinivasan S, Srinivasa A. Functional behavior of isotropic magnetorheological gels. Smart Materials and Structures. 2010;**19**: 085019

[74] Gong X, Chen L, Li J. Study of utilizable magnetorheological elastomers. International Journal of Modern Physics B. 2007;**21**:4875-4882

[75] See H, Chen R, Keentok M. The creep behaviour of a field-responsive fluid. Colloid and Polymer Science. 2004;**282**:423-428

[76] Xu Y, Gong X, Xuan S, Li X, Qin L, Jiang W. Creep and recovery behaviors of magnetorheological plastomer and its magnetic-dependent properties. Soft Matter. 2012;**8**:8483-8492

[77] Qi S, Yu M, Fu J, Li PD, Zhu M. Creep and recovery behaviors of magnetorheological elastomer based on polyurethane/epoxy resin IPNs matrix. Smart Materials and Structures. 2016;**25**: 015020

[78] Xu Y, Liu T, Liao G, Lubineau G. Magneto-dependent stress relaxation of magnetorheological gels. Smart Materials and Structures. 2017;**26**: 115005

[79] Bellan C, Bossis G. Field dependence of viscoelastic properties of MR elastomers. International Journal of Modern Physics B. 2002;**16**:2447-2453

[80] Varga Z, Filipcsei G, Zrinyi M. Magnetic field sensitive functional elastomers with tunable elastic modulus. Polymer. 2006;**47**:227-233

[81] Kallio M, Lindroos T, Aalto S, Jarvinen E, Karna T, Meinander T. Dynamic compression testing of a tunable spring element consisting of a magnetorheological elastomer. Smart Materials and Structures. 2007;**16**: 506-514

[82] Koo J, Khan F, Jang D, Jung H. Dynamic characterization and modeling of magneto-rheological elastomers under compressive loadings. Smart Materials and Structures. 2010;**19**: 117002

[83] De Vicente J, Ruiz-Lopez J, Andablo-Reyes E, Segovia-Gutierrez J, Hidalgo-Alvarez R. Squeeze flow magnetorheology. Journal of Rheology. 2011;**55**:753-779

[84] Guo C, Gong X, Xuan S, Qin L, Fan Y. Squeeze behavior of magnetorheological fluids under constant volume and uniform magnetic field. Smart Materials and Structures. 2013;**22**:045020

[85] Zhang X, Gong X, Zhang P, Wang Q. Study on the mechanism of the squeeze-strengthen effect in magnetorheological fluids. Journal of Applied Physics. 2004;**96**:2359-2364

[86] Xu Y, Liu T, Gong X, Xuan S. Squeeze flow behaviors of magnetorheological plastomers under constant volume. Journal of Rheology. 2014;**58**:659-679

[87] Wang Y, Wang S, Xu C, Xuan S, Jiang W, Gong X. Dynamic behavior of magnetically responsive shear-stiffening gel under high strain rate. Composites

Science and Technology. 2016;**127**: 169-176

[88] Bica I. Electroconductive magnetorheological suspensions: Production and physical processes. Journal of Industrial and Engineering Chemistry. 2009;**15**:233-237

[89] Kono A, Shimizu K, Nakano H, Goto Y, Kobayashi Y, Ougizawa T, et al. Positive-temperature-coefficient effect of electrical resistivity below melting point of poly(vinylidene fluoride) (PVDF) in Ni particle-dispersed PVDF composites. Polymer. 2012;**53**:1760-1764

[90] Mietta J, Ruiz M, Antonel P, PerezO E, Butera A, Jorge G, et al. Anisotropic magnetoresistance and piezoresistivity in structured Fe_3O_4-silver particles in PDMS elastomers at room temperature. Langmuir. 2012;**28**:6985-6996

[91] Kchit N, Lancon P, Bossis G. Thermoresistance and giant magnetoresistance of magnetorheological elastomers. Journal of Physics D-Applied Physics. 2009;**42**:105506

[92] Pang H, Xuan S, Liu T, Gong X. Magnetic field dependent electro-conductivity of the graphite doped magnetorheological plastomers. Soft Matter. 2015;**11**:6893-6902

[93] Wang Y, Xuan S, Ge L, Wen Q, Gong X. Conductive magnetorheological elastomer: Fatigue dependent impedance-mechanic coupling properties. Smart Materials and Structures. 2017;**26**:015004

[94] Xu Y, Gong X, Liu T, Xuan S. Magneto-induced microstructure characterization of magnetorheological plastomers using impedance spectroscopy. Soft Matter. 2013;**9**: 7701-7709

[95] Bednarek S. The giant magnetostriction in ferromagnetic composites within an elastomer matrix. Applied Physics A-Materials Science & Processing. 1999;**68**:63-67

[96] Ulicny J, Balogh M, Potter N, Waldo R. Magnetorheological fluid durability test-iron analysis. Materials Science and Engineering A-Structural Materials Properties Microstructure and Processing. 2007;**443**:16-24

[97] Rosensweig RE. Ferrohydrodynamics. New York: Dover Publications; 1997

[98] Klingenberg D, Ulicny J, Golden M. Mason numbers for magnetorheology. Journal of Rheology. 2007;**51**:883-893

[99] Bossis G, Lemaire E, Volkova O, Clercx H. Yield stress in magnetorheological and electrorheological fluids: A comparison between microscopic and macroscopic structural models. Journal of Rheology. 1997;**41**:687-704

[100] Ginder JM, Davis LC, Elie LD. Rheology of magnetorheological fluids: Models and measurements. International Journal of Modern Physics B. 1996;**10**:3293-3303

[101] Bossis G, Volkova O, Lacis S, Meunier A. Magnetorheology: Fluids, structures and rheology. Lecture Notes in Physics. 2002;**594**:202-230

[102] Jolly M, Carlson J, Munoz B, Bullions T. The magnetoviscoelastic response of elastomer composites consisting of ferrous particles embedded in a polymer matrix. Journal of Intelligent Material Systems and Structures. 1996;**7**:613-622

[103] Davis L. Model of magnetorheological elastomers. Journal of Applied Physics. 1999;**85**:3348-3351

[104] Shen Y, Golnaraghi M, Heppler G. Experimental research and modeling of magnetorheological elastomers. Journal of Intelligent Material Systems and Structures. 2004;**15**:27-35

[105] Chen L, Jerrams S. A rheological model of the dynamic behavior of magnetorheological elastomers. Journal of Applied Physics. 2011;**110**:013513

[106] Danas K, Kankanala S, Triantafykkidis N. Experiments and modeling of iron-particle-filled magnetorheological elastomers. Journal of the Mechanics and Physics of Solids. 2012;**60**:120-138

[107] Cantera M, Behrooz M, Gibson R, Gordaninejad F. Modeling of magneto-mechanical response of magnetorheological elastomers (MRE) and MRE-based systems: A review. Smart Materials and Structures. 2017;**26**: 023001

[108] Han Y, Hong W, Faidley L. Coupled magnetic field and viscoelasticity of ferrogel. International Journal of Applied Mechanics. 2011;**3**: 259-278

[109] Zubarev A. On the theory of the magnetic deformation of ferrogel. Soft Matter. 2012;**8**:3174-3179

[110] Liu T, Gong X, Xu Y, Xuan S. Simulation of magneto-induced rearrangeable microstructures of magnetorheological plastomers. Soft Matter. 2013;**9**:10069-10080

Chapter 5

Autowave Processes in Magnetic Fluid: Electrically Controlled Interference

Vladimir S. Chekanov, Natalya V. Kandaurova, Viktoria I. Drozdova, Galina V. Shagrova, Veniamin V. Romantsev and Mikhail Yu. Shevchenko

Abstract

The chapter considers autowaves that were observed in the thin near-electrode layer of concentrated magnetic fluid. Autowave process is a unique object for the study of self-organization. We observed pacemakers (leading centers), reverberators (spiral waves), and wave diffraction. A mechanism for the appearance of an autowave process has been developed; its mathematical model has been proposed and realized by means of computer simulation. As a basic method of observation, we used electrically controlled interference. This method watches the changes in the spectrum of reflected light from a two-layer structure with variable thickness: "conductive ITO coating—a layer of concentrated magnetic fluid" in an electric field.

Keywords: interference, magnetic fluid, reflection, thin membrane, near-electrode layer, autowaves, pacemaker, reverberator, self-organization

1. Introduction

In this chapter, we describe a new active excitable medium—a thin near-electrode layer of magnetic fluid. The uniqueness of this environment lies in the fact that its electrophysical and optical properties can be controlled using a weak electric field. In addition, in a layer of concentrated magnetic fluid, we were able to observe and investigate the autowave process. Also, with the help of external periodic effects, it was possible to achieve synchronization of autowaves. While working on this chapter, we studied the latest innovative work on ferrofluids and modeling the processes occurring in them [1–3].

A unique phenomenon that we observed in the near-electrode layer of a magnetic fluid is an autowave process. The best known is the autowave process in the Belousov-Zhabotinsky reaction, when the color of the solution changes periodically. In nature, the autowave process is a change of predator–prey populations. In human body, autowaves spread in the heart muscle and the retina. Interest in the study of autowave processes is primarily due to the fact that in chemical, biological systems, neural networks, and the human brain, they follow the same rules of propagation [4].

IntechOpen

We managed to observe an autowave process (Video 1 available at https://yadi.sk/i/rUBv-Mx12DqFLQ) in a thin near-electrode layer of a magnetic fluid (MF) placed between two electrodes in an electric field [5, 6].

The uniqueness of the experiments is that autowaves can be observed in laboratory conditions with the help of simple equipment, including transient processes and synchronization (Video 2 available at https://yadi.sk/i/-9l5aXD5aT3tKA). The purpose of this chapter is to study the autowave process in the near-surface layer of a magnetic fluid at the boundary with an electrode and to describe the physical model and the autowaves' appearance mechanism. Also, the goal of the chapter is to create a mathematical model and obtain its solution in the environment of modeling physical processes COMSOL Multiphysics 5.2.

2. Materials and methods

Magnetizing liquid media—magnetic fluids—is colloidal solutions of ferromagnetic particles in a liquid (kerosene, water). In our experiments, we used a liquid like "magnetite in kerosene" [7]. The typical particle size is 10 nm, which corresponds to the single-domain state in such particles and determines the superparamagnetic behavior of these systems. A surfactant (*HOl* in our experiments) is used to prevent particles from coagulating. Oleic acid creates a stabilizing layer that compensates for the dipole-dipole attraction between the magnetic particles. The thickness of the stabilizing layer is 1–2 nm. The magnetic fluid that we used in the experiments has a concentration of 3.2 vol.%, dielectric constant $\varepsilon = 2.1$, and conductivity $\sigma = 3.8 \times 10^{-7}$ $(\Omega{\cdot}\mathrm{m})^{-1}$ (measured at 1000 Hz).

The autowave process observation is described below. The experiment was carried out on the unit shown in **Figure 1**.

The magnetic fluid (1) is placed into a cell between two electrodes (2, 5), made of glass with a conductive transparent coating $InSnO_2$ (ITO). The area of electrode surface is $S = 30 \times 40$ mm^2 with the glass thickness of 4 mm and conductive coating thickness of $h_0 = (160 \pm 5)$ nm; the thickness of magnetic fluid first layer is 1–40 μm. A beam from a source of white light (8) falls on the surface of a cell with a magnetic fluid. The falling rays were reflected from the "electrode-magnetic liquid" border and were recorded by a camera (7). A constant or pulse voltage was applied to the electrodes (9). The prism (6) was used to separate the rays that reflected from

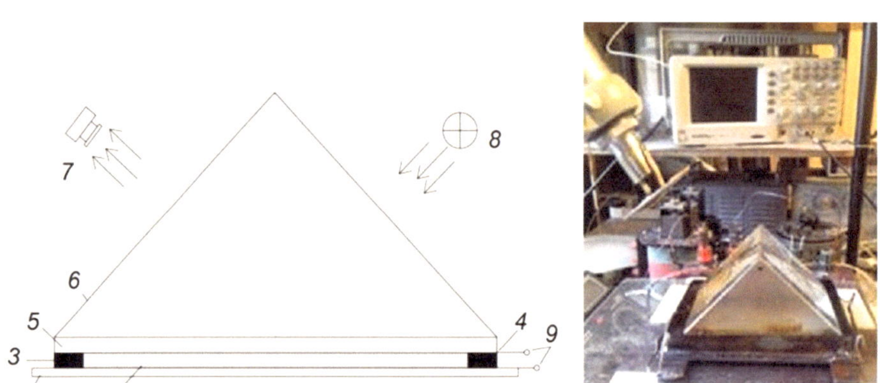

A B

Figure 1.
(A) Schematic drawing of the experimental unit: (1) "magnetite in kerosene" magnetic fluid; (2, 5) glass with conductive coating; (3, 4) insulating gaskets; (6) prism; (7) camera or photodiode; (8) lighting (white or monochromatic light); (9) electrodes. (B) Photo of the experimental unit.

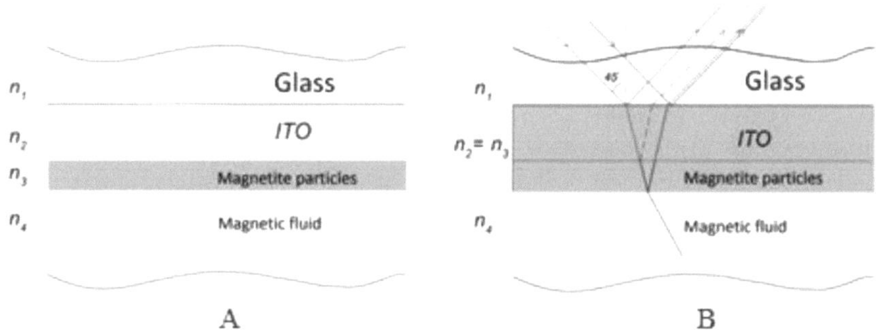

Figure 2.
Model of a multilayer structure at the cell surface. (A) Model of a multilayer structure "glass – ITO – layer of magnetite particles – magnetic fluid." (B) Model of three optical layers "glass – ITO + a layer of magnetite particles – magnetic fluid" with reflected rays.

the surface of the glass (5) and from the surface of the transparent ITO electrode. The complex refraction indices and thickness of the conductive coating (ITO) were measured by using a spectroscopic ellipsometer SE 800 SENTECH. The refraction index of glass is $n_1 = 1.52$, and the refraction index of conductive coating is $\hat{n}_2 = 1.76(1 + 0.04i)$. The refraction index of magnetic fluid with a concentration of 3.2% vol. is $\hat{n}_4 = 1.45(1 + 0.01i)$. The rays of light with a wide range of wavelengths (white light) fall on the ITO electrode and are reflected from its upper and lower borders. Since the thickness of ITO is ~200 nm, the interference of the reflected rays occurs. When voltage is applied to the electrodes, the magnetite particles move to the electrodes. Thin layers of a concentrated magnetic fluid (~27–30%) are formed near the electrodes, which corresponds to the dense packing of particles with a protective coating. The MF refraction index of such concentration is $\hat{n}_3 = 1.76(1 + 0.03i)$.

The refraction indices of ITO and near-electrode layer are close in value: $n_2 \approx n_3$. Therefore, the growth of near-electrode layer of concentrated magnetic fluid in the electric field is optically equivalent to the increase of conductive coating (ITO) thickness. The interference of falling rays is taking place in two-layer structure —"conductive ITO coating—a near-electrode layer." It can be visually observed by changing of the cell surface color from green to crimson. We called this electrooptical effect as electrically controlled interference (**Figure 2**).

Depending on the near-electrode layer thickness, the surface of a cell with a magnetic fluid changed its color. The visible color change of the layer is explained by the shift of spectrum maximum due to the increase of the structure "conductive ITO coating – near-electrode layer" optical thickness. If the voltage on electrodes is more than critical voltage (~12 V), particles in the near-electrode layer become unstable, and the autowave process (AW-process) starts [6].

In our experiments, we considered the temperature of the liquid to be the same and did not take into account the heating during the passage of electric current and heat transfer [8].

3. Experimental results

3.1 Excitation autowave spreading modes

3.1.1 "Fast" autowave mode

A technique for the calculation of the near-electrode layer thickness is presented in [9]. Depending on the cell voltage applied to the electrodes, the thickness of the layer varies up to 100 nm. The specific resistance of this thickness layer is several orders of magnitude greater than the specific resistance of the liquid in the cell (the thickness of the liquid layer in the cell is ≈40 μm). Hence, the field strength at a steady-state current in a layer of such thickness also will be several orders of magnitude greater. At some critical voltage on the electrodes, the tension in the layer becomes $\sim 10^7$ V/m, and it becomes conductive (Wien effect). Individual elements (ensembles of particles) get the same charge from the electrode and move away from it. A running wave is visible on the surface of the cell (**Figure 3**).

Note that complete information about the autowave structure evolution can be obtained by describing only the time evolution of the wave front position. This is the basis for the kinematic approach to the description of autowave structures [10]. Since the cell has dimensions of 3 × 4 cm, we can estimate the average velocity of the wave front motion. Approximately 0.25 s passes from the appearance of green color areas (**Figure 3A**) before filling the entire surface of the cell with this color. Such wave is called "fast"; the velocity of its movement is about 16 cm/s. It looks like a flame spreading over the steppe by setting fire in different places. This is a well-known problem solved by Zeldovich and Frank–Kamenetsky.

3.1.2 "Slow" autowave mode

Further, the layer elements (ensembles of particles) fall into the cell, break apart, and lose the same charge as the electrode. They get oppositely charged in the cell and begin their movement to the electrode due to electro- and dipolar phoresis. Behind the first wave follows the second one; on the surface of the cell, there appears a picture of the autowave process (**Figure 4**). The waves move with a velocity of ≈1 cm/s. These are the so-called phase ("slow") autowaves. That is, the MF near-electrode layer is an excitable medium with a restoration. An example of such a process can be watched on video (Video 3 available at https://yadi.sk/i/WR1vDK5IzGGxLg).

3.2 Spiral waves (reverberators)

We managed to observe the appearance and development of spiral waves (reverberators) in the near-electrode layer of magnetic fluid (Video 4 available at

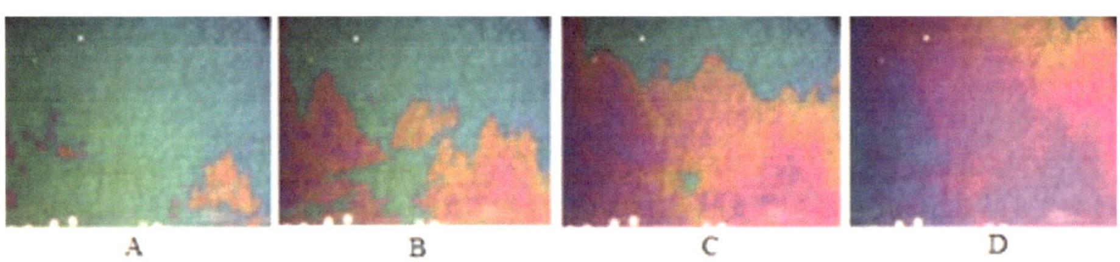

Figure 3.
"Fast" wave in the near-electrode layer of a cell with magnetic fluid.

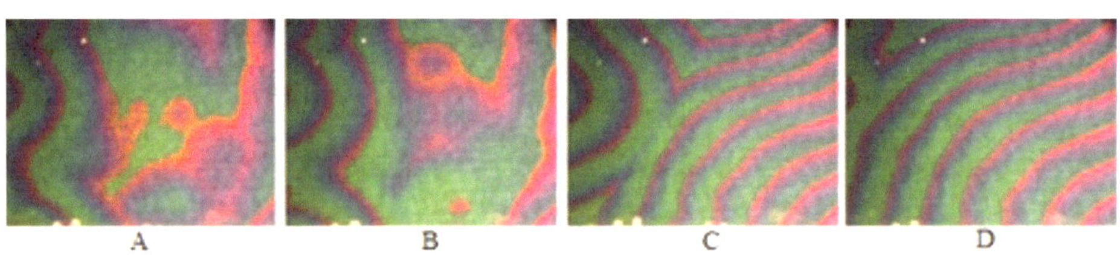

Figure 4.
Phase ("slow") autowaves in the near-electrode layer of a cell with magnetic fluid.

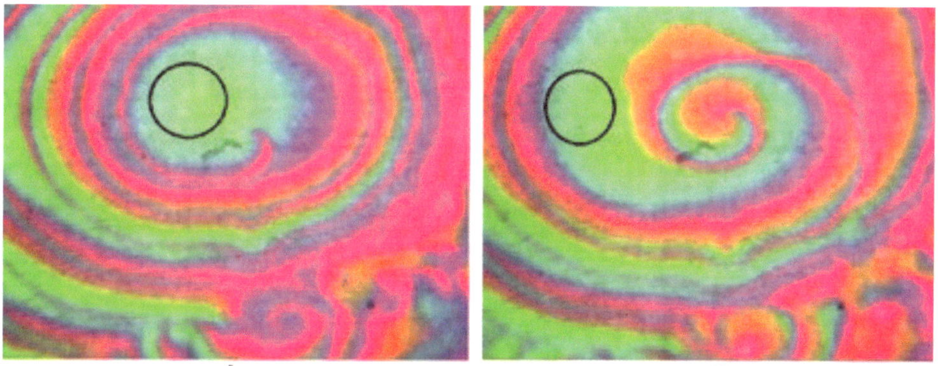

Figure 5.
Spiral waves: reverberators in the autowave process. (A) Reverberator at the beginning of the rotation period, and (B) reverberator at half of the rotation period. Singular domains are marked.

https://yadi.sk/i/8OcbP4y831nPMQ). These were single reverberators (**Figure 5**) and multiple ones (**Figure 6**). The reverberator's lifetime is limited to a few turns. If several reverberators appear in the observation field, then after 20 s, only one remains [11].

Figure 5 shows a photograph of the reverberator: (A) the beginning of the reverberator rotation period and (B) half of the period. As seen in the figure, there is a region in the center of the reverberator core, corresponding to the interference pattern of reflection from the electrode at a constant layer thickness. So in this region, there is no oscillation of the layer. This is the so-called singular domain. Its existence comes from the reasoning described in [12].

3.3 Leading centers (pacemakers)

Figure 7 shows the pacemaker (leading center), which we observed in the considered active medium. Pacemakers can emit waves with different periods, as can be seen in **Figure 7A** and **B**. The larger the radius of curvature of the cylindrical wave, the lower the speed of the wave. So in **Figure 7A**, the period of the wave emitted by the pacemaker is 0.31 s. Since all characteristics in the autowave process are determined by the system itself, the period of the waves emitted by each

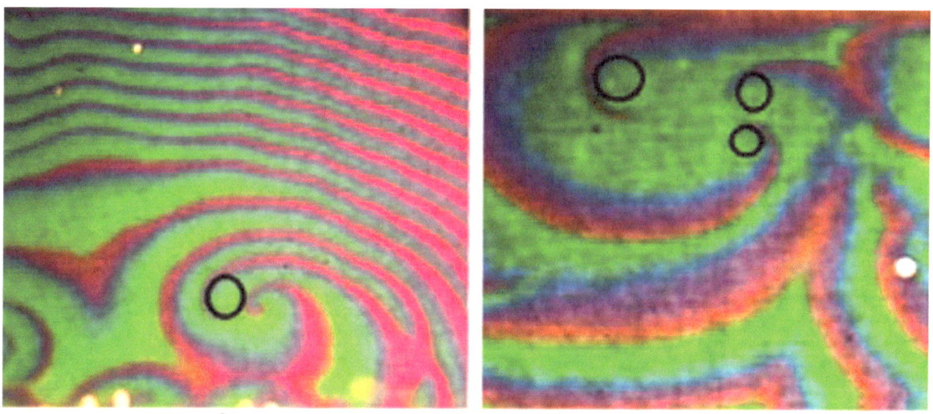

Figure 6.
(A) Two-arm reverberator and (B) three-arm reverberator. Singular domains are marked.

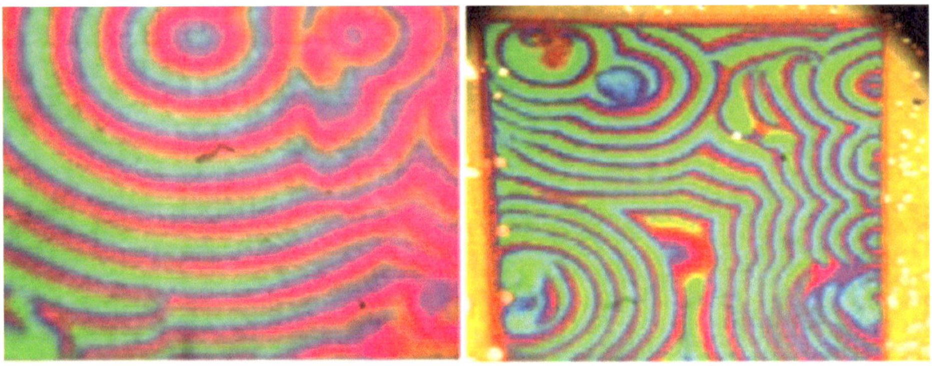

Figure 7
Experiment: pacemakers with different periods. Frame width, 1.2 cm.

pacemaker is determined only by its own properties and cannot be adjusted by external influence.

3.4 Autowave diffraction on an obstacle

Autowave is a special kind of wave that does not have standard wave properties. Autowaves do not interfere and do not transfer energy. For autowaves, the principle of superposition is not valid; they mutually destroy each other in a collision (a wave with a smaller period "destroys" a slower wave). The only common property of autowaves with conservative waves is diffraction or rounding of the obstacle. The bending of autowave near the obstacle in a cell with magnetic fluid is shown in **Figure 8**.

3.5 Propagation of autowaves in a medium with local heterogeneity: simulation of cardiac pathology

Local heterogeneity was created in the near-electrode layer of the magnetic fluid cell, as shown in **Figure 9**. This figure also shows the propagation of the autowave, which shows that there is a region around the heterogeneity and the oscillations which occur with a phase different from the oscillations of the cell surface. In this case, the current waveform has the form shown in **Figure 10**.

Figure 8.
Experiment: the autowave diffraction around the obstacle. Frame width, 1.2 cm.

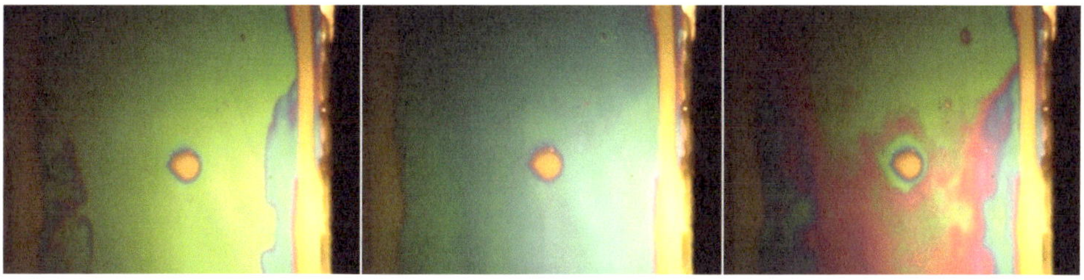

Figure 9.
Propagation of autowaves in a medium with local heterogeneity.

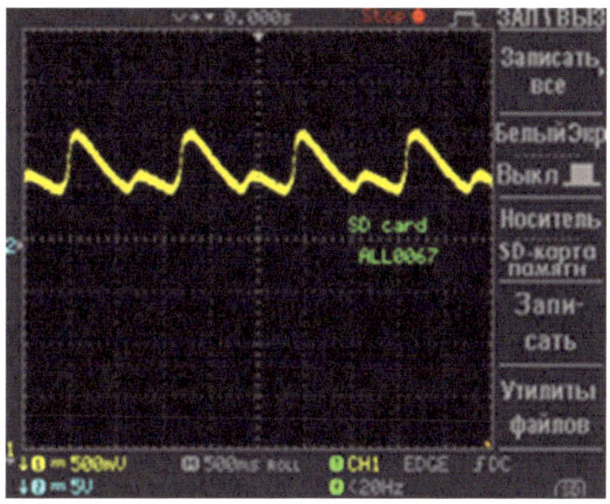

Figure 10.
Oscillogram of relaxation oscillations of a current in a cell with local heterogeneity.

We assume that the region around the local inhomogeneity has a refractoriness time R_1, different from the refractoriness time R_2 of the medium. Therefore, the wave front reaches the section with refractoriness R_1 and gets broken. Thus, arrhythmia occurs in oscillations between the surface of the cell as a whole and the area around the local heterogeneity.

If there is a heterogeneity in the myocardium (muscle middle layer) of the heart, for example, a heart attack, which means that there is no blood supply in some area, then a similar rhythm transformation may occur in the heart muscle.

This experiment can be used as a model for processes occurring in the myocardium.

4. The physical model of the autowave process

The physical mechanism of the autowave process in the near-electrode layer of a magnetic fluid is the following: as mentioned earlier, in an electric field, the charged particles of magnetite due to electro- and dipolar phoresis move to the nearest electrode and form a close-packed layer 0–170 nm thick around it (**Figure 11**).

A thin layer—the so-called structural-mechanical barrier—appears between the layer of particles and the electrode. It consists of a dielectric—a mixture of kerosene and oleic acid molecules (surfactant surrounding the magnetite particles). The thickness of this barrier is about 10 nm, but it prevents the particle discharge because of the contact with the electrode. With the growth of the near-electrode layer, the intensity increases in the barrier. At its critical value $E \sim 10^7$ V/m, the dielectric becomes conductive and magnetite particles get recharged and start moving from the electrode—a single flat autowave passes.

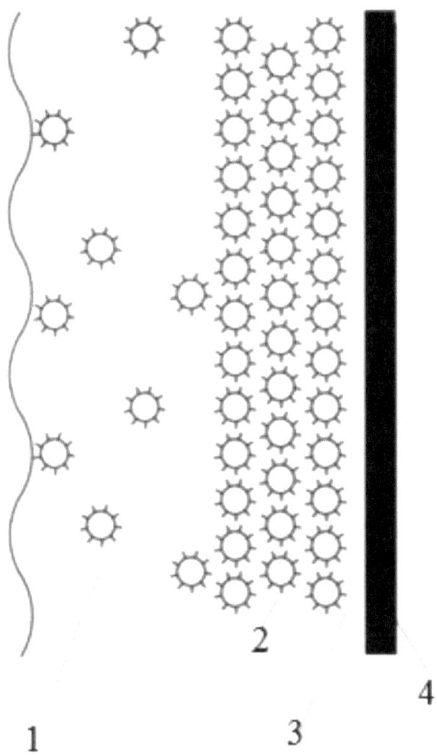

Figure 11.
Schematic of a cell with a magnetic fluid with a formed near-electrode layer. (1) Magnetic fluid in the cell, (2) a layer of magnetite particles, (3) dielectric (structural-mechanical barrier), and (4) electrode (ITO).

We assume that the medium is excitable and that it consists of a set of elementary areas of near-electrode layer. According to definition, each such site is an autonomous source of energy. Energy is accumulated due to the so-called polarization capacitance of near-electrode layer of particles [13]. The elements of an excitable medium interact with each other, transferring the energy of electrical and hydrodynamic interaction from one to another. This is how the excitation pulse transmits. This process is similar to the interaction of nonlinear coupled oscillators with a short coupling, where each oscillator is associated with only several adjacent ones. Thus, the physical meaning of the autowave process mechanism considered by us is a system of coupled nonlinear oscillators.

An excitable element has only one stable stationary state. An external impact that exceeds the threshold level can take an element out of a steady state and make it take action before it returns to that state. During the action, the active element can affect the elements associated with it and, in turn, remove them from the stationary state. As a result, an excitation wave propagates in such a medium. This is the most common type of autowaves in biological systems such as nerve tissue or heart muscle. In our medium, each section of the near-electrode layer affects the neighboring areas, transferring the charge and thereby removing them from the equilibrium state.

The element of an excitable medium can be in three states—rest, excitation, and refractoriness.

The local elements that make up the active medium under study (elements of the near-electrode layer) have “the only distinguished state of rest that is stable with respect to fairly weak external influences” [13]. This is a state of the medium, when the electrode layer has already formed, (the voltage on the electrodes is less than the critical one), but the autowave process has not yet begun.

When the voltage on the electrodes exceeds critical, the element gets excited and becomes active, and a discharge occurs in the elementary capacitor. This is the state of excitation or the so-called active transition, after which the elementary segment of the layer disappears, because the layer in this local region is destroyed. In this case, a single wave front passes over the surface.

Refractoriness is a state that comes after excitation when the medium accumulates energy and cannot produce a new impulse. The medium must restore its properties with the energy coming from outside and prepare for the next impulse. The energy from the external source continues to flow, and the magnetite particles move to the nearest electrode, forming a system of successively connected capacitors: first electrode – structural-mechanical barrier – the layer of close-packed particles – magnetic fluid of low concentration – the layer of close-packed particles – structural-mechanical barrier – second electrode. The period of refractoriness corresponds to the charging of capacitors, one plate of which is an electrode and the other is near-electrode layer.

5. Synchronization of the autowave process under the influence of a periodic electric signal

In this section, we describe the synchronization of an autowave process by an external influence in a thin near-electrode layer of a magnetic fluid.

One of the main trends in the living world is the tendency to achieve a common rhythm in collective behavior—a tendency toward synchronization. An example of synchronization in nature is a colony of simultaneously flashing fireflies; these are simultaneously flapping birds flying in a flock. In technics, synchronization is important when designing computers with parallel architecture.

It was also noted that the most famous example of an active autowave environment is the Belousov-Zhabotinsky chemical reaction. Petrov et al. [14] experimented with the photosensitive form of such a reaction using periodic optical effects.

Earlier we noted that the autowave process in the near-electrode layer of magnetic fluid is considered from the standpoint of oscillations of coupled oscillators. Each element of the layer (an ensemble of particles) interacts with its nearest neighbors or, in a more complicated case, with several neighbors.

Each oscillator oscillates with its own frequency, but also we could see oscillator clusters that oscillated with the same frequency.

We affected the self-oscillation medium (oscillators) with an external periodic force (pulsed electric field) and achieved synchronization. As the synchronization criterion, the conditions of frequency and/or phase adjustment are used. It is considered that the oscillator 1:1 is synchronized by an external periodic effect, if its own frequency f0 becomes equal to the frequency of the external signal f_e.

In addition to the criterion "frequency capture," another criterion is used —"phase capture." This criterion means that for any moment of time t, the condition $|f_e\ (t)\ \text{-}\ f_0(t)\text{-}\ const| < 2\pi$ is satisfied.

We affected on the observed self-oscillations in a thin near-electrode layer of magnetic fluid by a periodic pulsed electric field and selected its frequency f_e in such a way that it matched with the natural oscillation frequency of the layer elements. Thus, with a frequency capture of 1:1, we obtained a picture of autowave synchronization as shown in **Figure 12**. This picture allows us to determine the natural oscillation frequency f_0 of the elements of the near-electrode layer of magnetic fluid.

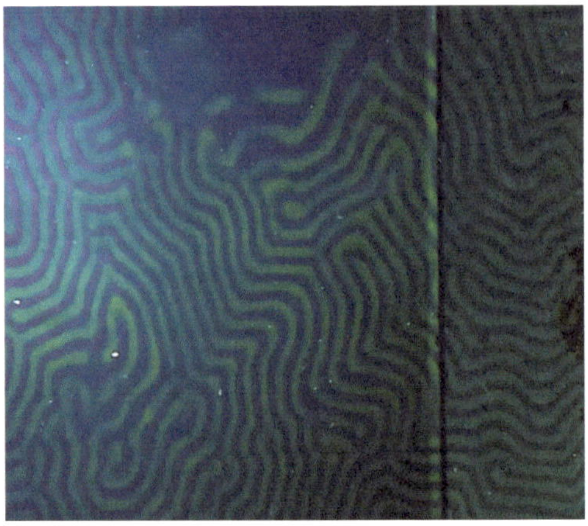

Figure 12.
Synchronization of autowaves in a thin layer of concentrated magnetic fluid by an external pulsed electric field. The ratio of the natural oscillation frequency of the elementary section of the layer f_o *and the frequency of the pulsed electric field* f_e *is 1:1.*

Figure 13.
The formation of breaks and labyrinth structures with the ratio of frequencies: $f_e \sim 2f_o$.

If the ratio of the frequency of a pulsed electric field and the natural frequency are 2:1, then a few time after the beginning of the effect, we observed breaks and labyrinth structures, as shown in **Figure 13**, or two clouds, which oscillate with a phase shift of π, as shown in **Figure 14**.

Since the near-electrode layer of magnetic fluid is a heterogeneous medium, it can be assumed that the synchronization mechanism is similar to the formation of clusters in the intestine: oscillators with similar frequencies tend to group together.

With a ratio of 1:3, three uniformly oscillating clouds were observed (**Figure 15**).

The synchronization mechanism is the following: each point tends to synchronize with external force, as well as with neighboring points. If there were no interaction in the medium, then, with a frequency ratio of about 1:*m*, neighboring points would have an equal probability of the phase difference $2\pi/m\ i$, where i = 1, ..., m^{-1}. Because of the interaction, the points tend to have the same phases as the neighbors, and a compromise is achieved through the formation of clouds. Obviously, the phase difference between the clouds is $2\pi/m\ i$ [15].

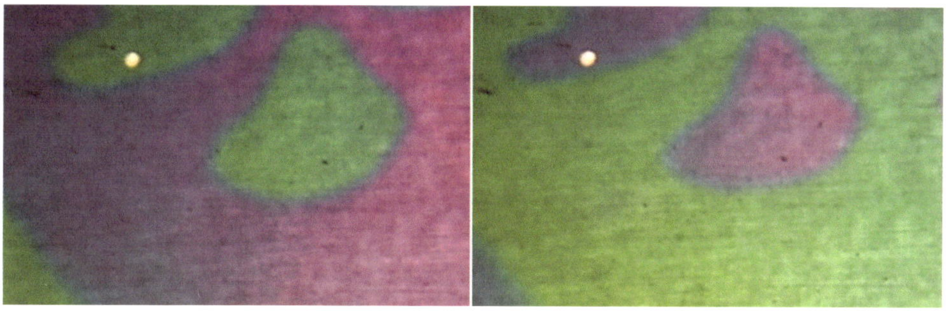

Figure 14.
Two clouds of synchronously oscillating points. The shift of phase between oscillations is π.

Figure 15.
Three uniformly oscillating clouds.

6. Mathematical model of the autowave process

To simulate spiral waves, pacemakers, and synchronization, we use the basic FitzHugh-Nagumo model (FHN). It describes an excitable medium and is named after Richard FitzHugh (1922–2007), who proposed the system in 1961, and Nagumo, who proposed a similar scheme the next year. It allows to adjust the characteristics of the autowave process widely and change the parameters of its behavior. Model solution can be obtained using the finite element method [16].

The model consists of two equations; the first equation describes a "fast" process—a change in the intensity in the near-electrode layer after the increase in the conductivity of the structural-mechanical barrier.

The second equation describes a slow process—the change in charge of the near-electrode layer, when charged particles of magnetite accumulate in it.

$$\frac{\partial V}{\partial t} + \frac{\partial^2 \mathrm{V}}{\partial \mathrm{t}^2} = D\Delta V - \mathrm{V}^3 + \mathrm{V} - \mathrm{I}$$
$$\frac{\partial I}{\partial t} = \varepsilon(V + \alpha - bV) \tag{1}$$

where V is the function, depending on the field intensity in a thin near-electrode layer of a magnetic fluid (activator), I is the function associated with a change in charge (inhibitor), D is the diffusion coefficient of the activator, and ε is a small parameter; presumably this is the ratio of the time of single pulse passage and the time of the near-electrode layer formation.

Depending on the values of the parameters α and β, the medium element can be either in the oscillatory mode or in the excitable mode.

The proposed model of the autowave process was implemented by means of the COMSOL Multiphysics 5.2. The solutions that were obtained for spiral waves, pacemakers, and obstacle diffraction correspond to full-scale experiments and confirm the adequacy of the proposed model (1).

The simulation results and comparison with experimental data are shown in **Figures 16–18**.

We were able to show that the diffraction of autowaves (bending the obstacles) is a consequence of the existence of a region with nonlinear characteristics near the obstacles. One of the characteristics is the deceleration parameter φ.

The coefficient φ depends only on the distance to the nearest point of the obstacle and is calculated by the following formula:

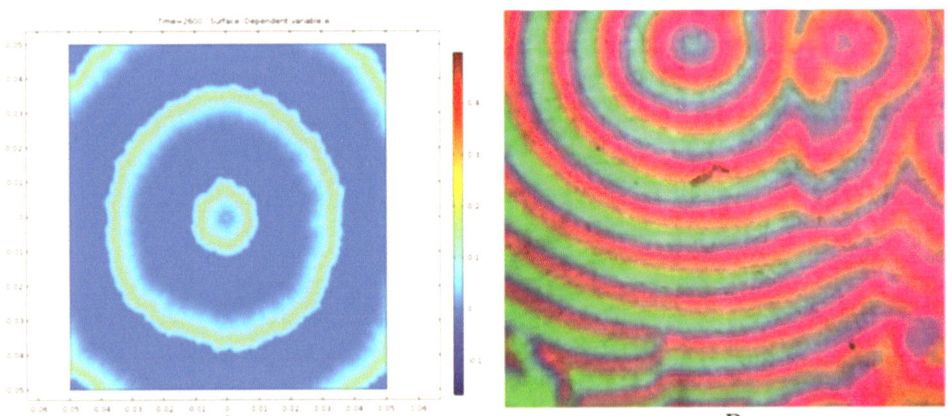

A B

Figure 16.
Comparison of the results of pacemaker simulation and experiment. (A) The result of modeling a single pacemaker; (B) the result of a full-scale experiment.

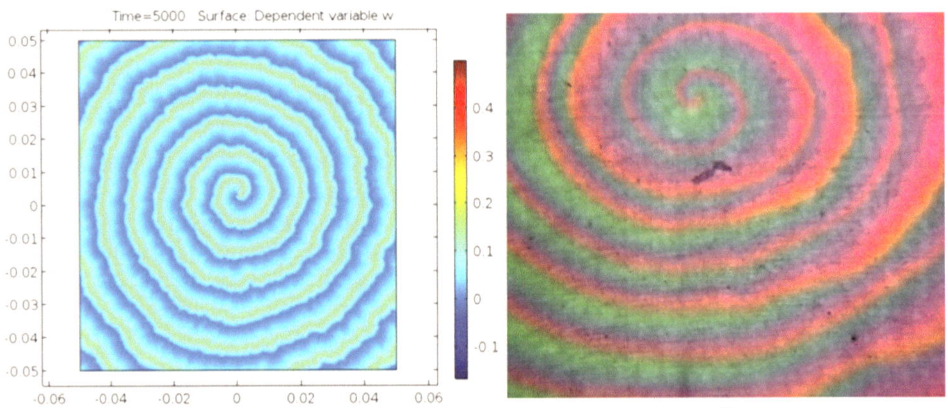

A B

Figure 17.
Comparison of the results of spiral wave simulation and experiment. (A) The result of modeling the development of a one-arm reverberator; (B) the result of a full-scale experiment.

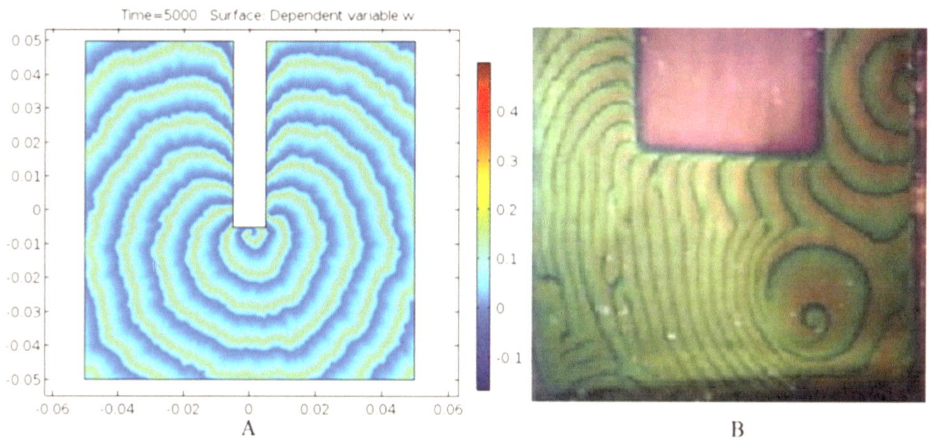

Figure 18.
Comparison of the autowave diffraction simulation results and experiment: (A) modeling the rounding of an obstacle; (B) the result of a full-scale experiment.

$$\varphi = \left(\frac{d}{r}\right)^4 * H(d < r) \tag{2}$$

where *d* is the distance to the obstacle, *r* is the range of the deceleration function, and *H* is the Heaviside function.

7. Conclusions

This chapter describes the study of the autowave process in a layer of concentrated magnetic fluid that forms at the electrode in an electric field. The main modes of the autowave process are investigated: pacemakers and reverberators (spiral waves) and obstacle rounding (autowave diffraction). All visualizations are obtained by the author's method of electrically controlled interference in a thin layer.

The uniqueness of the medium we study is that the autowave process is controlled by a weak electric field. It can be observed indefinitely, and this process is easily reproduced in the laboratory. Also synchronization by external influence was studied in the electrically controlled medium.

We consider very important to investigate and control heterogeneities (obstacles) in the active medium. Heterogeneities are the main reason that causes the development of cardiac arrhythmias and heart attacks in the human heart. By controlling the movement of a wave around a heterogeneity, arrhythmia can be eliminated. This is important for practical use.

A mathematical model of the autowave process, based on the FitzHugh-Nagumo basic model, is created for the near-electrode thin layer of a magnetic fluid. The model is implemented in the interactive environment for physical processes simulation—COMSOL Multiphysics 5.2. The coefficients in the equations of the modified FHN system are found; using these coefficients, we can simulate different autowave modes.

A criterion for the adequacy of the model is a visual correspondence to the full-scale experiment.

All the features of the formation and propagation of reverberators, pacemakers, and diffraction in the autowave medium observed in the natural experiment are repeated in computer simulation, which will allow us to find further methods for controlling the development of autowave process. This has practical application either in fundamental field or in practical purposes.

Acknowledgements

This paper was written as part of the state task Project No 3.5385.2017/8.9 for the implementation of the project on the topic: "Experimental research and mathematical modeling of interphase and near-surface phenomena in a thin membrane of nanostructured magnetic fluid" in MIREA—Russian Technological University.

Thanks

Authors give their great thanks to Prof. Chekanov V.V. (RIP) for the invaluable help in preparing the materials and experimental results.

Author details

Vladimir S. Chekanov[1], Natalya V. Kandaurova[1*], Viktoria I. Drozdova[2], Galina V. Shagrova[2], Veniamin V. Romantsev[3] and Mikhail Yu. Shevchenko[4]

1 MIREA—Russian Technological University, Stavropol Branch, Stavropol, Russia

2 North-Caucasian Federal University, Stavropol, Russia

3 St. Petersburg State Electrotechnical University "LETI" named after V.I. Ulyanov (Lenin), St. Petersburg, Russia

4 Wonder Technologies LLC, Moscow, Russia

*Address all correspondence to: candaur18@yandex.ru

References

[1] Sheikholeslami M, Barzegar Gerdroodbary M, Moradi R, Shafee A, Li Z. Application of neural network for estimation of heat transfer treatment of Al_2O_3-H_2O nanofluid through a channel. Computer Methods in Applied Mechanics and Engineering. 2019;**344**: 1-12

[2] Sheikholeslami M. Numerical approach for MHD Al_2O_3-water nanofluid transportation inside a permeable medium using innovative computer method. Computer Methods in Applied Mechanics and Engineering. 2019;**344**:306-318

[3] Sheikholeslami M. New computational approach for exergy and entropy analysis of nanofluid under the impact of Lorentz force through a porous media. Computer Methods in Applied Mechanics and Engineering. 2019;**344**:319-333

[4] Zaikin AN, Zhabotinsky AM. Concentration wave propagation in two-dimensional liquid-phase self-oscillating system. Nature. 1970;**225**: 535-537

[5] Chekanov VV, Iljuch PM, Kandaurova NV, Bondarenko EA. Autowaves in near-surface layer of magnetic fluid. Journal of Magnetism and Magnetic Materials. 2005;**289**: 107-109

[6] Chekanov VV, Kandaurova NV, Chekanov VS. Phase autowaves in the near-electrode layer in the electrochemical cell with a magnetic fluid. Journal of Magnetism and Magnetic Materials. 2017;**431**:38-41

[7] Rosensweig RE. Ferrohydrodynamics. Cambridge: Univ. Press; 1985. 344p

[8] Sheikholeslami M, Mehryan SAM, Shafee A, Sheremet MA. Variable magnetic forces impact on magnetizable hybrid nanofluid heat transfer through a circular cavity. Journal of Molecular Liquids. 2019;**277**:388-396

[9] Chekanov VV, Kandaurova NV, Chekanov VS. Thickness calculation of thin transparent conductive membrane on the border with a magnetic fluid. Journal of Nano-and Electronic Physics. 2016;**8**:143-148

[10] Davydov VA, Zykov VS, Mikhailov AS. Kinematics of autowave structures in excitable media. Advances in Physical Sciences. 1991;**161**:45-83

[11] Chekanov VV, Kandaurova NV, Chekanov VS. Experimental study of the properties of autowave sources (reverberators) in the magnetic fluid near-electrode layer using reflected light interference. Magnetohydrodynamics. 2017;**53**(3):495-500

[12] Ivanitsky GR, Krinsky VI, Selkov EE. Mathematical Biophysics of the Cell. Moscow: Science Publ. House; 1978. p. 308

[13] Kandaurova NV, Chekanov VV, Chekanov VS. Effect of a near-surface nanolayer formation on the magnetic fluid electrical properties. Acta Tech. 2018;**63**(4):555-562

[14] Petrov V, Ouyang Q, Swinney HL. Resonant pattern formation in a chemical system. Nature. 1997;**388**: 655-657

[15] Pikovsky A, Rosenblum M, Kurths J. Synchronization. A Universal Concept in Nonlinear Sciences. Cambrige, Great Britain: Cambridge University Press. pp. 167-169

[16] Sheikholeslami M, Haq R-u, Shafee A, Li Z. Heat transfer behavior of nanoparticle enhanced PCM solidification through an enclosure with V shaped fins. International Journal of Heat and Mass Transfer. 2019;**130**:1322-1342

Chapter 6

Synthesis and Application of Porous Kaolin-Based ZSM-5 in the Petrochemical Industry

Ebrahim Mohiuddin, Yusuf Makarfi Isa,
Masikana M. Mdleleni and David Key

Abstract

Zeolites are advanced chemical materials that play a significant role in many petrochemical applications. In recent years, research interest in improving and enhancing the effectiveness of ZSM-5 as a catalyst has grown immensely. In particular, finding cheaper, environmentally friendly alternative starting materials for the synthesis of ZSM-5 has gained much attention. Kaolin has been widely investigated as a zeolite precursor as it comprises the required constituents for an aluminosilicate zeolite material; ubiquitous nature and its benefit in synthesising zeolites are well known as an inexpensive way of obtaining catalysts. This chapter deals with the factors affecting ZSM-5 synthesis when utilising a kaolin precursor. The effects of kaolin crystallinity, kaolinite content and synthesis parameters on ZSM-5 formation and its physicochemical properties are discussed. The potential of kaolin-based ZSM-5 as an oligomerisation catalyst is investigated. Pure, crystalline ZSM-5 could be successfully synthesised from a kaolin precursor. Physicochemical properties such as morphology, porosity and acidity are affected by the kaolin precursor and optimum synthesis conditions are required for synthesis of ZSM-5 from particular kaolin. Kaolin-based ZSM-5 catalyst showed good activity and selectivity to valuable fuel range hydrocarbons.

Keywords: kaolin, ZSM-5, synthesis, oligomerisation, cracking

1. Introduction

Access to a variety of energy sources has been fundamental in driving human development. Fossil fuels have been a major source of energy for mankind for more than 5000 years. Today, crude oil continues to be a significant contributor to the energy sector; it accounts for a large percentage of the world's energy consumption. The production of chemicals has also continued to play a pivotal role in our daily activities. Interestingly, the amount of chemicals produced and used for both domestic and industrial purposes is very much related to the growth in global population. However, energy sources such as crude oil are non-renewable sources of fuel and current estimations show that world oil supplies will be depleted in the next century. Apart from the uncertainties in crude oil reserves, a major cause for concern is the impact crude oil extraction and its refining has on the environment; the combustion of fossil fuels leads to a net increase in greenhouse gases (GHG) leading to global warming. These concerns among others have led researchers in the recent past to explore alternative energy sources to the traditional crude oil for the production of fuels and petrochemicals. Various practises such as the use of efficient catalysts and augmented reactor technology are currently being employed towards ensuring that production technologies are eco-friendly and sustainable.

Nanoporous materials are a large class of materials which consist of either an organic or inorganic framework structure containing ordered porous networks. They are generally classified by having pores sizes less than 100 nanometres and may be subdivided into three categories i.e. Microporous (<2 nm), Mesoporous (2–50 nm) and Macroporous (>50 nm). Their ability to interact or discriminate molecules depending on size has granted them scientific and technological importance. Research interest in nanoporous materials continues to grow as researches attempt to understand structure–property relations and design materials tailored for certain applications. Dependent on the properties of the nanoporous materials, applications range from purification and separation, sorption and drug delivery to energy storage, solar and fuel cells as well as electronic and magnetic devices. Typical examples include activated carbon, metal organic frameworks, ceramics, various polymers, aerogels, silicates and zeolites to name a few. Zeolites are microporous aluminosilicate materials that possess a 3-dimensional pore structure and play a prominent role in the petrochemical industry as ion exchangers, adsorbents, in separation and catalysis [1–4]. Their shape selective properties permit control of product distribution in chemical reactions and as such have become indispensable catalysts in many petrochemical processes [5]. Of particular importance to the petrochemical industry is zeolite ZSM-5. ZSM-5 because of its unique channel structure, acidity, and hydrothermal stability has been used as a shape selective catalyst in isomerization, alkylation, oligomerisation and catalytic cracking reactions [6–9]. It is conventionally synthesised using chemical sources such as sodium silicate solutions or silica gels and aluminium salts as the starting materials. Commercial synthetic zeolites are preferred over their naturally occurring analogues due to higher purity and uniform particle size which makes them more suitable for scientific and industrial applications [10]. However, zeolite synthesis using conventional methods leads to large amounts of waste being produced and chemical sources may be expensive, leading to high costs of zeolite production which limit commercialisation and use in many industrial applications [11]. Recently there have been increased efforts to explore the use of more affordable, natural raw materials possessing the necessary requirements for the synthesis of zeolites. ZSM-5 has been synthesised from natural silica and alumina sources such as rice husk ash [12–14], expanded perlite [15], palygorskite [16], fly ash [17], and kaolinite [18–21]. The main drive to utilise these rich aluminosilicate minerals is their relative abundances, cost effectiveness and overall more environmentally friendly synthetic procedures. Although many natural minerals and manufacturing wastes have been utilised to synthesise a wide variety of zeolite structures, this chapter will focus on kaolin-based ZSM-5 synthesis and possible application in the petrochemical industry.

Kaolin is a white clay composed mainly of kaolinite, a hydrous aluminosilicate mineral containing silica and alumina in a 1:1 ratio as well as impurities such as quartz and mica. Kaolin may require beneficiation to remove impurities depending on its application. Due to its low Si/Al ratio it has been extensively use in the synthesis of low silica zeolites [22, 23]. High silica zeolites such as ZSM-5 have also been synthesised with the addition of supplementary silica sources as well as through dealumination of kaolin via acid treatment [24–26]. Of the extensive range of aluminosilicate minerals used as zeolite precursors, kaolin has been favoured due to its ubiquitous nature. However, the studies of kaolin from different areas are significant since kaolin varies in composition depending on its geological occurrence. The chemical compositions of materials affect their properties and variations in the kaolin structure and composition can thus affect its subsequent chemical reactivity [27, 28].

In this chapter the synthesis of kaolin-based ZSM-5 and the factors affecting synthesis are discussed. The work presented will focus on the synthesis of ZSM-5 using kaolin of South African origin. Most studies of kaolin derived ZSM-5 is performed on commercial kaolin. Only few have been done using raw kaolin. Chemical reactivities of kaolins obtained from different geological areas and the need to optimise synthesis conditions tailored to particular kaolin are highlighted. The effects of kaolinite content and synthesis parameters such as crystallisation time and temperature are discussed. The work is extended to include the effects of silica to alumina ratio on the physicochemical properties of ZSM-5 and is the main focus of this chapter. Furthermore, the application of kaolin-based ZSM-5 in important petrochemical processes such as the oligomerisation of olefins to fuel range hydrocarbons is evaluated.

2. Kaolin in zeolite synthesis

In the search for cheaper and more environmentally friendly alternatives to chemical sources, much research has been conducted on the feasibility of kaolin. By converting kaolin to the more reactive metakaolin via thermal activation and subjecting it to hydrothermal treatment in a NaOH medium, zeolite is produced. The use of kaolin as a source of silica and alumina was reported by Barrer [29] after it was calcined between 700 and 1000°C to form metakaolin. However due to the variations in kaolin composition and structure its subsequent chemical reactivity may be affected. Synthesis of zeolites from kaolin is affected by factors such as degree of crystallinity of the kaolin [23], kaolin composition [30], mineralogical impurities [23, 31], calcination temperature of kaolin [32], specific surface area of kaolin [33], synthesis parameters such as crystallisation temperature, time [30] and silica alumina ratio.

2.1 Factors affecting kaolin-based zeolite synthesis

2.1.1 Kaolin crystallinity

Many studies have been performed employing kaolin as the starting material for zeolite synthesis [34]. Investigations on the effects of kaolin crystallinity are contradictory as some researchers have shown that differences in the reaction kinetics of zeolite formation are observed for kaolin with different crystallinities or structural ordering [23] whereas others have reported that no significant differences were established when synthesising zeolites from kaolin of different crystallinity and reported that reactions of metakaolinites are independent of defects in the original crystal structure [22]. When two kaolins of South African origin from different geological areas i.e. Grahamstown (BK) and Fishoek (SK) were analysed it was shown that they differed in their structural order as well as composition and SK was more crystalline than BK [35]. The ZSM-5 synthesised using the two kaolin precursors resulted in differences in the crystallisation kinetics. The more disordered kaolin (BK) showed faster crystallisation kinetics than the more ordered SK. The physical and chemical properties of the reactive metakaolin of BK and SK were compared. The morphology obtained from SEM analysis shown that BK was composed of highly disordered loose kaolin plates compared to SK which possessed highly ordered stacked layers. The ordering remained even after calcination to form metakaolin. It was suggested that the highly disordered kaolin dissolved at a faster rate into the gel solution compared to SK in which the stacking layers were preserved. The chemical properties of the kaolin samples were analysed using ^{27}Al and ^{29}Si NMR MAS spectroscopy. The different coordination's of aluminium and the relative amounts of each coordination type were correlated with the differences in reactivity of the two kaolins. These were determined for the kaolin before and after thermal activation, in the starting gels and products in order to observe the transformation of the Al environment as the zeolite was formed. After calcination of the precursors to form metakaolin, three distinct coordination's of Al i.e. tetrahedral (AlO_4), penta (AlO_5) and octahedral (AlO_6) species were observed however the relative amount of each differed between the two kaolins. The AlO_6 in kaolin was converted to more reactive AlO_5 and AlO_4 sites via dehydroxylation of kaolinite [36]. Metakaolin SK contained a larger amount of AlO_5 coordinated Al whereas metakaolin BK contained a larger amount of tetrahedrally coordinated Al at higher chemical shifts indicating the Aluminium existed in different environments of AlO_4 coordination. The peak occurring at higher chemical shift in ^{27}Al MAS NMR suggested the Al atoms are surrounded by different neighbouring atoms. Both q^4 and q^3 groups were identified. The q^4 group has 4 neighbouring Si atoms and the q^3 Al has 3 Si and 1 OH neighbouring group and is a highly reactive species [37]. It had been reported that metakaolins containing larger amounts of AlO_4 species due to metakaolinisation at the optimal thermal activation temperature showed increased chemical reactivity. The optimum thermal activation temperature however differed for kaolins of different structural order [38]. In the case of BK and SK the thermal activation temperatures were identical. Therefore, the results suggested that the amount of reactive q^3 AlO_4 species in particular may influence kaolin chemical reactivity and that chemical reactivity of the kaolin is indeed related to its crystallinity when thermally activated at the same temperature. ^{29}Si NMR also showed that Si atoms existed in different environments. Q^2 and Q^4 groups were identified. The Q^n groups refer to the number of neighbouring Si atoms [39, 40]. The relative amount of Q^2 and Q^4 groups showed that metakaolin of SK possessed more Q^4 (74%) species compared to metakaolin BK having more Q^2 (79%) species. This indicated that the Al and Si environments of BK and SK were different after calcination at the same temperature. The higher amount of neighbouring Si atoms in SK confirmed a higher degree of networking between Si atoms. This may therefore be related to a higher degree in structural order which may affect dissolution of metakaolin into the gel and the hence the crystallisation kinetics. When the dried gels of the zeolite batch mixtures were analysed by ^{29}Si NMR it was noticed that the gels were very similar and both exhibited a high degree of networking between Si atoms. However the ^{27}Al NMR differed as shown in **Figure 1**.

The dried gel spectra of BK and SK both show broad peaks in the tetrahedral coordination range with peak maxima at ~55 ppm corresponding to q^4 Al atoms. Dried gel BK however, possesses a shoulder peak at ~65 ppm indicating the presence of the reactive q^3 Al species. Therefore the q^3 Al from the calcined kaolin dissolves into the gel mixture and provides more reactive aluminium species to the starting gel of BK. The results also indicate that since the Si NMR spectra are similar the difference in chemical reactivity is mainly due to reactive alumina. The presence of the q^3 Al species may quickly form aluminosilicate species and govern the incorporation of Al into the framework and is most likely responsible for the faster crystallisation kinetics observed in ZSM-5 derived from BK.

The difference in crystallinity of kaolins and hence the difference in crystallisation kinetics leads to variations in the physicochemical properties of the zeolite. The synthesis parameters therefore have to be optimised to obtain a highly crystalline ZSM-5 zeolite. In this case for the more ordered SK a longer crystallisation time of 96 h compared to 48 h for BK was required to obtain a crystalline ZSM-5 with well-developed micropores. Furthermore, the acidic properties are also affected

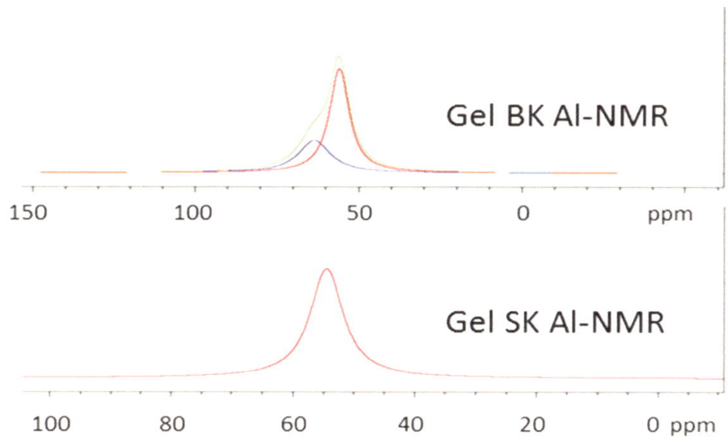

Figure 1.
^{27}Al MAS NMR spectra of dried gels of BK and SK showing the presence of different tetrahedral species [35].

and ZSM-5 was noticed to increase in acidity with an increase in crystallisation time when the more ordered SK was used a precursor.

2.1.2 Kaolinite content

The composition of kaolin varies greatly depending on its formation process and also can affect the chemistry of the clay. The clay comprises mainly kaolinite as well as impurities such as quartz, muscovite and feldspars. Other contaminants such as iron oxide and titania may also be present. The purity of kaolin determines its use in a range of applications. Kaolin if properly processed could be utilised in the production of whiteware ceramics, paper and filler pigment [41, 42] as well as in catalysis and cement production [43]. Highly contaminated kaolins such as those containing high amounts of iron oxide are use in the manufacture of bricks. While most studies have focused on the use of pure commercial kaolin in the synthesis of zeolites, only a few studies on the use of raw or virgin kaolins have been reported. Synthesis of ZSM-5 and zeolite A from kaolin of Nigerian origin containing a high amount of quartz has been reported [20, 44]. The respective zeolites could be synthesised after using beneficiation techniques (i.e. extensive settling and flocculation) or a modified autoclave to separate impurities from the synthesis gel. However, the ZSM-5 final product still contained quartz and mordenite impurities attributed to unreacted metakaolin and similarities in the synthesis conditions for both ZSM-5 and mordenite. The zeolite A purity was affected by colloidal impurities in the dispersion and the 'virgin' kaolin still required some treatment before use.

ZSM-5 was synthesised using highly impure kaolin with a high quartz content of large and finely grained quartz and muscovite originating from Grahamstown, South Africa [30]. Synthesis of ZSM-5 from raw (RK) and beneficiated kaolin (BK) was performed and the effect of the kaolinite content was investigated. Beneficiation was used to remove the majority of quartz and muscovite impurities and increase the kaolinite content. The composition of the kaolin was shown to play a significant effect on the formation of ZSM-5. From the beneficiated kaolin a highly pure crystalline ZSM-5 could be obtained. A higher kaolinite content results in more active silicate and aluminate species originating from the metakaolinite being present, which easily dissolve into the gel medium and form the primary building units necessary for nucleation and crystal growth in a shorter time period [21]. The results suggested higher kaolinite content shortens the induction period, increases the nucleation rate and hence the crystallisation of ZSM-5. However, from the raw kaolin ZSM-5 could only be synthesised under optimum synthesis conditions i.e. crystallisation temperature and time which was established for the particular kaolin during the study. Even under optimum conditions ZSM-5 synthesised from RK still contained small quartz impurities. The quartz impurities are difficult to dissolve and are therefore undesirable in kaolin [45]. Substantial amounts of alumina may also be required when kaolin is used in zeolite synthesis [46]. Thus, utilising kaolin with low kaolinite contents (i.e. alumina source) and high quartz content may result in the hindrance of ZSM-5 formation under certain crystallisation conditions. Furthermore, the results from this study suggested that beneficiation is a necessary step for the synthesis of pure ZSM-5.

2.1.3 Crystallisation temperature and time

The synthesis parameters of crystallisation temperature and time are critical in controlling the phase purity of ZSM-5. Temperature is a major factor in the formation of zeolites due to its strong effects on nucleation and crystal growth [47]. The effect of crystallisation temperature on ZSM-5 formation from BK was studied by holding the crystallisation time constant at shorter (24 h) medium (48 h) and longer times (96 h) and varying the temperature from 120 to 190°C [30]. The XRD diffractograms showed a large amorphous peak corresponding to amorphous aluminosilicate in the gel mixture is present at 120°C. As the synthesis temperature was increased to 150°C the amorphous gel is converted into pure, crystalline ZSM-5 phase as the nucleation rate increased. At 190°C the metastable ZSM-5 re-dissolves into the gel and the intensity of the ZSM-5 peaks decreases as a more thermodynamically stable quartz phase crystallises. At a high temperature of 190°C zeolite synthesis follows the Ostwald's law of successive reactions. The initial metastable phase is replaced successively by more stable phases, in this case quartz. This trend is also observed at shorter and longer crystallisation times. Therefore at all crystallisation times studied, the optimum temperature for obtaining a highly crystalline pure ZSM-5 was determined to be 150°C. Similar results were reported by for the synthesis of ZSM-5 from Ahoko Nigerian Kaolin [48]. Although we might expect increased nucleation and crystal growth with an increase in temperature to 190°C, thermodynamic effects predominate over kinetic effects and favour the formation of the thermodynamically more stable quartz phase rather than growth of metastable ZSM-5 at 190°C.

Crystallisation time is critical in controlling the crystallinity of the synthesised ZSM-5. It was also shown to have a major impact on the physicochemical properties such as porosity and morphology of ZSM-5. The relative crystallinity determined from XRD increased with time at optimum crystallisation temperature of 150°C. SEM results showed different morphologies and crystal sizes could be obtained by varying the crystallisation times between 24 and 96 h. ZSM-5 with high external surface area and both micro and mesopores were obtained at 150°C 24 h using BK and that micro-pore area increased with time as the relative crystallinity increased. The studies on the effects of crystallisation temperature and time clearly highlighted the need to determine optimum synthesis conditions to obtain pure well crystallised ZSM-5.

2.1.4 SiO_2/Al_2O_3 ratio

In the synthesis of ZSM-5 the Si/Al ratio is known to affect physical properties such as crystal size and morphology as well as chemical properties such as acidity. The acid site density, type and strength are affected by the presence of aluminium and can be controlled by adjusting the Si/Al ratio. The acidity of the zeolite is

Zeolite	Atomic % Si	Atomic % Al	Si/Al ratio
Si-Al 150 RK	11.67	0.30	39.0
Si-Al 70 RK	97.2	2.77	35.1
Si-Al 42 RK	91.51	6.48	14.1
Si-Al 150 BK	11.97	0.30	39.9
Si-Al 70 BK	42.18	2.51	16.8
Si-Al 42 BK	37.2	5.04	7.4

Table 1.
EDS results showing atomic composition (%) of synthesised ZSM-5 with different Si/Al ratios using RK and BK.

important as ZSM-5 is used in many acid catalysed petrochemical reactions such as oligomerisation of olefins, cracking, isomerisation and alkylation. In this study the Si/Al ratios were varied and its effects on the physicochemical properties of the ZSM-5 are investigated. Raw kaolin (RK) and beneficiated kaolin (BK) are used as alumina source. The study shows the effects of the kaolin compositions when the Si/Al ratios are varied on the formation and properties of the zeolites. The ZSM-5 zeolites were synthesised using RK and BK with molar SiO_2/Al_2O_3 ratios of 150, 70 and 42 in the batch mixture. The Si/Al ratios in the final product were determined using EDS atomic analysis. (It must be noted that the term Si/Al is used interchangeably but SiO_2/Al_2O_3 is meant when referring to the batch ratio and Si/Al is meant when referring to the ratio in the product as determined by EDS analysis.) The atomic percentages of Si and Al are shown in **Table 1**.

The Si/Al ratios of the final ZSM-5 product decrease as the initial Si/Al ratios in the batch mixtures decrease for both RK and BK. The Si/Al ratios of ZSM-5 synthesised from batch mixtures 70 and 42 BK are lower, approximately half that synthesised from RK of the same starting ratios respectively. This suggests that as the Si/Al ratio decreases less silica may be incorporated into ZSM-5 formed from BK compared to RK.

The effect of Si/Al ratio on the structure of ZSM-5 was confirmed by x-ray diffraction. The diffractograms are shown in **Figure 2(a)** and **(b)**. All patterns exhibited the characteristic ZSM-5 peaks at 7.94° (011), 8.90° (020), 23.10° (051), 24.0° (033) and 24.35° (313) 2θ. All samples synthesised from RK also showed a reflection at 26.67° 2θ corresponding to the presence of quartz. The intensity of this peak

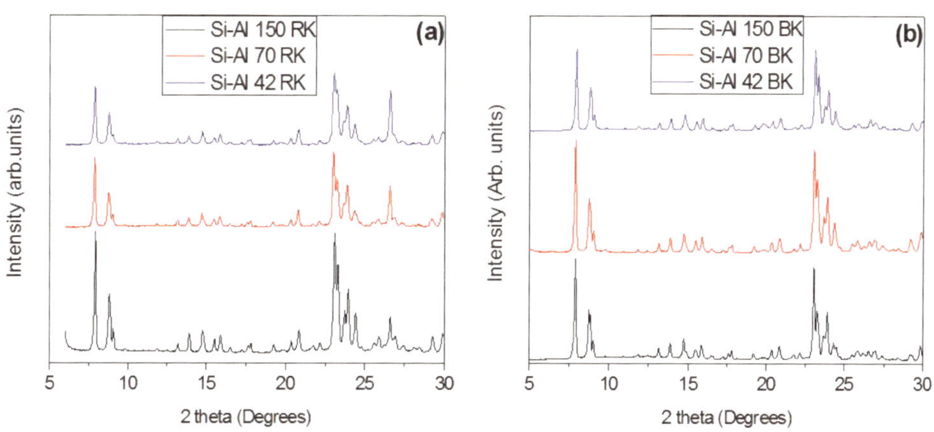

Figure 2.
XRD powder patterns of ZSM-5 synthesised from (a) RK and (b) BK with different Si/Al ratios.

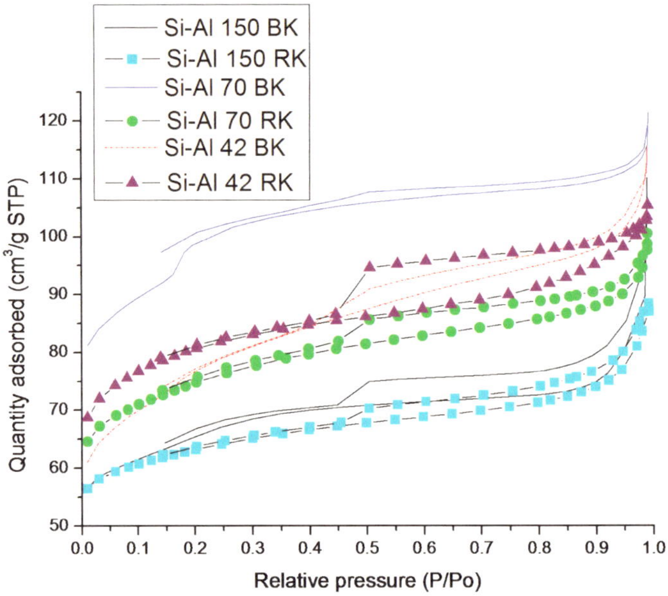

Figure 3.
N_2 adsorption-desorption isotherms of the representative samples.

increased as the Si/Al ratio decreased. RK used as the aluminium source contains a significant amount of quartz. Samples synthesised with lower Si/Al ratios thus contained a larger amount of quartz which remained unreacted hence the increase in intensity of the quartz phase. As the Si/Al ratio in the RK samples decreased the intensity of the (011) and (020) reflections diminished and peaks shifted to slightly lower angles. This is due to a change in the crystal size and an increase in Al substitution in the framework structure. Peak intensity however increased from sample Si-Al 150 BK to Si-Al 70 BK as the material became more crystalline but then decreased in sample Si-Al 42 BK.

The specific surface area and porosity characteristics of the samples are shown in **Table 2** and N_2 adsorption-desorption isotherms in **Figure 3**. From **Table 2** it is noticed that the specific surface area increases as the Si/Al ratio decreases for the RK samples. The micropore, external surface area as well as the micropore volume increase as the Si/Al ratio decreases.

Zeolite	BET s/a m^2/g	Micropore area m^2/g	External area m^2/g	Avg. pore vol. m^3/g	Micropore vol. m^3/g	Avg. pore size nm
Si-Al 150 RK	214	159	55	0.125	0.073	5.8
Si-Al 70 RK	253	171	82	0.143	0.079	4.6
Si-Al 42 RK	274	189	85	0.155	0.087	4.8
Si-Al 150 BK	223	137	86	0.133	0.063	6.8
Si-Al 70 BK	340	162	178	0.174	0.072	3.1
Si-Al 42 BK	265	121	144	0.162	0.055	4.2

Table 2.
Specific surface area and porosity characteristics of RK and BK samples with different Si/Al ratios.

This indicates that the increase in aluminium content in the samples leads to a well-developed internal microporous structure. The increase in external surface area with lower Si/Al ratios is likely due to the decrease in crystal size and correlates with the decreased intensity of the XRD reflections. The presence of more quartz as the Si/Al ratio decrease may also result in a higher external surface area. All samples possess a type 1 adsorption isotherm indicative of the microporous nature of the synthesised ZSM-5. The samples also exhibit a H4 type hysteresis loop in the relative partial pressure range 0.4–1.0, indicating capillary condensation occurred. This has been attributed to the presence of a mesoporous phase associated with the slit shaped voids between packed crystals [49]. A large uptake in the P/Po range 0.9–1.0 suggests the presence of larger macropores. This however decreases with decrease in Si/Al ratio. The hysteresis loop becomes wider as the Si/Al ratio decreases. Similar results have been shown by Liu et al. [50] for ZSM-5 synthesised from gel mixtures of varying polymerisation degrees. The more depolymerised gel formed uniform aggregates of fine crystals. Therefore as the Si/Al ratio decreased the gel became more depolymerized forming aggregates of tiny crystals which when tightly packed form mesopores between them [51]. Si-Al 42 RK also showed a broad peak centred at approximately 12.4 nm in the pore size distribution curves shown in **Figure 4(a)** not seen in the other RK samples which is indicative of mesopores and hence the broader hysteresis loop. Samples from BK had a lower micropore area and volume and higher external surface area compared to those synthesised from RK. The crystallisation process is different when using RK and BK as the compositions differ. This may suggest a difference in the degree of polymerisation of silica and alumina in RK and BK. As suggested, the N_2 isotherms and SEM images confirm the presence of aggregated nanocrystals in RK samples which form well depolymerized gel mixtures forming tetrahedral silica and alumina building blocks that form highly networked aluminosilicate species leading to a highly crystalline microporous structure. Zhang et al. [52] suggests a destruction of the microporous structure of ZSM-5 occurs with isomorphous substitution of Si with Al. This is possible as there is a higher content of aluminium in the BK samples compared to RK for each particular ratio. It is more likely however that extra-framework aluminium is present which blocks the micropores since Si-Al 70 BK has a higher micropore area and volume than Si-Al 150 BK but Si-Al 42 BK containing the highest aluminium content has the lowest microporosity. Therefore there is an optimum Si/Al ratio

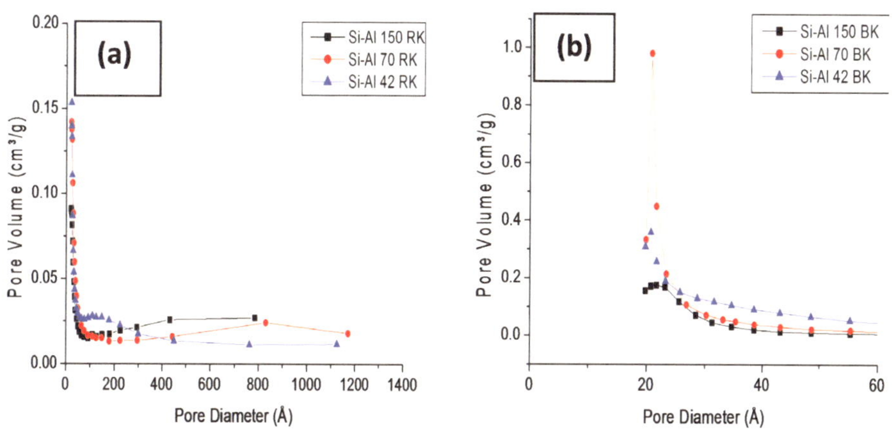

Figure 4.
Pore size distribution calculated using the BJH method for the representative samples synthesised from (a) RK and (b) BK.

when synthesising ZSM-5 using BK. Interesting to note is the shape of the Si-Al 70 BK isotherms. It is a combination of a type I isotherm typical of ZSM-5 with a microporous framework and a type IV isotherm representative of mesoporous materials. The hysteresis loop in the 0.4–1.0 P/Po range is much smaller than other samples probably due to less space between packed crystals. However, the steep uptake at approximately 0.2 P/Po with a second small hysteresis loop may be due to the presence of uniform pores in the 2 nm range as shown in the pore size distribution graph for BK samples in **Figure 4(b)**. Isotherms of this type have been reported in the literature for hierarchical zeolites [49, 53–55] and are ascribed to filling of mesopores with a narrow pore size distribution between 3 and 4 nm. The pore size in this sample is smaller and similar to the size of supermicropores described by Yang et al. [56]. From **Figure 4(b)** it is noticed that the presence of 2 nm sized pores increase with decreasing Si-Al ratio and exist even for Si-Al 42 BK which has the lowest Si-Al ratio although the amount decreased probably due to the blockage of pores by extra-framework aluminium. We consider that these are real pores and not physical phenomena observed in gas adsorption such as fluid to crystalline transitions observed in MFI zeolite structures on the basis that Si-Al 70 BK has a low Si/Al ratio. Fluid to crystalline effects are usually observed on high Si/Al ratio MFI and silicalite-1 materials due to the energetically homogenous surface created by the large amount of Si atoms which result in a well-pronounced sub-step in a narrow P/Po range and an increase in Al content induces energetic heterogeneity [57].

HRSEM studies revealed that the morphology and particle size is affected by the change in Si/Al ratio. The images for the RK and BK samples are shown in **Figure 5**. Si-Al 150 RK has rounded boat shaped crystals that are highly inter-grown and agglomerated. There is also the presence of some amorphous material with no particular morphology.

As the Si/Al ratio decreases the crystal size also decreases as shown in **Figure 5(a–c)**. This agrees well with results reported in literature [24] and correlates with the increase in surface area and decrease of intensity of reflections as shown by BET and XRD results respectively. The material also becomes less intergrown and aggregates with a uniform size distribution are noticed. Looking at the insets of **Figure 5(b) and (c)** it can be clearly seen that the aggregates are made up of nano-sized crystals that interlock and form an overall cross discus shape. ZSM-5 with similar morphology has been synthesised by Yue et al. [58] in the synthesis of hierarchical zeolites from kaolin and rectorite using a nanoscale depolymerisation-reorganisation approach and by Liu et al. [50] from highly depolymerised gel mixtures which also show similar trends in gas adsorption analysis to our samples. Thus an explanation of the change in morphology to aggregates of nano-crystals with a decrease in Si/Al ratio is that the gel mixture becomes more depolymerised with increased amount of the aluminium source. This trend is only observed in RK samples and not BK. The formation of the aggregates is due to the difference in composition of RK and BK in which the former contains a high quantity of quartz. As the aluminium amount is increased to obtain lower Si/Al ratios the quartz content in the gel mixture also increases and is present as shown in XRD. The quartz is difficult to dissolve which means that the alkalinity may be higher for the more soluble species i.e. that obtained from kaolinite and additional water-glass which is readily consumed. As crystallisation proceeds it becomes more depolymerized and leads to formation of fine crystals [50].

Si-Al 150 BK had larger crystals than Si-Al 70 and 42 BK and followed the trend of decreasing crystal size with lower Si/Al ratios. As noticed in **Figure 5(d)** the morphology is hexagonally shaped crystals. Twinning is also apparent as the 100 face protrudes from the 010 face and is commonly seen in ZSM-5 crystal twinning.

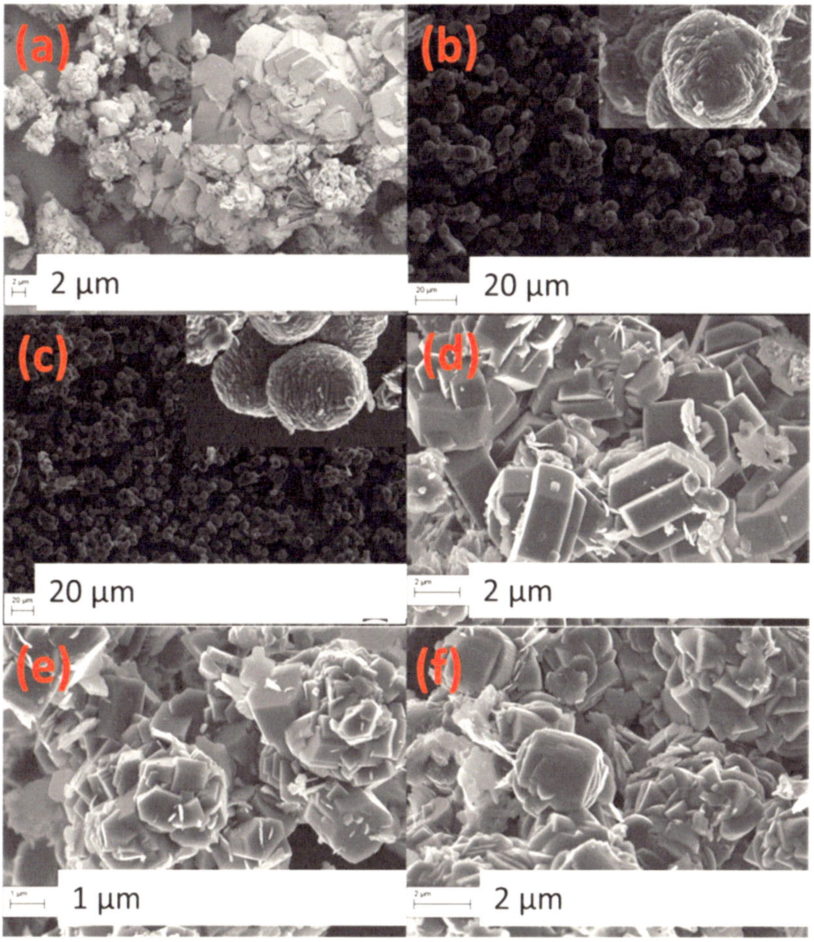

Figure 5.
High resolution SEM images showing morphology and crystal size of (a) Si-Al 150 RK, (b) Si-Al 70 RK, (c) Si-Al 42 RK, (d) Si-Al 150 BK, (e) Si-Al 70 BK and (f) Si-Al 42 BK.

Si-Al 70 BK and Si-Al 42 BK both have wide crystal size distributions of both micrometre and sub-micrometre crystals that are highly intergrown with the former possessing smooth hexagonal crystals and the latter more rounded and rough edged crystals. Thus the Si/Al ratio also has an effect on morphology which agrees with work that has been reported in the literature [24].

The amount and strength of acid sites were determined by the NH_3 detected during desorption and the peak maximum in the desorption profile respectively. The results are summarised in **Table 3**.

It is clearly seen from **Table 3** that the total amount of acid sites increase with decreasing Si/Al ratio for RK samples as the higher the Si/Al ratio the lesser amount of total acid sites is [59]. This is true for the BK samples with the exception of Si/Al 42 BK which has the lowest Si/Al ratio but has lesser amount of acid sites than Si-Al 70 BK. This is most likely due to a larger amount of extra-framework aluminium and thus less Bronsted acid sites. This is supported by the fact that it has the least amount of strong acid sites (206 µmol/g), which are mainly due to NH_3 desorption from Bronsted sites. The decrease in the amount and strength of acid sites when a batch ratio of 42 is used may be due to the aluminium atom being in close proximity to each other forming Al pairs which reduce the amount of acid sites. Both the amounts of weak and strong acid sites in general increase with decreasing Si/Al

Samples	Peak (°C)		Acidity distribution (μmol NH_3/g)		
	Low temp	High temp	Weak	Strong	Total
Si-Al 150 RK	191	395	230	225	455
Si-Al 70 RK	192	408	269	279	548
Si-Al 42 RK	200	421	536	500	1036
Si-Al 150 BK	200	413	354	279	633
Si-Al 70 BK	210	440	689	519	1208
Si-Al 42 BK	203	423	584	206	790

Table 3.
Distribution of acidity on the ZSM-5 samples as determined by NH_3-TPD.

ratio. The RK samples have an almost 1:1 ratio of weak and strong acid sites. The BK samples however possess more weak acid sites than strong ones which may be due to NH_3 desorption from the silanols caused by the defects in the crystalline structure. The strength of the acid sites varies with aluminium content and seems to increase with decreasing Si/Al ratio. The maxima of the peaks shift with decreasing Si/Al ratio to higher temperatures indicating that the NH_3 molecule is more strongly bound to the acid sites. The maximum peak temperatures as shown in **Table 3** indicate that ZSM-5 synthesised from BK have stronger acid sites when compared to those synthesised from RK with similar aluminium content. This may be due to the difference in aluminium environment as those synthesised from RK are more crystalline whereas those synthesised from BK contain more defects.

It is well known that the aluminium content is directly related to the acidic properties of aluminosilicates. The presence of different Al species or coordination types leads to the formation of both Bronsted and Lewis acid sites. Therefore ZSM-5 samples with different Si/Al ratios synthesised from RK and BK were studied in order to determine the aluminium state, coordination, stability and degree of incorporation. The data obtained from ^{27}Al MAS NMR investigations are presented in **Table 4**.

Two peaks are commonly noticed in the ^{27}Al NMR spectra of ZSM-5 zeolites. The major peak occurring at a chemical shift of approximately 55 ppm corresponding to the tetrahedrally coordinated Al in the zeolite framework and another smaller peak at approximately 0 ppm is related to the presence of octahedrally coordinated Al and is usually referred to as extra-framework aluminium [60]. This notion has been disputed, however, as evidence for octahedrally coordinated Lewis Al present

Sample	Al framework (% Al_{fr})	Al extra-framework (% Al_{efr})	PW at 1/2 max (~55 ppm Al_{fr})
Si-Al 150 RK	100	n.d	4.9
Si-Al 70 RK	92	8	5.5
Si-Al 42 RK	92	8	5.7
Si-Al 150 BK	93	7	5.3
Si-Al 70 BK	89	11	5.8
Si-Al 42 BK	83	17	5.6

Table 4.
Framework and extra-framework Al contents and line widths at half height for the representative ZSM-5 samples.

as Al-$(OSi)_3(H_2O)_3$ in the framework has been confirmed by Woolery et al. [61]. The relative amounts of the two species have been obtained by integration of the two peaks and are shown in **Table 4**. It is noticed that Si-Al 150 RK contains only tetrahedrally coordinated framework aluminium (Al_{fr}) and any presence of extra-framework aluminium (Al_{efr}) is negligible. It is common for highly siliceous ZSM-5 to possess only Al_{fr}. Si-Al 150 BK which has a similar Si-Al ratio as Si-Al 150 RK as shown in **Table 1** does have a small amount of Al_{efr}. This may be due to the difference in the composition and hence solubility of RK and BK. A slight increase in Al_{efr} is seen as the aluminium content increases for RK samples although there is not much difference between Si-Al 70 and Si-Al 42 RK. The increase in Al_{efr} with increase in aluminium content however, is drastically enhanced for the BK samples. EDS analysis shows that Al content is almost double in Si-Al 70 and Si-Al 42 BK compared to that synthesised from RK with the same staring gel ratios confirming that more Al is incorporated into ZSM-5 from BK due to a more abundant supply of reactive aluminium. However, NMR suggests that the Al is not necessarily all in framework positions and Al available from RK is better incorporated into the framework structure as it has a greater amount (92%) compared to Si-Al 70 BK and Si-Al 42 BK only having 89 and 83% Al_{fr} respectively. This result correlates with that obtained from N_2 physisorption and SEM analysis which showed a larger micropore area and nano crystal aggregates respectively for the RK samples compared to BK. This suggests a well-structured crystalline material made up of tetrahedrally coordinated Si and Al. It must be noted that quantitatively the BK samples form more Al-rich ZSM-5 even though the Al_{efr} amounts are greater. BK samples had a large external surface area greater than its micropore area due to mesoporosity. Larger amounts of aluminium may therefore be present at the external surface of the zeolite in a less stable octahedral form as suggested by Serrano et al. [53] for materials with more than 1 type of porosity such as hierarchical zeolites. They may also be present in different environments with more than 1 type of neighbouring atom as the resonances observed are broad and Si-Al 42 BK also shows a shoulder peak at ~6 ppm as shown in **Figure 6**.

From the ^{27}Al NMR spectra of the RK and BK samples, The RK samples and Si-Al 150 BK all have a peak max at ~55.7 ppm. Si-Al 70 BK shows a slight shift to a lower value ~54 ppm and Si-Al 42 BK a larger shift to ~58.3 ppm. The peak width values shown in **Table 4** are an indication of the uniformity of the Al environments in zeolite materials [53, 62] and crystallinity [63]. Si-Al 150 RK has the smallest peak width indicating high crystallinity and correlates with the XRD results and a more uniform Al environment. The peak width increases for all other samples indicating different environments of Al. The base of the peak begins at ~50 ppm and ends at ~62 ppm for all samples except Si-Al 42 BK which shows a shift of approximately ~4 ppm towards higher ppm values. Previous studies by Dedecek et al. [37] have shown that different environments for AlO_4 tetrahedra exist after deconvolution and simulation of the tetrahedral peak. Peaks observed at 50, 53, and 58 ppm correspond to Al atoms with only Si neighbours, while resonances at 62 ppm corresponds to Al atoms with 3 Si and 1 neighbouring OH group which are highly reactive. Thus the resonances fall within the range of the peak observed for the ZSM-5 samples. The resonances at 50, 53 and 58 ppm reflect the difference in environment of the AlO_4 tetrahedra in the sample which is due to the effect of the vicinity of second Al atom in the sample [37]. The closeness of the next Al atom in sequences such as Al-O-$(Si\text{-}O)_2$-Al can change the observed shift of the Al atom by ~4 ppm. Thus the peak observed for Si-Al 42 BK at 58.3 ppm suggests that the Al atoms are in close proximity to each other and has been reported for Al rich zeolites [37, 64]. Si-Al 42 BK also more likely has more Al atoms with a neighbouring OH which are highly reactive as the resonance shifts towards 62 ppm. Therefore it is

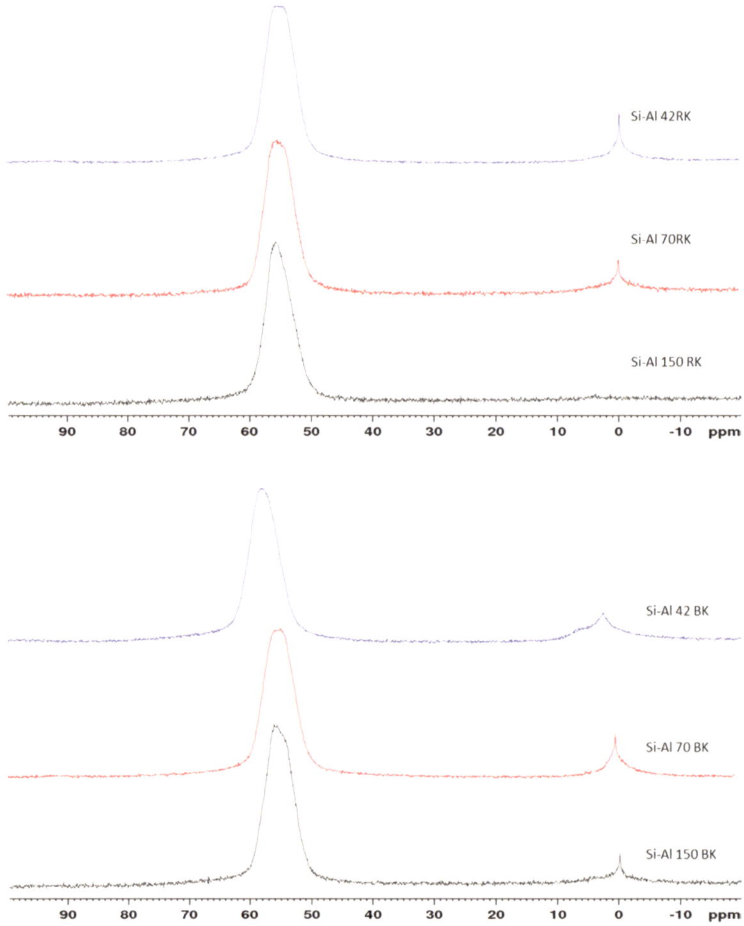

Figure 6.
^{27}Al MAS NMR spectra of the RK and BK samples with different Si/Al ratios.

possible that BK forms more Al rich ZSM-5 than RK for two reasons. The first being that RK causes a higher degree of depolymerisation due to the presence of quartz (as discussed above) of the remaining Na-silicate which is broken down to silicate species. This then forms more networked silicate and aluminosilicate and has less formation of close Al atoms resulting in higher ratios. Second for the BK samples, which were prepared with a smaller amount of additional Na-silicate, was not well depolymerised due to a lower alkalinity i.e. the silicate remained non-transformed and balanced by the Na^+ and only a much lower concentration of Si was available for the highly reactive Al species resulting in aluminosilicate with a much lower Si/Al ratio having also more Al_{efr} and in the case of Si-Al 42 BK, in a high probability of aluminosilicate species with close Al atoms [37].

3. Potential application of kaolin-based ZSM-5

3.1 Catalytic oligomerisation

The performance of the zeolites as catalysts was tested in the transformation of 1-hexene, used as a model compound for the oligomerisation of alkenes; reaction was performed at the following conditions: T = 350°C, pressure = 1 atm, WHSV = 8 h^{-1}. The activities of the catalysts were determined by assuming all

1-hexene present in the product was due to unconverted feed. The graph showing activity as a function of time on stream is shown in **Figure 7**.

The catalytic tests were conducted over a time period of 420 min. All catalysts showed some deactivation over the total time on stream although some were more rapidly deactivated than others. All catalysts had a conversion not less than 90% in the first hour on stream. The highest conversion achieved was approximately 98%. It is clearly noticed that the conversion of the catalysts is highly dependent on the acidity, in particular the Bronsted acidity. Si-Al 42 RK and Si-Al 70 BK with similar acidities and possessing the highest acidities of the catalysts had the highest conversions and both showed only slight deactivation as the conversion remained above 90% for the duration of the reaction. Si-Al 70 RK and Si-Al 150 BK having identical amounts of strong acid sites and tetrahedral alumina (**Tables 3** and **4** respectively) have the same conversion and deactivation for the time on stream. Si-Al 42 BK which had the least amount of Bronsted sites showed the lowest conversion and the most rapid deactivation. Interestingly, Si-Al 150 RK having a much lower acidity compared to Si-Al 42 RK and Si-Al 70 BK has a very similar activity and shows even better stability (**Figure 7**). This was attributed to the effect of quartz coating the surface of the ZSM-5 which prevents deactivation. Therefore it is seen that this catalyst functions better than catalysts with more than double its acidity and further validates the influence of quartz on the performance of the catalyst.

The catalysts tested in the transformation of 1-hexene all possess a wide product distribution. The products are grouped accordingly: (C_2–C_5) range, (C_6–C_9) gasoline range and (C_{10}+) diesel range. This indicates multiple types of reactions may occur at the set reaction conditions such as oligomerisation, cracking, isomerisation and alkylation to name a few. The selectivities to these ranges however change with time on stream and it is noticed that small changes in conversion can have large changes in selectivity to products. The selectivities of the catalysts are discussed further and correlated with their physicochemical properties.

Si-Al 150 BK and Si-Al 70 RK are compared to each other since they have similar acidities as well as conversions over time on stream as shown in **Figure 7**. The selectivities however differ from each other. The selectivities to gasoline (C_6–C_9) and diesel (C_{10}+) ranges are shown in **Figure 8(a)** and **(b)** respectively.

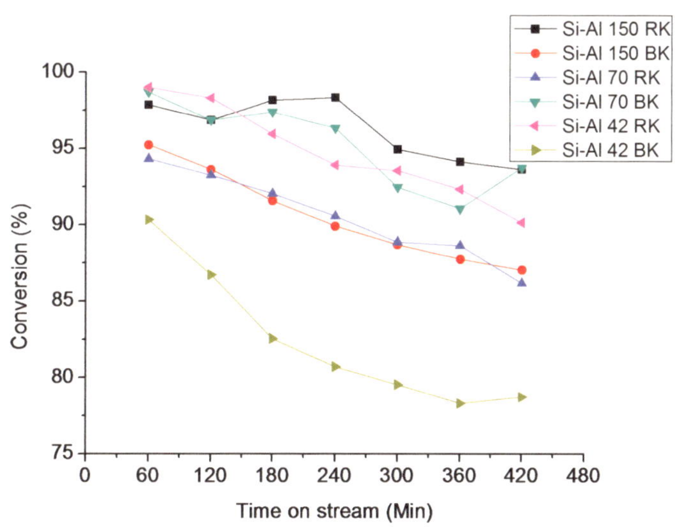

Figure 7.
Graph showing hexane conversion over different catalysts (P = 1 atm, T = 350°C, WHSV = 8 h^{-1}).

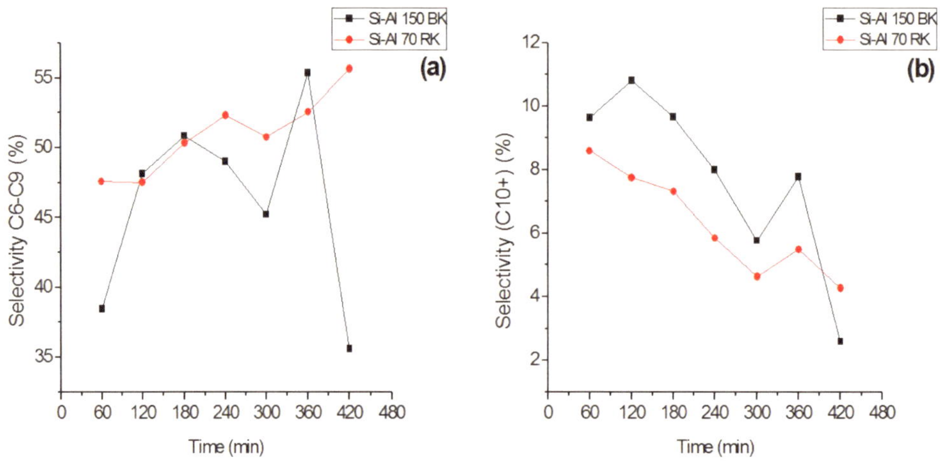

Figure 8.
Selectivity of Si-Al 150 BK and Si-Al 70 RK to gasoline (C_6–C_9) (a) and diesel (C_{10}+) (b) range products over time on stream.

Both catalysts have a similar trend to selectivity of diesel range products as noticed in the graph and decrease with time as the conversion decreased. However Si-Al 150 BK has a slightly higher selectivity throughout the reaction. Both show good selectivity (40–55%) to gasoline products with the selectivity of Si-Al 70 RK possessing a steady increase over the reaction time. Si-Al 150 BK was more unstable and the selectivity varied over time and was slightly less than Si-Al 70 RK. It was thought Si-Al 70 RK was more selective to reactions pertaining to chain growth such as oligomerisation and alkylation but a closer look at the product selectivity of the gasoline range revealed that Si-Al 70 RK was highly selective to C_6 hydrocarbons as compared to Si-Al 150 BK as shown in **Figure 9**. An increase from ~12% to above 30% was observed. The C_2–C_5 selectivity trend is also shown and decreases over time for the Si-Al 70 RK catalyst but shows an overall increase in the C_2–C_5 products

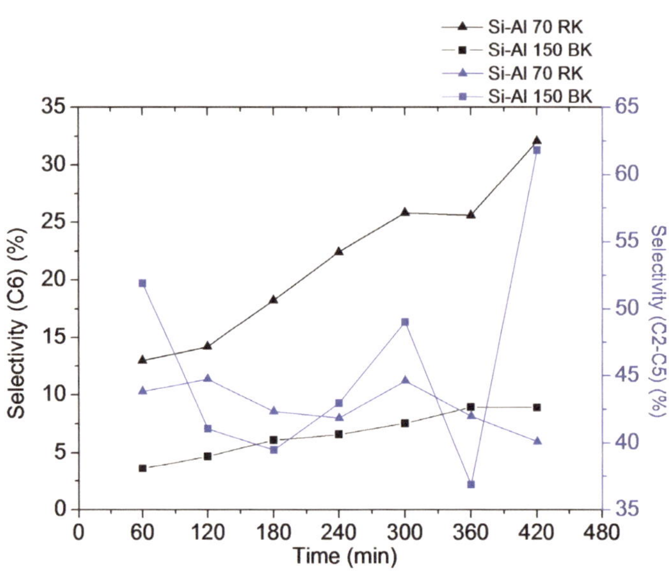

Figure 9.
Selectivity of Si-Al 150 BK and Si-Al 70 RK to C_6 (black curves) and C_2–C_5 (blue curves) range products over time on stream.

for Si-Al 150 BK. Thus there may be a greater selectivity to isomerisation reactions over cracking reactions for the Si-Al 70 RK catalyst whereas Si-Al 150 BK has a better selectivity to chain growth reactions as its total selectivity to C_7+ hydrocarbons is higher. This may suggest a slight difference in strength of acid sites available as oligomerisation and cracking occur on stronger acid sites than isomerisation reactions which occur on sites of intermediate acidity [65]. The main difference between the two catalysts however is the crystal morphology and size. The Si-Al 70 RK possessing nanocrystals as shown in HRSEM, this may reduce the effects of diffusion limitation. As shown by Buchanan et al. [66] larger crystals showed a higher selectivity to C_3/C_4 products and a lower isomerization/cracking of olefin ratio due to diffusion limitation. Therefore it is possible that as the strength of the acidic sites are reduced over time due to deactivation which happens more in Si-Al 70 RK, isomerization reactions which occur faster than cracking reactions [66], increase and due to the nanocrystals, the isomers which form are able to diffuse out before any secondary cracking leading to more C_6 isomers, less cracked C_2–C_5 products and a better gasoline selectivity. Therefore the physical properties of the catalyst synthesised from RK are shown to affect its catalytic performance.

Figure 10 shows the selectivity of Si-Al 150 RK, Si-Al 42 RK, Si-Al 70 BK and Si-Al 42 BK to diesel and gasoline range hydrocarbons. The three catalysts with similar conversions i.e. Si-Al 150 RK, Si-Al 42 RK and Si-Al 70 BK can be compared in terms of selectivity.

From **Figure 10** it is noticed that after the first hour on stream the selectivity to C_{10}+ is the highest for the catalyst Si-Al 42 RK showing ~16% selectivity. This may indicate that oligomerisation to long chain hydrocarbons largely depends on the acidity since Si-Al 42 RK which has a high Bronsted acidity had the highest selectivity. The more acidic catalysts have better selectivity over the first 3 h on stream. As the reaction proceeds however, all catalysts show a large decrease in selectivity to C_{10}+ hydrocarbons except Si-Al 150 RK which shows a greater stability. The decrease for Si-Al 42 RK is almost linear over the reaction time and drops to ~3%. The decrease in selectivity to C_{10}+ suggests the deactivation of certain acid sites which are most likely strong Bronsted acid sites as oligomerisation reactions are favoured at these catalytic centres [65]. Si-Al 70 BK which has similar acidity to Si-Al 42 RK shows a slightly better selectivity to C_{10}+ hence less deactivation of Bronsted sites. This is most likely due to the higher surface area and increased mesoporosity which may inhibit the build-up of carbonaceous deposits usually responsible for deactivation and hence further highlights the effects of the physicochemical properties of the zeolites on its catalytic behaviour. As mentioned in our previous work [30], the effect of quartz deposits on

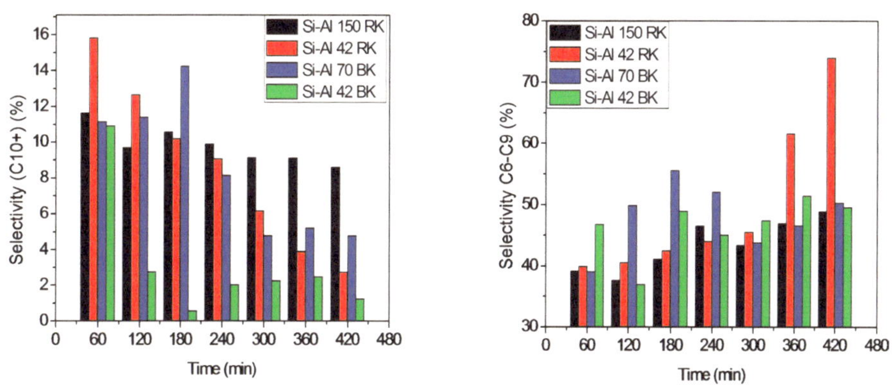

Figure 10.
Selectivities to diesel range (C_{10}+) and gasoline range (C_6–C_9) hydrocarbons of catalysts Si-Al 150 RK, Si-Al 42 RK, Si-Al 70 BK and Si-Al 42 BK as a function of time.

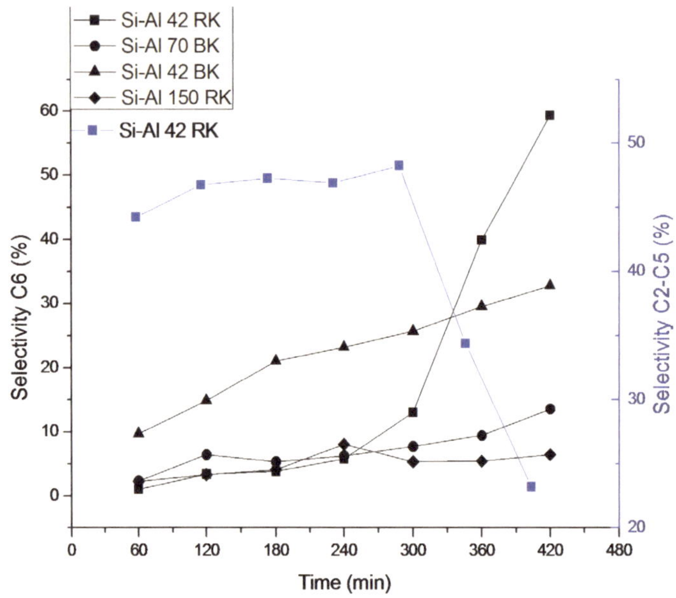

Figure 11.
Selectivity to C_6 hydrocarbons and C_2–C_5 selectivity of Si-Al 42 RK (blue curve) as a function of time.

the external surface of the crystals in Si-Al 150 RK plays a major role in inhibiting the formation of carbonaceous material causing deactivation. Thus even when compared to catalysts with greater acid site density and strength (almost double its acidity), Si-Al 150 RK possesses greater stability and selectivity is almost double that of the more acidic catalysts. This clearly suggests that the acid sites are prevented from being deactivated by the deposition of quartz and continue to catalyse oligomerisation reactions for the duration of the reaction. Si-Al 42 BK had the lowest selectivity to C_{10}+. This is most likely due to its low Bronsted acidity but also the high content of extra-framework aluminium may cause pore blockage and prevent access to acid sites for oligomerisation. The closeness of Al atoms in the structure may also have an effect on the acidity and types of catalytic reactions that are favoured.

All catalysts showed good selectivity to the gasoline range. Si-Al 42 RK shows the opposite trend for selectivity to gasoline as this increase over the reaction time. The selectivity spikes in the last 2 h from 45 to 74%. **Figure 11** shows the selectivity to C_6 and C_2–C_5 products.

The spike in gasoline activity is again due to an increase in C_6 selectivity. This further confirms our observations with the Si-Al 70 RK catalyst. The decrease in C_{10}+ selectivity indicates deactivation of strong acid sites. Thus isomerisation reactions are then favoured when only less acidic site are available and due to Si-Al 70 RK and Si-Al 42 RK possessing the same morphology the isomerisation products diffuse out before cracking. **Figure 11** clearly shows the contrasting selectivities between isomerisation (C_6) and cracking reactions (C_2–C_5) for the Si-Al 42 RK catalyst. Si-Al 42 BK also shows good isomerisation activity which further indicates the reaction occurring on weaker acid sites.

4. Conclusion

In conclusion, this chapter discusses some of the key factors affecting the synthesis of kaolin-based ZSM-5 such as kaolin crystallinity, kaolinite content, crystallisation parameters and Si/Al ratio. Additionally, the catalytic performance of the ZSM-5 derived from kaolin was evaluated. These factors are important considerations when attempting to synthesise ZSM-5 with high purity and crystallinity. However, ZSM-5 can also be synthesised from impure kaolin sources and these impurities may act as poisons or promoters during synthesis and catalytic application. The requirements to develop synthesis conditions that are optimised for specific sources of kaolin are established. The physicochemical properties such as porosity, morphology and acidity of ZSM-5 can be controlled by choosing the right synthesis procedures. Kaolin-based ZSM-5 zeolites are promising as catalysts for petrochemical reactions such as oligomerisation. High activity and selectivity to gasoline and diesel range hydrocarbons was attainable. The activity of the catalysts correlated well with the acidity of ZSM-5 samples. The catalytic performance of the zeolites also correlated well with the physical properties such as morphology and surface area which were shown to influence selectivity to certain products by favouring isomerisation and oligomerisation reactions respectively. Impurities in the kaolin precursor may also have positive effects on catalytic performance, in this case quartz deposition on ZSM-5 inhibiting deactivation and increasing catalyst stability. Therefore, ZSM-5 zeolites can be successfully synthesised from cheaper, more environmentally friendly alternative starting materials that have satisfactory performances in catalytic application.

Acknowledgements

The authors would like to thank the Petroleum, Oil and Gas Corporation of South Africa (PetroSA) for their financial support and technical discussions, the National Research Foundation (NRF) for granting Ebrahim Mohiuddin a scholarship, Makana Brick for supplying the raw kaolin, the electron microscope unit, Physics department, University of the Western Cape for the SEM images, Mrs. E. Antunes, Chemistry department, University of the Western Cape and Dr. D.J. Brand, University of Stellenbosch for the solid state NMR work and Ithemba labs for the XRD work.

Author details

Ebrahim Mohiuddin[1], Yusuf Makarfi Isa[1,2], Masikana M. Mdleleni[1*] and David Key[1]

1 PetroSA Synthetic Fuels Innovation Centre, South African Institute for Advanced Materials Chemistry, University of the Western Cape, Bellville, South Africa

2 Chemical Engineering Department, Durban University of Technology, Durban, South Africa

*Address all correspondence to: masikana.mdleleni@petrosa.co.za

References

[1] Chen CY, Zones SI. Post-synthetic treatment and modification of zeolites. In: Cejka J, Avelino C, Stacey Z, editors. Zeolites and Catalysis: Synthesis, Reactions and Applications. Weinheim: Wiley-VCH Verlag; 2010. pp. 155-170

[2] Townsend RP. Ion exchange in zeolites: Some recent developments in theory and practice. Pure and Applied Chemistry. 1986;**58**(10):1359-1366

[3] Ackley M. Application of natural zeolites in the purification and separation of gases. Microporous and Mesoporous Materials. 2003;**61**(1-3):25-42

[4] Weitkamp J. Zeolites and catalysis. Solid State Ionics. 2000;**131**:175-188

[5] Chen NY, Garwood WE. Industrial application of shape-selective catalysis. Catalysis Reviews. 1986;**28**(2-3):185-264

[6] Xiaoning W, Zhen Z, Chunming X, Aijun D, Li Z, Guiyuan J. Effects of light rare earth on acidity and catalytic performance of HZSM-5 zeolite for catalytic cracking of butane to light olefins. Journal of Rare Earths. 2007;**25**(20373043):321-328

[7] De Klerk A. Oligomerization of 1-hexene and 1-octene over solid acid catalysts. Industrial and Engineering Chemistry Research. 2005;**44**(11):3887-3893

[8] Mohiuddin E, Mdleleni MM, Key D. Catalytic cracking of naphtha: The effect of Fe and Cr impregnated ZSM-5 on olefin selectivity. Applied Petrochemical Research. 2018;**8**:119

[9] Aboul-Gheit AK, SM A-H, El-Desouki DS. Catalytic para-xylene maximization. Applied Catalysis A: General. 2001;**209**(1-2):179-191

[10] Abdullahi T, Harun Z, Othman MHD. A review on sustainable synthesis of zeolite from kaolinite resources via hydrothermal process. Advanced Powder Technology. 2017;**28**(8):1827-1840

[11] Yaping Y, Xiaoqiang Z, Weilan Q, Mingwen W. Synthesis of pure zeolites from supersaturated silicon and aluminum alkali extracts from fused coal fly ash. Fuel. 2008;**87**(10-11):1880-1886

[12] Kordatos K, Gavela S, Ntziouni a, Pistiolas KN, Kyritsi a, Kasselouri-Rigopoulou V. Synthesis of highly siliceous ZSM-5 zeolite using silica from rice husk ash. Microporous and Mesoporous Materials. 2008;**115**(1-2):189-196

[13] Mohamed MM, Zidan FI, Thabet M. Synthesis of ZSM-5 zeolite from rice husk ash: Characterization and implications for photocatalytic degradation catalysts. Microporous and Mesoporous Materials. 2008;**108**(1-3):193-203

[14] Panpa W, Jinawath S. Synthesis of ZSM-5 zeolite and silicalite from rice husk ash. Applied Catalysis B: Environmental. 2009;**90**:389-394

[15] Wang P, Shen B, Shen D, Peng T, Gao J. Synthesis of ZSM-5 zeolite from expanded perlite/kaolin and its catalytic performance for FCC naphtha aromatization. Catalysis Communications. 2007;**8**(10):1452-1456

[16] Jiang J, Duanmu C, Yang Y, Gu X, Chen J. Synthesis and characterization of high siliceous ZSM-5 zeolite from acid-treated palygorskite. Powder Technology. 2014;**251**:9-14

[17] Chareonpanich M, Namto T, Kongkachuichay P, Limtrakul J.

Synthesis of ZSM-5 zeolite from lignite fly ash and rice husk ash. Fuel Processing Technology. 2004;**85**(15):1623-1634

[18] Feng H, Li C, Shan H. Effect of calcination temperature of kaolin microspheres on the In situ synthesis of ZSM-5. Catalysis Letters. 2008;**129**(1-2):71-78

[19] Khatamian M. Preparation and characterization of nanosized ZSM-5 zeolite using kaolin and investigation of kaolin content, crystallization time and temperature changes on the size and crystallnity of products. Journal of the Iranian Chemical Society. 2009;**6**(1):187-194

[20] Kovo AS, Hernandez O, Holmes SM. Synthesis and characterization of zeolite Y and ZSM-5 from Nigerian Ahoko Kaolin using a novel, lower temperature, metakaolinization technique. Journal of Materials Chemistry. 2009;**19**(34):6207-6212

[21] Pan F, Lu X, Wang Y, Chen S, Wang T, Yan Y. Organic template-free synthesis of ZSM-5 zeolite from coal-series kaolinite. Materials Letters. 2014;**115**(1):5-8

[22] Gualtieri A, Norby P, Artioli G, Hanson J. Kinetics of formation of zeolite Na-A [LTA] from natural kaolinites. Physics and Chemistry of Minerals. 1997;**24**:191-199

[23] Murat M, Amokrane A, Bastide J, Montanaro L. Synthesis of zeolites from thermally activated kaolinite. Some observations on nucleation and growth. Clay Minerals. 1992;**27**(1):119-130

[24] Shirazi L, Jamshidi E, Ghasemi MR. The effect of Si/Al ratio of ZSM-5 zeolite on its morphology, acidity and crystal size. Crystal Research and Technology. 2008;**43**(12):1300-1306

[25] Caballero I, Colina FG, Costa J. Synthesis of X-type zeolite from dealuminated kaolin by reaction with sulfuric acid at high temperature. Industrial and Engineering Chemistry Research. 2007;**46**(4):1029-1038

[26] Lafi AAF, Matam SK, Hodali HA. New synthesis of ZSM-5 from high silica kaolinite and its use in vapor phase conversion of 1-phenylethanol to styrene new synthesis of ZSM-5 from high silica kaolinite and its use in vapor phase conversion of 1-phenylethanol to styrene. Industrial and Engineering Chemistry Research. 2015;**54**(15):3754-3760

[27] Bergaya F, Lagaly G. Chapter 1: General introduction: Clays, clay minerals, and clay science. In: Developments in Clay Science. Netherlands: Elsevier B.V. 2006. pp. 1-18

[28] Bloodworth AJ, Highley DE, Mitchell CJ. Industrial Minerals Laboratory Manual: Kaolin. BGS TechnicalReport WG/93/1, 2014. 1993

[29] Barrer RM. Hydrothermal Chemistry of Zeolites. London, New York: Academic Press; 1982

[30] Mohiuddin E, Isa YM, Mdleleni MM, Sincadu N, Key D, Tshabalala T. Synthesis of ZSM-5 from impure and beneficiated Grahamstown kaolin: Effect of kaolinite content, crystallisation temperatures and time. Applied Clay Science. 2016;**119**:213-221

[31] Belviso C, Cavalcante F, Lettino A, Fiore S. A and X-type zeolites synthesised from kaolinite at low temperature. Applied Clay Science. 2013;**80-81**:162-168

[32] Chandrasekhar S. Influence of metakaolinization temperature on the formation of zeolite 4A from kaolin. Clay Minerals. 1996;**31**(2):253-261

[33] Granizo ML, Blanco-Varela MT, Palomo A. Influence of the starting kaolin on alkali-activated materials based on metakaolin. Study of the reaction parameters by isothermal conduction calorimetry. Journal of Materials Science. 2000;**35**(24):6309-6315

[34] Johnson EBG, Arshad SE. Hydrothermally synthesized zeolites based on kaolinite: A review. Applied Clay Science. 2014;**97-98**:215-221

[35] Mohiuddin E, Isa YM, Mdleleni MM, Key D. Effect of kaolin chemical reactivity on the formation of ZSM-5 and its physicochemical properties. Microporous and Mesoporous Materials. 2017:237

[36] Chandrasekhar S, Pramada PN. Microwave assisted synthesis of zeolite A from metakaolin. Microporous and Mesoporous Materials. 2008;**108**(1-3):152-161

[37] Dedecek J, Balgová V, Pashkova V, Klein P, Wichterlová B. Synthesis of ZSM—5 zeolites with defined distribution of Al atoms in the framework and multinuclear MAS NMR analysis of the control of Al distribution. Chemistry of Materials. 2012;**24**:3231-3239

[38] Maia AÁB, Angélica RS, de Freitas Neves R, Pöllmann H, Straub C, Saalwächter K. Use of ^{29}Si and ^{27}Al MAS NMR to study thermal activation of kaolinites from Brazilian Amazon kaolin wastes. Applied Clay Science. 2014;**87**:189-196

[39] Magusin PCMM, Zorin VE, Aerts A, Houssin CJY, Yakovlev AL, Kirschhock CEA, et al. Template-aluminosilicate structures at the early stages of zeolite ZSM-5 formation. A combined preparative, solid-state NMR, and computational study. The Journal of Physical Chemistry. B. 2005;**109**(48):22767-22774

[40] Eilertsen EA, Haouas M, Pinar AB, Hould ND, Lobo RF, Lillerud KP, et al. NMR and SAXS analysis of connectivity of aluminum and silicon atoms in the clear sol precursor of SSZ-13 zeolite. Chemistry of Materials. 2012;**24**:571-578

[41] Murray LJ, Heckroodt RO. South African kaolins. In: International Clay Conference. 1978. pp. 601-608

[42] Saikia NJ, Bharali DJ, Sengupta P, Bordoloi D, Goswamee RL, Saikia PC, et al. Characterization, beneficiation and utilization of a kaolinite clay from Assam, India. Applied Clay Science. 2003;**24**(1-2):93-103

[43] Panda AK, Mishra BG, Mishra DK, Singh RK. Effect of sulphuric acid treatment on the physico-chemical characteristics of kaolin clay. Colloids and Surfaces A: Physicochemical and Engineering Aspects. 2010;**363**(1-3):98-104

[44] Holmes SM, Alomair AA, Kovo AS. The direct synthesis of pure zeolite-A using “virgin” kaolin. RSC Advances. 2012;**2**(30):11491

[45] Velho JAGL, SF GC d. Characterization of Portuguese kaolins for the paper industry: Beneficiation through new delamination techniques. Applied Clay Science. 1991;**6**(2):155-170

[46] Mignoni ML, Petkowicz DI, Fernandes Machado NRC, Pergher SBC. Synthesis of mordenite using kaolin as Si and Al source. Applied Clay Science. 2008;**41**(1-2):99-104

[47] Yu J. Synthesis of zeolites. In: Cejka J, van Bekkum H, Corma A, Schuth F, editors. Introduction to Zeolite Science and Practice. 3rd Revise ed. Netherlands: Elsevier B.V. 2007. pp. 39-103

[48] Kovo SA. Development of zeolites and zeolite membranes from Ahoko Nigerian Kaolin. [doctoral thesis].

England: The University of Manchester; 2011

[49] Na J, Liu G, Zhou T, Ding G, Hu S, Wang L. Synthesis and catalytic performance of ZSM-5/MCM-41 zeolites with varying mesopore size by surfactant-directed recrystallization. Catalysis Letters. 2013;**143**(3):267-275

[50] Liu C, Kong D, Guo H. The morphology control of zeolite ZSM-5 by regulating the polymerization degree of silicon and aluminum sources. Microporous and Mesoporous Materials. 2014;**193**:61-68

[51] Gobin OC, Reitmeier SJ, Jentys A, Lercher JA. Comparison of the transport of aromatic compounds in small and large MFI particles. Journal of Physical Chemistry C. 2009;**113**(47):20435-20444

[52] Zhang R, Zhang B, Shi Z, Liu N, Chen B. Catalytic behaviors of chloromethane combustion over the metal-modified ZSM-5 zeolites with diverse SiO_2/Al_2O_3 ratios. Journal of Molecular Catalysis A: Chemical. 2015;**398**:223-230

[53] Serrano DP, García RA, Vicente G, Linares M, Procházková D, Čejka J. Acidic and catalytic properties of hierarchical zeolites and hybrid ordered mesoporous materials assembled from MFI protozeolitic units. Journal of Catalysis. 2011;**279**(2):366-380

[54] Wang S, Dou T, Li Y, Zhang Y, Li X, Yan Z. A novel method for the preparation of MOR/MCM-41 composite molecular sieve. Catalysis Communications. 2005;**6**(1):87-91

[55] Tang Q, Xu H, Zheng Y, Wang J, Li H, Zhang J. Catalytic dehydration of methanol to dimethyl ether over micro-mesoporous ZSM-5/MCM-41 composite molecular sieves. Applied Catalysis A: General. 2012;**413-414**:36-42

[56] Yang ZX, Xia YD, Mokaya R. Zeolite ZSM-5 with unique supermicropores synthesized using mesoporous carbon as a template. Advanced Materials. 2004;**16**(8):727-732

[57] Groen JC, Peffer LAA, Pérez-Ramírez J. Pore size determination in modified micro- and mesoporous materials. Pitfalls and limitations in gas adsorption data analysis. Microporous and Mesoporous Materials. 2003;**60**(1-3):1-17

[58] Yue Y, Liu H, Yuan P, Li T, Yu C, Bi H, et al. From natural aluminosilicate minerals to hierarchical ZSM-5 zeolites: A nanoscale depolymerization–reorganization approach. Journal of Catalysis. 2014;**319**:200-210

[59] Rodríguez-González L, Hermes F, Bertmer M, Rodríguez-Castellón E, Jiménez-López A, Simon U. The acid properties of H-ZSM-5 as studied by NH_3-TPD and ^{27}Al-MAS-NMR spectroscopy. Applied Catalysis A: General. 2007;**328**(2):174-182

[60] Cabral de Menezes SM, Lam YL, Damodaran K, Pruski M. Modification of H-ZSM-5 zeolites with phosphorus. 1. Identification of aluminum species by ^{27}Al solid-state NMR and characterization of their catalytic properties. Microporous and Mesoporous Materials. 2006;**95**(1-3):286-295

[61] Woolery GL, Kuehl GH, Timken HC, Chester a W, Vartuli JC. On the nature of framework Brønsted and Lewis acid sites in ZSM-5. Zeolites. 1997;**19**(4):288-296

[62] Oldfield E, Haase J, Schmitt KD, Schramm SE. Characterization of zeolites and amorphous silica—Aluminas by means of aluminum-27 nuclear magnetic resonance spectroscopy: A multifield, multiparameter investigation. Zeolites. 1994;**14**(2):101-109

[63] Prasad JV, Rao KV, Bhat YS, Halgeri AB. MAS NMR studies of ZSM-5 zeolites: Correlation to para selectivity and SEM observations. Catalysis Letters. 1992;**14**(3-4):349-357

[64] Smaihi M, Barida O, Valtchev V. Investigation of the crystallization stages of LTA-type zeolite by complementary characterization techniqnes. European Journal of Inorganic Chemistry. 2003;**24**:4370-4377

[65] Van Donk S, Bitter JH, De Jong KP. Deactivation of solid acid catalysts for butene skeletal isomerisation: On the beneficial and harmful effects of carbonaceous deposits. Applied Catalysis A: General. 2001;**212**(1-2):97-116

[66] Buchanan JS, Olson DH, Schramm SE. Gasoline selective ZSM-5 FCC additives: Effects of crystal size, SiO_2/Al_2O_3, steaming, and other treatments on ZSM-5 diffusivity and selectivity in cracking of hexene/octene feed. Applied Catalysis A: General. 2001;**220**(1-2):223-234

Chapter 7

Thermal Radiation and Thermal Diffusion for Soret and Dufour's Effects on MHD Flow over Rotating Infinite Disk

Gamal M. Abdel-Rahman and Faiza M.N. El-fayez

Abstract

In general, the thermal radiation and thermal diffusion effects over an electrically conducting, Newtonian fluid in a steady laminar magnetohydrodynamic convective flow over a porous rotating infinite disk with the consideration of heat and mass transfer in the presence of Soret and Dufour's diffusion effects have been obtained and studied numerically. The governing continuity, momentum, energy and concentration equations are converted into a system of non-linear ordinary differential equations by means of similarity transformation. The resulting system of coupled non-linear ordinary differential equations is solved numerically. In this chapter, numerical results were presented for velocity (radial, axial and tangential), temperature, concentration and pressure profiles for different parameters of the problem Also, the effects of the pertinent parameters on the radial and tangential skin friction, the rate of heat and mass transfer are obtained and discussed numerically and illustrated graphically.

Keywords: MHD, heat and mass transfer, thermal radiation, slip flow, porous medium, rotating disk

1. Introduction

The flow due to rotating disks is of great interest in many practical and engineering aspects. Rotating disk flows of electrically conducting fluids have practical applications in many areas, such as rotating machinery, lubrication, oceanography, computer storage devices, viscometer and crystal growth processes etc. Also, the study is interesting from the mathematical point of view. During the last two decades, research on renewable energy sources, as for example, solar, wind energy or energy from hydro-power and the preparation of oxygenated additives to blend diesel fuel, has been intensified.

Pioneering study of fluid flow due to an infinite rotating disk was carried by authors [1–3]. Chemical reactions usually accompany a large amount of exothermic and endothermic reactions. These characteristics can be easily seen in a lot of industrial processes, it has been realized that it is not always permissible to neglect the convection effects in porous constructed chemical reactors [4]. The reaction produced in a porous medium was extraordinarily in common, such as the topic of

PEM fuel cells modules and the polluted underground water because of discharging the toxic substance, etc.

Fourier's law, for instance, described the relation between energy flux and temperature gradient. In other aspects, Fick's law was determined by the correlation of mass flux and concentration gradient. Moreover, it was found that energy flux can also be generated by composition gradients, pressure gradients, or body forces. The energy flux caused by a composition gradient was discovered in 1873 by Dufour and was correspondingly referred to the Dufour effect.

It was also called the diffusion-thermo effect. On the other hand, mass flux can also be created by a temperature gradient, as was established by Soret. This is the thermal-diffusion effect. In general, the thermal-diffusion and the diffusion-thermo effects were of a smaller order of magnitude than the effects described by Fourier's or Fick's law and were often neglected in heat and mass transfer processes. There were still some exceptional conditions. The thermal-diffusion effect has been utilized for isotope separation and in mixtures between gases with very light molecular weight (H_2, He) and of medium molecular weight (N_2, air), the diffusion-thermo effect was found to be of a magnitude such that it may not be neglected in certain conditions [5].

The first traceable interest in magnetohydrodynamics (MHD) flow was in 1907, when Northrop built an MHD pump prototype [6, 7]. Since then, analysis of the effects of both rotation and magnetic fields on fluid flows has been an active area of research. While technology expanded in many directions, the subject of MHD has developed in the use of magnetic fields and the range of fluid and thermal processes by [8–13]. This study considers the effect of slip as a result of rarefied effect, a type of flow commonly encountered in many engineering tasks such as high altitude flight, micro-machines, vacuum technology, aerosol reactors, etc. In this study, the slip and no-slip regimes that lie in the range $0.1 > K_n > 0$ are considered. A completely different extension of von Karman's one-disk problem is the analysis of Sparrow et al. [14]. They considered the flow of a Newtonian fluid due to the rotation of a porous-surfaced disk and for that purpose replaced the conventional no-slip boundary conditions at the disk surface with a set of linear slip flow conditions. A substantial reduction in torque then occurred as a result of surface slip. Recently Frusteri and Osalusi [15] studied the effects of variable properties on MHD and slip flow over a porous rotating disk.

In all these studies Soret and Dufour effects were assumed to be negligible. Such effects are significant when density differences exist in the flow regime. For example, when species are introduced at a surface in fluid domain, with different (lower) density than the surrounding fluid, both Soret (thermo-diffusion) and Dufour (diffusion-thermal) effects can be influential. An analytical study of convection along a horizontal cylinder for a Helium-air system was reported subsequently by Sparrow et al. [16]. In view of the importance of above mentioned effects, Maleque [17] studied Soret effect on convective heat and mass transfer past a rotating porous disk and he neglected the Dufour effect. Ahmed [18] investigated the Dufour and Soret effects on free convective heat and mass transfer over a stretching surface considering suction or injection. Recently, numerical study of free convection magnetohydrodynamic heat and mass transfer due to a stretching surface under saturated porous medium with Soret and Dufour effects was also discussed by Anwar Beg et al. [19].

In these two papers [15, 20] they have studied the effect of the magnetic field on the equations of motion that I found them used in the cylindrical coordinate, although the magnetic field parameters are in the Cartesian coordinate, which is wrong.

2. Mathematical formulation

Consider the steady, axially symmetric, incompressible flow of an electrically conducting fluid with heat and mass transfer flow due to a rotating porous disk in the presence of radiation has been considered. Assume that the fluid is infinite in extent in the positive z-direction. Let (r, φ, z) be the set of cylindrical polar coordinates and let the disk rotate with constant angular velocity Ω and is placed at $z = 0$. The components of the flow velocity are (u, v, w) in the directions of increasing (r, φ, z), respectively. p, T and C are the pressure, the temperature and the concentration distribution, respectively. The surface of the rotating disk is maintained at a uniform temperature T_w and uniform concentration C_w. Far away from the surface, the free stream is kept at a constant temperature T_∞, at a constant concentration C_∞ and at a constant pressure p_∞. The fluid is assumed to be gray, emitting and absorbing heat, but not scattering medium and is assumed to be Newtonian. The physical model and geometrical coordinates are shown in **Figure 1**.

The MHD body forces $\overline{J} \times \overline{B}$ the Maxwell's equations:

$$div\,\overline{B} = 0,\; Curl\,\overline{B} = \mu_m \overline{J} \text{ and } div\,\overline{E} = 0$$

where $\overline{J}$ is the electric current density, $\overline{B} = B + b$ is the total magnetic field, μ_m is the magnetic permeability and b is the induced magnetic field. The external uniform magnetic field B is imposed in the direction normal to the surface of the disk which is assumed unchanged by taking small magnetic Reynolds number, so that the flow induction distortion of the applied magnetic field can be neglected as in the case with most of conducting fluids. In addition, a uniform suction is applied at the surface of the disk for the entire range.

The magnetic body force $\overline{J} \times \overline{B}$ takes the form $\sigma(\overline{V} \times \overline{B}) \times \overline{B}$, therefore, $\sigma(\overline{V} \times \overline{B}) \times \overline{B} = -\sigma B^2 \overline{V}/r$, where σ is the electrical conductivity of fluid and $\overline{V}$ is

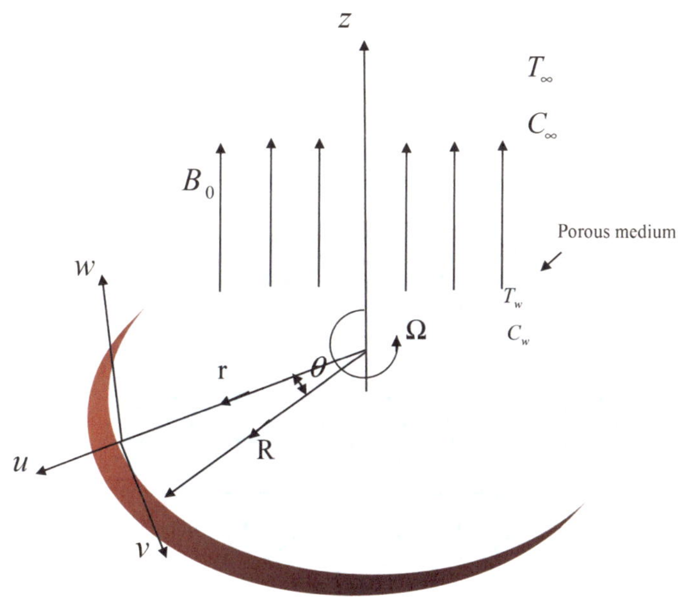

Figure 1.
Schematic diagram of the problem.

velocity vector $\overline{V} = (u, v, w)$ and $\overline{B} = (0, 0, B)$. The Lorentz force (MHD body force) has two components:

$$F_r = -\sigma B^2 u/r, F_\theta = -\sigma B^2 v/r.$$

Under these assumptions, the governing equations for the continuity, momentum, energy and concentration in laminar incompressible flow can be written as follows:

$$\frac{\partial u}{\partial r} + \frac{u}{r} + \frac{\partial w}{\partial z} = 0 \,, \tag{1}$$

$$\begin{aligned} u\frac{\partial u}{\partial r} - \frac{v^2}{r} + w\frac{\partial u}{\partial z} = -\frac{1}{\rho}\frac{\partial p}{\partial r} + \nu\left(\frac{\partial^2 u}{\partial r^2} + \frac{1}{r}\frac{\partial u}{\partial r} - \frac{u}{r^2} + \frac{\partial^2 u}{\partial z^2}\right) \\ - \frac{\sigma B^2}{\rho r}u - \frac{\nu}{K_1^*}u + g\beta_T(T - T_\infty) + g\beta_C(C - C_\infty), \end{aligned} \tag{2}$$

$$u\frac{\partial v}{\partial r} + \frac{uv}{r} + w\frac{\partial v}{\partial z} = \nu\left(\frac{\partial^2 v}{\partial r^2} + \frac{1}{r}\frac{\partial v}{\partial r} - \frac{v}{r^2} + \frac{\partial^2 v}{\partial z^2}\right) - \frac{\sigma B^2}{\rho r}v - \frac{\nu}{K_1^*}v, \tag{3}$$

$$u\frac{\partial w}{\partial r} + w\frac{\partial w}{\partial z} = -\frac{1}{\rho}\frac{\partial p}{\partial z} + \nu\left(\frac{\partial^2 w}{\partial r^2} + \frac{1}{r}\frac{\partial w}{\partial r} + \frac{\partial^2 w}{\partial z^2}\right), \tag{4}$$

$$\begin{aligned} u\frac{\partial T}{\partial r} + w\frac{\partial T}{\partial z} = \frac{k}{\rho c_p}\left(\frac{\partial^2 T}{\partial r^2} + \frac{1}{r}\frac{\partial T}{\partial r} + \frac{\partial^2 T}{\partial z^2}\right) - \frac{1}{\rho c_p}\frac{\partial q_r}{\partial z} + \frac{D k_T}{c_s c_p}\left(\frac{\partial^2 C}{\partial r^2} + \frac{1}{r}\frac{\partial C}{\partial r} + \frac{\partial^2 C}{\partial z^2}\right) \\ + \frac{\sigma B^2}{\rho c_p r}u^2 + \frac{Q}{\rho c_p}(T - T_\infty), \end{aligned} \tag{5}$$

$$u\frac{\partial C}{\partial r} + w\frac{\partial C}{\partial z} = D\left(\frac{\partial^2 C}{\partial r^2} + \frac{1}{r}\frac{\partial C}{\partial r} + \frac{\partial^2 C}{\partial z^2}\right) + \frac{D k_T}{T_m}\left(\frac{\partial^2 T}{\partial r^2} + \frac{1}{r}\frac{\partial T}{\partial r} + \frac{\partial^2 T}{\partial z^2}\right) - k_1(C - C_\infty) \tag{6}$$

When the mean free path of the fluid particle is comparable to the characteristic dimensions of the flow field domain, Naiver-Stokes equations break down since the assumption of continuum media fails. In the range of $0.1 < k_n < 10$ Knudsen Number, the higher order continuum equation, for example, Burnett equation should be used. For the range of $0.001 \le k_n \le 0.1$, no slip boundary conditions cannot be used and should be replaced with the following expression (Gad-el-Hak [20]):

$$U_t = \frac{2 - \xi}{\xi}\lambda\frac{\partial U_t}{\partial n}, \tag{7}$$

Where U_t is the tangential velocity, n is the normal direction to the wall, ξ is the tangent momentum accommodation coefficient and λ is the mean free path. For $k_n < 0.001$, the no-slip boundary condition is valid, therefore, the velocity at the surface is equal to zero. In this study the slip and the no-slip regimes of the Knudsen number that lies in the range $0.1 > k_n > 0$ is considered. By using Eq. (7), the appropriate boundary conditions for the flow induced by an infinite disk ($z = 0$) which rotates with constant angular velocity Ω subjected to uniform suction w_0 through the disk are given by

$$\begin{aligned} z = 0: \quad & u = \frac{2 - \xi}{\xi}\lambda\frac{\partial u}{\partial z}, \quad v = \Omega r + \frac{2 - \xi}{\xi}\lambda\frac{\partial v}{\partial z}, \quad w = w_0, \quad T = T_w, \quad C = C_w, \\ z \to \infty: \quad & u \to 0, v \to 0, \quad T \to T_\infty, \quad C \to C_\infty, \quad p \to p_\infty. \end{aligned} \tag{8}$$

where k is the thermal conductivity, $\nu = \mu/\rho$ is the kinematic viscosity of the ambient fluid, σ is the electrical conductivity, K_1^* is the permeability of the porous medium, g is the gravitational acceleration, β_T and β_C are the expansion coefficients of temperature and concentration, ρ, μ and c_p are the density, dynamic viscosity and the specific heat at constant pressure, respectively, Q is the volumetric heat generation/absorption rate, D is the molecular diffusion coefficient, k_T, c_s, T_m and q_r are the thermal-diffusion rate, concentration susceptibility, the mean fluid temperature and the radiative heat flux. Using the Rosseland approximation (Rashed [21]), the radiative heat flux q_r could be expressed by

$$q_r = -\frac{4\sigma^*}{3k^*}\frac{\partial T^4}{\partial z} \tag{9}$$

Where the σ^* represents the Stefan-Boltzman constant and k^* is the Rosseland mean absorption coefficient.

Assuming that the temperature difference within the flow is sufficiently small such that T^4 could be approached as the linear function of temperature

$$T^4 \cong 4T_\infty^3 T - 3T_\infty^4 \tag{10}$$

For the flow under study, it is relevant to assume that the applied magnetic field $B(r)$ has the form cobble [22] $B = B_0\sqrt{r}$, where B_0 is constant magnetic flux density.

To obtain the non-dimensional form of the above equations, the following dimensionless variables are introduced.

$$\begin{gathered}\overline{R} = \frac{r}{L},\ \overline{Z} = \frac{z}{L},\ \overline{U} = \frac{u}{\Omega L},\ \overline{V} = \frac{v}{\Omega L},\ \overline{\lambda} = \frac{\lambda}{L},\ \overline{W} = \frac{w}{\Omega L},\ \overline{P} = \frac{p - p_\infty}{\rho\Omega^2 L^2},\\ \overline{\nu} = \frac{\nu}{\Omega L^2},\ \overline{k_1} = \frac{k_1}{L^2},\ \overline{T} = \frac{T - T_\infty}{T_w - T_\infty},\ \overline{C} = \frac{C - C_\infty}{C_w - C_\infty},\ k_1^* = \frac{\overline{k_*}}{L^2}\end{gathered} \tag{11}$$

Substituting Eqs. (9)–(11) in Eqs. (1)–(6), we obtain the following dimensionless equations:

$$\frac{\partial \overline{U}}{\partial \overline{R}} + \frac{\overline{U}}{\overline{R}} + \frac{\partial \overline{W}}{\partial \overline{Z}} = 0\ , \tag{12}$$

$$\begin{aligned}\overline{U}\frac{\partial \overline{U}}{\partial \overline{R}} - \frac{\overline{V^2}}{\overline{R}} + \overline{W}\frac{\partial \overline{U}}{\partial \overline{Z}} &= -\frac{\partial \overline{P}}{\partial \overline{R}} + \overline{\nu}\left(\frac{\partial^2 \overline{U}}{\partial \overline{R^2}} + \frac{1}{\overline{R}}\frac{\partial \overline{U}}{\partial \overline{R}} - \frac{\overline{U}}{\overline{R^2}} + \frac{\partial^2 \overline{U}}{\partial \overline{Z^2}}\right) - \frac{\sigma B_0^2}{\rho\Omega}\overline{U} - \frac{\overline{\nu}}{\overline{k_1}}\overline{U}\\ &+ \frac{g\beta_T(T_w - T_\infty)}{\Omega^2 L}\overline{T} + \frac{g\beta_c(C_w - C_\infty)}{\Omega^2 L}\overline{C},\end{aligned} \tag{13}$$

$$\overline{U}\frac{\partial \overline{V}}{\partial \overline{R}} - \frac{\overline{U}\,\overline{V}}{\overline{R}} + \overline{W}\frac{\partial \overline{V}}{\partial \overline{Z}} = \overline{\nu}\left(\frac{\partial^2 \overline{V}}{\partial \overline{R^2}} + \frac{1}{\overline{R}}\frac{\partial \overline{V}}{\partial \overline{R}} - \frac{\overline{V}}{\overline{R^2}} + \frac{\partial^2 \overline{V}}{\partial \overline{Z^2}}\right) - \frac{\sigma B_0^2}{\rho\Omega}\overline{V} - \frac{\overline{\nu}}{\overline{k_1}}\overline{V}, \tag{14}$$

$$\overline{U}\frac{\partial \overline{W}}{\partial \overline{R}} + \overline{W}\frac{\partial \overline{W}}{\partial \overline{Z}} = -\frac{\partial \overline{P}}{\partial \overline{Z}} + \overline{\nu}\left(\frac{\partial^2 \overline{W}}{\partial \overline{R^2}} + \frac{1}{\overline{R}}\frac{\partial \overline{W}}{\partial \overline{R}} + \frac{\partial^2 \overline{W}}{\partial \overline{Z^2}}\right), \tag{15}$$

$$\begin{aligned}\overline{U}\frac{\partial \overline{T}}{\partial \overline{R}} + \overline{W}\frac{\partial \overline{T}}{\partial \overline{Z}} &= \frac{k}{\rho c_p \Omega L^2}\left(\frac{\partial^2 \overline{T}}{\partial \overline{R^2}} + \frac{1}{\overline{R}}\frac{\partial \overline{T}}{\partial \overline{R}} + \frac{\partial^2 \overline{T}}{\partial \overline{Z^2}}\right) + \frac{16\sigma^* T_\infty^3}{3\rho c_p \Omega k^* L^2}\frac{\partial^2 \overline{T}}{\partial \overline{Z^2}}\\ &+ \frac{D k_T(C_w - C_\infty)}{c_s c_p \Omega L^2 (T_w - T_\infty)}\left(\frac{\partial^2 \overline{C}}{\partial \overline{R^2}} + \frac{1}{\overline{R}}\frac{\partial \overline{C}}{\partial \overline{R}} + \frac{\partial^2 \overline{C}}{\partial \overline{Z^2}}\right) + \frac{\sigma B_0^2 \Omega L}{\rho c_p (T_w - T_\infty)}\frac{\overline{U^2}}{\overline{R}} + \frac{Q}{\rho c_p \Omega}\overline{T},\end{aligned} \tag{16}$$

$$\overline{U}\frac{\partial\overline{C}}{\partial\overline{R}}+\overline{W}\frac{\partial\overline{C}}{\partial\overline{Z}}=\frac{D}{\Omega L^2}\left(\frac{\partial^2\overline{C}}{\partial\overline{R}^2}+\frac{1}{\overline{R}}\frac{\partial\overline{C}}{\partial\overline{R}}+\frac{\partial^2\overline{C}}{\partial\overline{Z}^2}\right)$$
$$+\frac{Dk_T(T_w-T_\infty)}{T_m\Omega L^2(C_w-C_\infty)}\left(\frac{\partial^2\overline{T}}{\partial\overline{R}^2}+\frac{1}{\overline{R}}\frac{\partial\overline{T}}{\partial\overline{R}}+\frac{\partial^2\overline{T}}{\partial\overline{Z}^2}\right)-\frac{\overline{k}}{\Omega L^2}\overline{C}. \tag{17}$$

The boundary conditions (8) are reduced to

$$\begin{aligned}&\overline{Z}=0: \quad \overline{U}=\frac{2-\xi}{\xi}\overline{\lambda}\frac{\partial\overline{U}}{\partial\overline{Z}},\quad \overline{V}=\overline{R}+\frac{2-\xi}{\xi}\overline{\lambda}\frac{\partial\overline{V}}{\partial\overline{Z}},\quad \overline{W}=\frac{w_0}{\Omega L},\quad \overline{T}=1,\quad \overline{C}=1,\\ &\overline{Z}\to\infty: \quad \overline{U}\to 0, \overline{V}\to 0,\quad \overline{T}\to 0,\quad \overline{C}\to 0,\quad \overline{P}\to 0.\end{aligned} \tag{18}$$

The governing equations are obtained by introducing a dimensionless normal distance from the disk, $\eta=\overline{Z}/\sqrt{\overline{\nu}}$ along with the von-Karman transformations,

$$\begin{aligned}&\overline{U}=\overline{R}F(\eta),\quad \overline{V}=\overline{R}G(\eta),\quad \overline{W}=\sqrt{\overline{\nu}}H(\eta).\\ &\overline{T}=\theta(\eta),\quad \overline{C}=\varphi(\eta),\quad \overline{P}=\overline{\nu}\overline{P}(\eta)\end{aligned} \tag{19}$$

Where F, G, H, θ, φ and $\overline{P}$ are non-dimensionless functions in terms of vertical co-ordinate η. Substituting these transformations into Eqs. (12)–(17) gives the nonlinear ordinary differential equations, expressed as

$$2F+H^{/}=0, \tag{20}$$

$$F^{//}-HF^{/}-F^2+G^2-(M+S)F+\alpha\theta+N\varphi=0, \tag{21}$$

$$G^{//}-HG^{/}-2FG-(M+S)G=0, \tag{22}$$

$$H^{//}-HH^{/}-\overline{P^{/}}=0, \tag{23}$$

$$\frac{1}{P_r}\left(1+\frac{4}{3R_d}\right)\theta^{//}-H\theta^{/}+D_u\varphi^{//}+JF^2+\delta\theta=0, \tag{24}$$

$$\frac{1}{S_c}\varphi^{//}-H\varphi^{/}+S_0\theta^{//}-\beta\varphi=0. \tag{25}$$

With the appropriate boundary conditions:

$$\begin{aligned}&\eta=0; F(0)=\gamma F^{/}(0),\ G(0)=1+\gamma G^{/}(0), H(0)=W_s,\ \ \theta(0)=1,\ \ \varphi(0)=1,\\ &\eta\to\infty; F(\infty)=0, G(\infty)=0,\ \theta(\infty)=0,\ \ \varphi(\infty)=0, \overline{P}(\infty)=0.\end{aligned} \tag{26}$$

Where $\gamma=\left[(2-\xi)\lambda\sqrt{\Omega}\right]/\xi\sqrt{\overline{\nu}}$ the slip is factor and $W_s=w_0/\sqrt{\nu\Omega}$ represents a uniform suction $(W_s<0)$ at the disk surface. The boundary conditions given by Eq. (26) imply that the radial (F), the tangential (G) components of velocity, temperature and concentration vanish sufficiently far away from the disk, whereas the axial velocity component (H) is anticipated to approach a yet unknown asymptotic limit for sufficiently large η-values.

3. Skin-friction coefficient, Nusselt number and Sherwood number

The parameters of engineering interest for the present problem are the local skin-friction coefficients and the local rates of heat and mass transfer to the surface are calculated. The radial shear stress and tangential shear stress are given by:

$$\tau_{\varphi z} = \mu \left(\frac{\partial v}{\partial z} + \frac{1}{r} \frac{\partial w}{\partial \varphi} \right)_{z=0},$$

$$\tau_{zr} = \mu \left(\frac{\partial u}{\partial z} + \frac{\partial w}{\partial r} \right)_{z=0},$$

The tangential and radial skin-friction coefficient are, respectively, given by, Eq. (27):

$$\begin{aligned} C_{f_1} &= \tau_{\varphi z} / \rho \Omega^2 L^2 = \sqrt{R_e} G^{/}(0), \\ C_{f_2} &= \tau_{zr} / \rho \Omega^2 L^2 = \sqrt{R_e} F^{/}(0), \end{aligned} \tag{27}$$

Now the heat flux (q_w) and the mass flux (J_w) at the surface are given by.

$$q_w = -k \left(\frac{\partial T}{\partial z} \right)_{z=0}, \text{ and } M_w = -D \left(\frac{\partial C}{\partial z} \right)_{z=0}.$$

Hence the Nusselt number, N_u and Sherwood number, S_h are obtained as Eqs. (28) and (29), respectively:

$$N_u = \frac{L q_w}{k (T_w - T_\infty)(\overline{\upsilon})^{-1/2}} = -\theta^{/}(0), \tag{28}$$

And

$$S_h = \frac{L M_w}{D (C_w - C_\infty)(\overline{\upsilon})^{-1/2}} = -\varphi^{/}(0). \tag{29}$$

Where $R_e = \overline{R^2 / \overline{\upsilon}}$ is the rotational Reynolds number.

4. Numerical results and discussion

The system of non-linear ordinary differential Eqs. (20)–(25) together with the boundary conditions (26) are locally similar and solved numerically by using the Keller-box method. In order to gain physical insight, the velocity (radial, axial and tangential), temperature, concentration and pressure profiles have been discussed by assigning numerical values to the parameter, encountered in the problem which the numerical results are tabulated and displayed with the graphical illustrations.

In order to verify the accuracy of our present method, we have compared our results with those of Frusteri and Osalusi [15], Osalusi and Sibanda [23] and Maleque and Sattar [24]. **Table 1** shows the values of $F^{/}(0)$, $-G^{/}(0)$ and $-\theta^{/}(0)$ for various values of W_s. The comparisons in all above cases are found to be excellent and agreed, so it is good.

Figures 2, **3** and **11a–d** display the velocity (radial, axial and tangential), temperature, concentration and pressure profiles under the magnetic field parameter, porosity parameter and Schmidt number. The (radial, axial and tangential) components of the velocity and pressure profile decrease with increase of magnetic field due to the inhibiting influence of the Lorentz force and increasing of all porosity parameter and Schmidt number, while the temperature and the concentration profiles increase with increasing of all magnetic field parameter, porosity parameter and Schmidt number. In **Figures 4** and **5a–d**, it is clear that the (radial and axial)

W_s	$F^{/}(0)$				$-G^{/}(0)$				$-\theta^{/}(0)$			
	Frusteri et al. [15]	Osalusi et al. [23]	Maleque et al. [24]	Present study	Frusteri et al. [15]	Osalusi et al. [23]	Maleque et al. [24]	Present study	Frusteri et al. [15]	Osalusi et al. [23]	Maleque et al. [24]	Present study
	$\gamma = 0, \varepsilon = 0$											
0.0	0.4241	0.4241	0.5102	0.4627	0.6514	0.6514	0.6160	0.6241	0.5387	0.5387	0.3258	0.3171
−2.0	0.2324	0.2324	0.2425	0.2435	2.0687	2.0687	2.0391	2.0452	1.5196	1.5196	1.4421	1.4746
−4.0	0.1246	0.1246	0.1248	0.1245	4.0065	4.0065	4.0054	4.0052	2.8520	2.8520	2.8447	2.8432
−5.0	0.0999	0.0999	0.0999	0.0998	5.0029	5.0029	5.0031	5.0032	3.5541	3.5541	3.5541	3.5539
−10.0	0.0499	0.0499	0.0506	0.0502	10.0003	10.0003	10.0017	10.0019	7.1002	7.1002	7.1020	7.1018

Table 1.
Comparison of the current and recent numerical values of the radial and tangential skin-friction coefficients and the rate of heat transfer coefficient for various of W_s with $S = 0.0, \alpha = 0.0, N = 0.0, J = 0.0, \delta = 0.0, P_r = 0.71, \beta = 0.0, R_d = 10^9, D_u = 0.0, S_0 = 0.0, S_c = 1.0$ and $M = 0$.

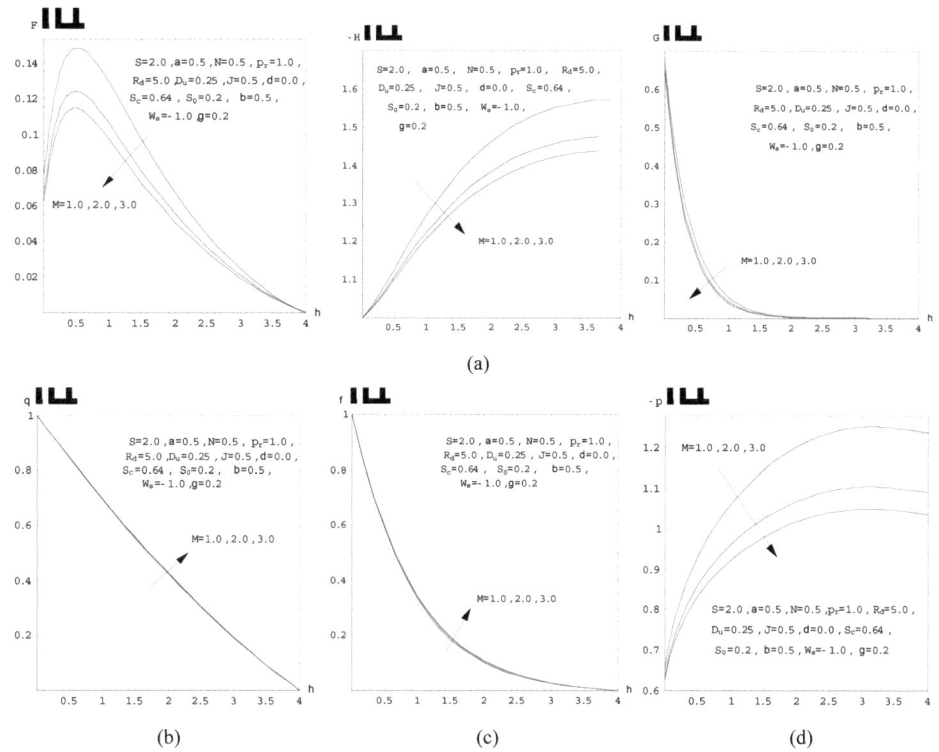

Figure 2.
Effect of magnetic field parameter on (a) the velocity (radial, axial and tangential) profile, (b) the temperature profile, (c) the concentration and (d) the pressure profile.

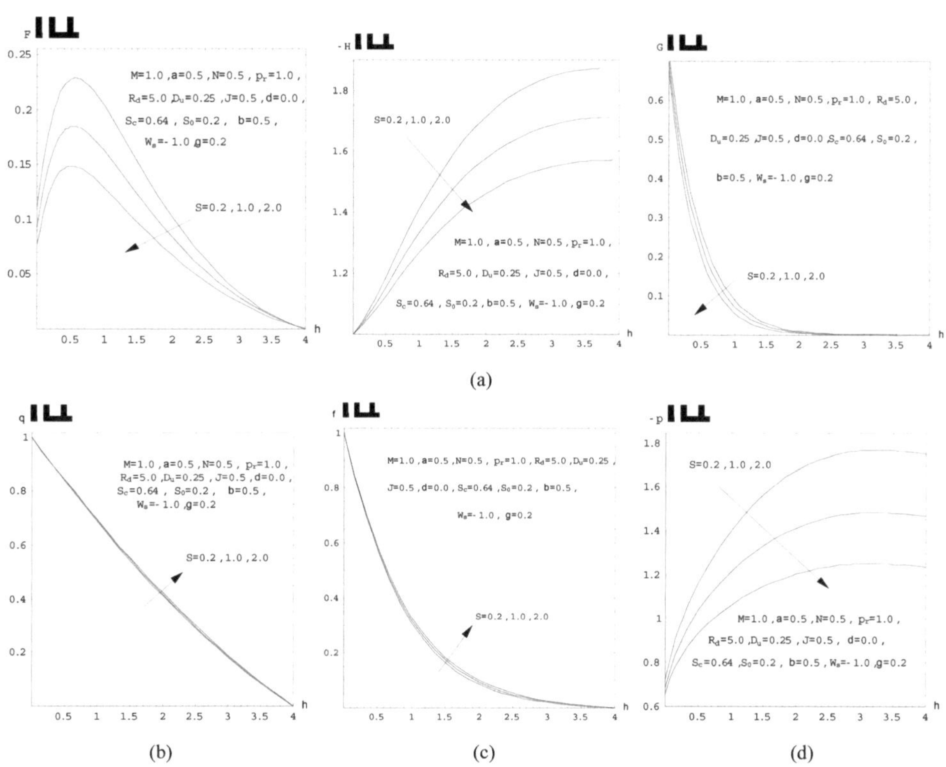

Figure 3.
Effect of porosity parameter on (a) the velocity (radial, axial and tangential) profile, (b) the temperature profile, (c) the concentration and (d) the pressure profile.

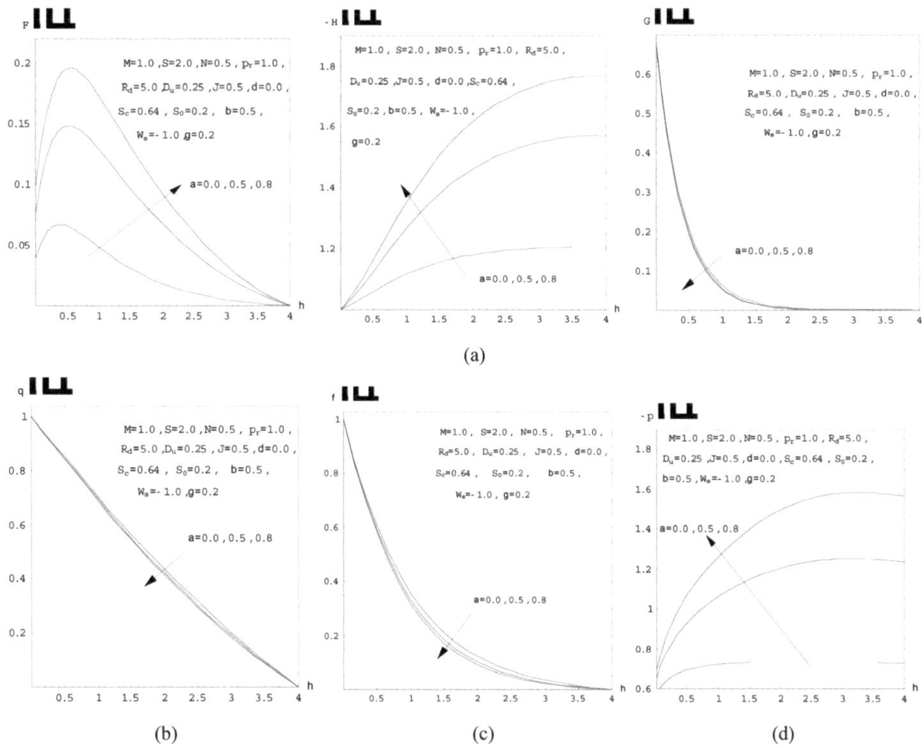

Figure 4.
Effect of temperature buoyancy parameter on (a) the velocity (radial, axial and tangential) profile, (b) the temperature profile, (c) the concentration and (d) the pressure profile.

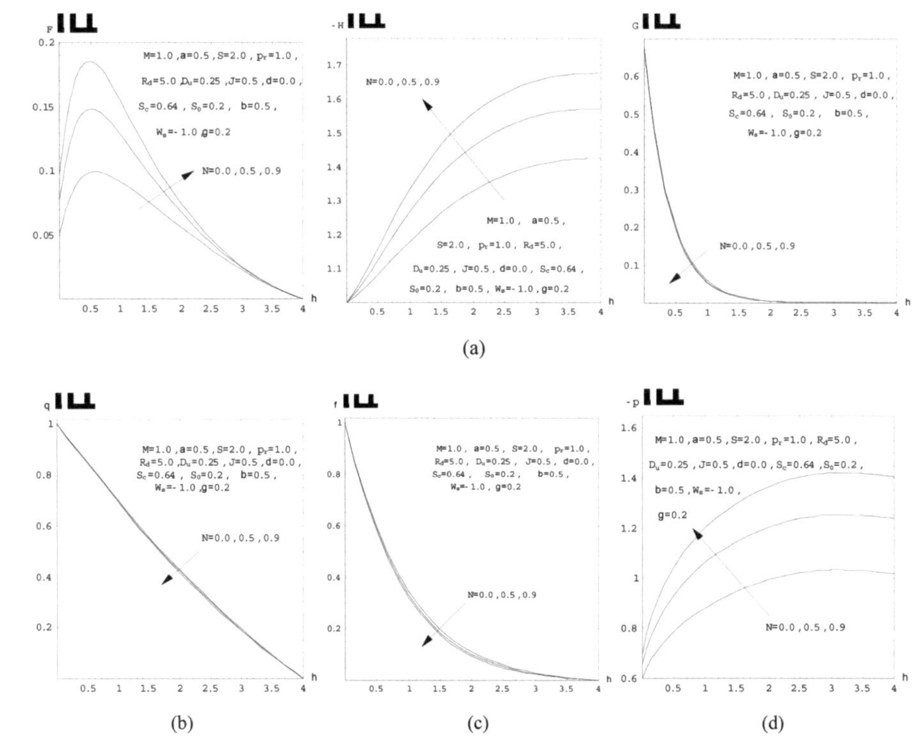

Figure 5.
Effect of concentration buoyancy parameter on (a) the velocity (radial, axial and tangential) profile, (b) the temperature profile, (c) the concentration and (d) the pressure profile.

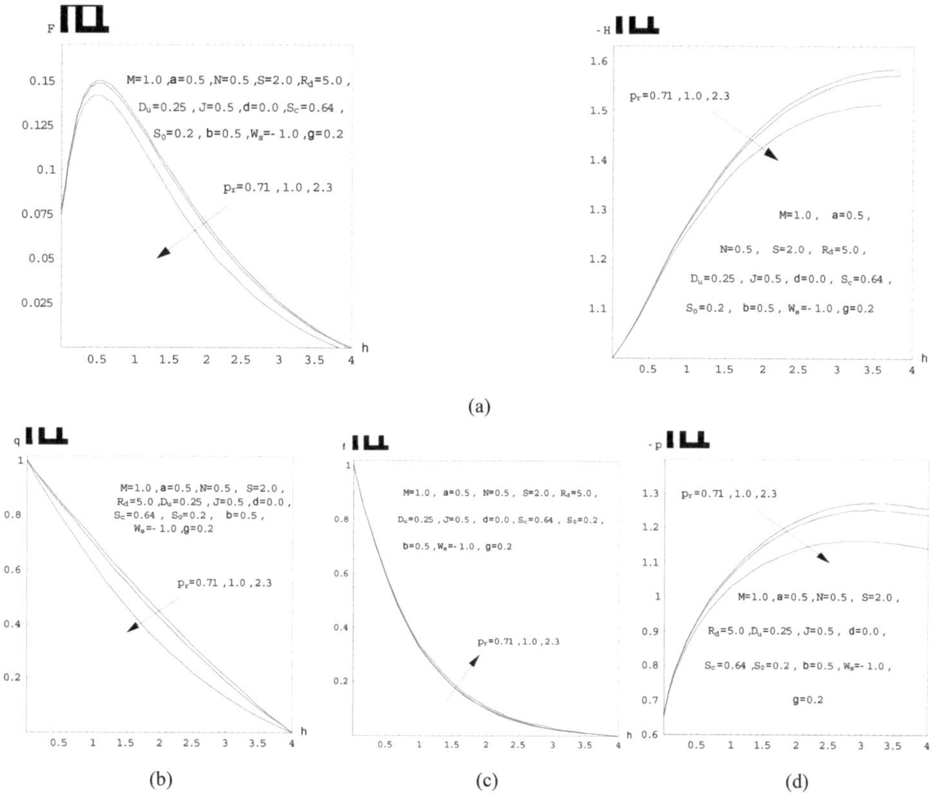

Figure 6.
Effect of Prandtl number on (a) the velocity (radial and axial) profile, (b) the temperature profile, (c) the concentration and (d) the pressure profile.

components of the velocity and pressure profiles increase with increase of temperature buoyancy parameter and increasing concentration buoyancy parameter, while the tangential component of the velocity, the temperature and the concentration profiles decrease with increasing temperature buoyancy parameter and concentration buoyancy parameter.

The effects of P_r on the (radial and axial) components of the velocity, the temperature, concentration and pressure profiles are shown in **Figure 6a–d**, respectively. It is observed that both the (radial and axial) components of the velocity, the temperature and pressure profiles decrease with the increase of Prandtl number. While, the concentration profile increase. Physically, it means that thermal boundary layer thickness gets decreased. In fact, it is well known that the thermal boundary layer thickness is inversely proportional to the square root of Prandtl number. Hence, the decrease of temperature profile with increasing P_r is straightforward.

Figures 7 and **10a–d** show the effects of radiation parameter and heat source parameter on the velocity (radial and axial), temperature, concentration and pressure profiles, also, we found that the (radial and axial) components of the velocity, temperature and pressure profiles increase with the increase of radiation parameter and increasing heat source parameter. While the concentration profile decrease. The effects of D_u and S_0 on the velocity (radial and axial), temperature, concentration and pressure profiles, are shown in **Figure 8a–d**, respectively. It is observed that the (radial and axial) components of the velocity, temperature, concentration and pressure profiles increase with the decreasing the Dufour's number and increasing Soret number.

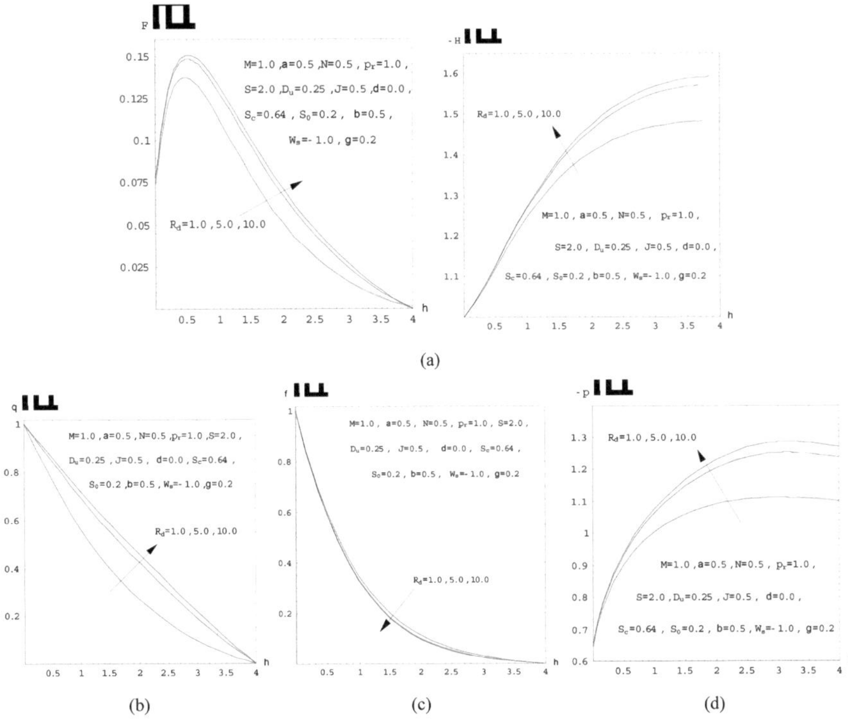

Figure 7.
Effect of radiation parameter on (a) the velocity (radial and axial) profile, (b) the temperature profile, (c) the concentration and (d) the pressure profile.

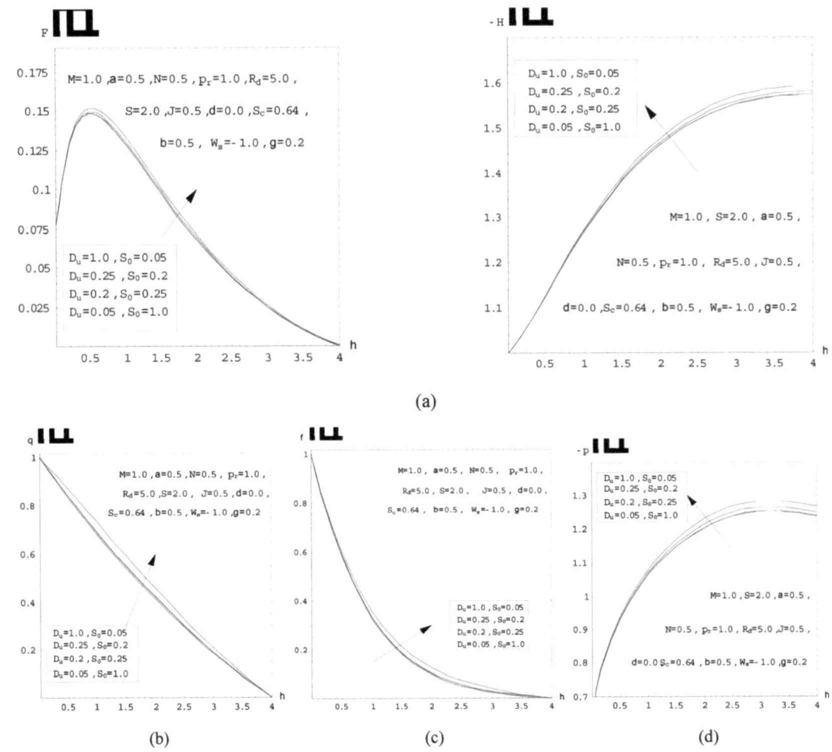

Figure 8.
Effect of Soret and Dufour's number on (a) the velocity (radial and axial) profile, (b) the temperature profile, (c) the concentration and (d) the pressure profile.

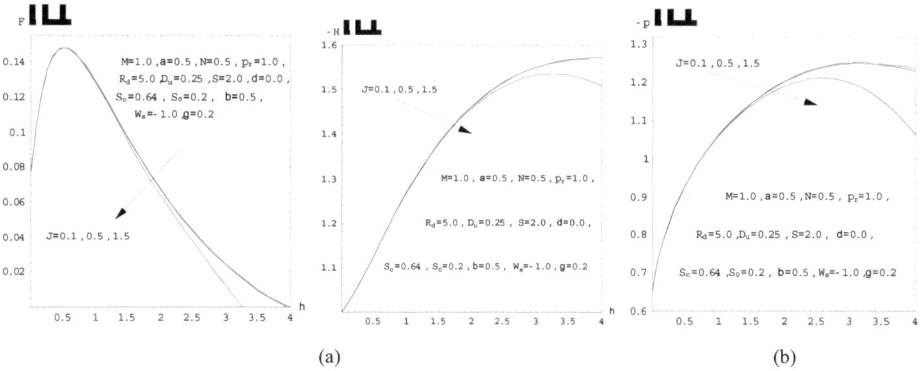

Figure 9.
Effect of Joule heating parameter on (a) the velocity (radial and axial) profile, and (b) the pressure profile.

The effects of J on the velocity (radial and axial) and pressure profiles, are shown in **Figure 9a** and **b**, respectively. We found that the (radial and axial) components of the velocity and pressure profiles decrease with the increase Joule heating parameter. In **Figure 12a–c**, it is clear that the (radial and axial) components of the velocity, concentration and pressure profiles decrease with increase of chemical reaction parameter. **Figure 13a–d** displays the velocity (radial, axial and tangential), temperature, concentration and pressure profiles under the suction

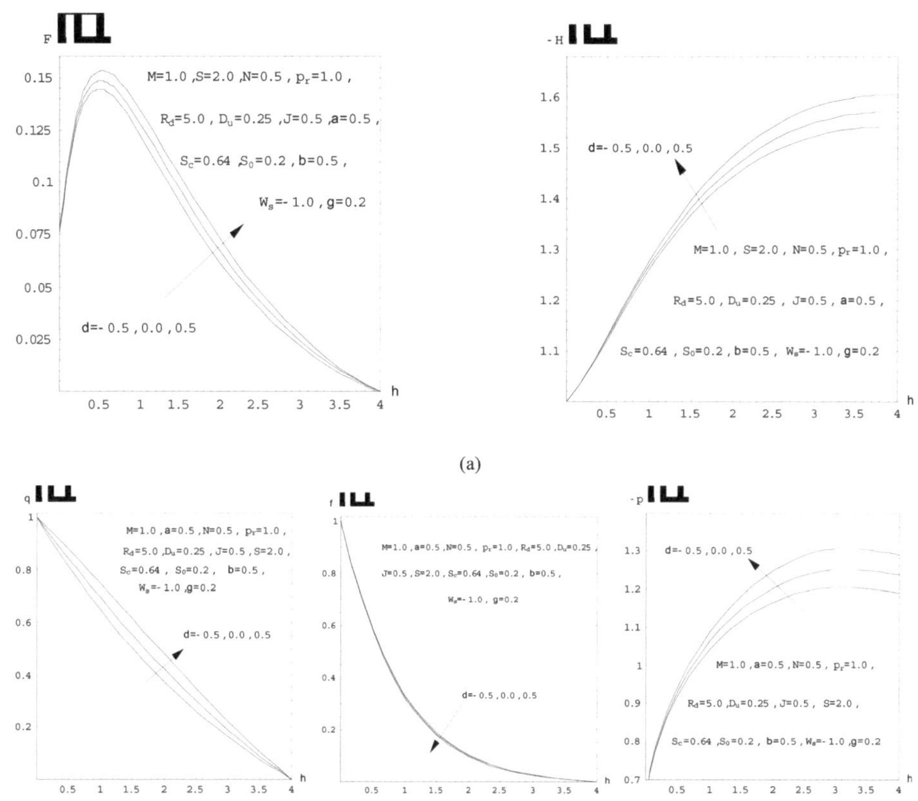

Figure 10.
Effect of heat source parameter on (a) the velocity (radial and axial) profile, (b) the temperature profile, (c) the concentration and (d) the pressure profile.

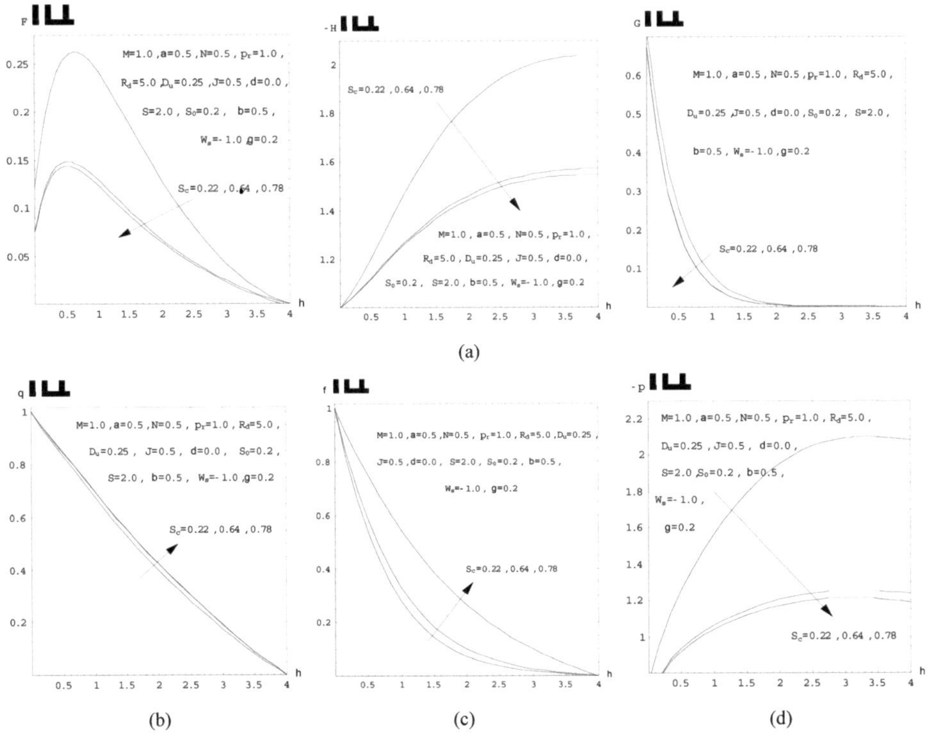

Figure 11.
Effect of Schmidt number on (a) the velocity (radial, axial and tangential) profile, (b) the temperature profile, (c) the concentration and (d) the pressure profile.

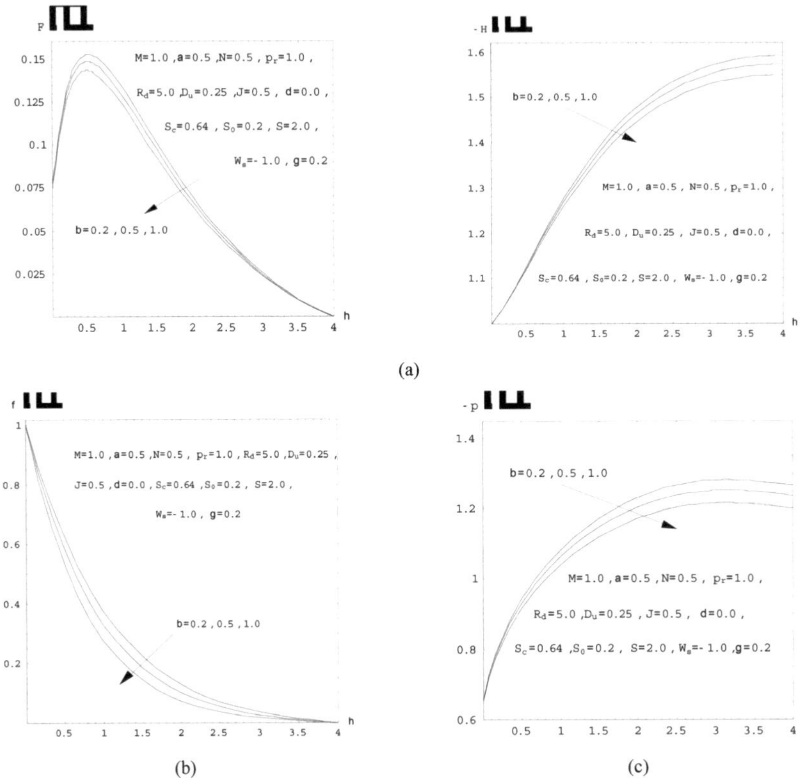

Figure 12.
Effect of chemical reaction parameter on (a) the velocity (radial and axial) profile, (b) the concentration and (c) the pressure profile.

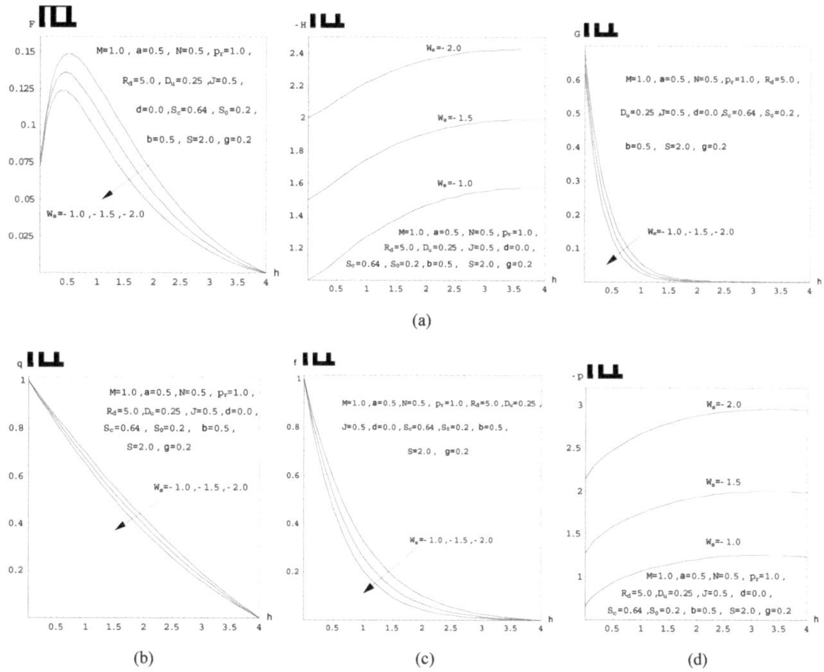

Figure 13.
Effect of suction parameter on (a) the velocity (radial, axial and tangential) profile, (b) the temperature profile, (c) the concentration and (d) the pressure profile.

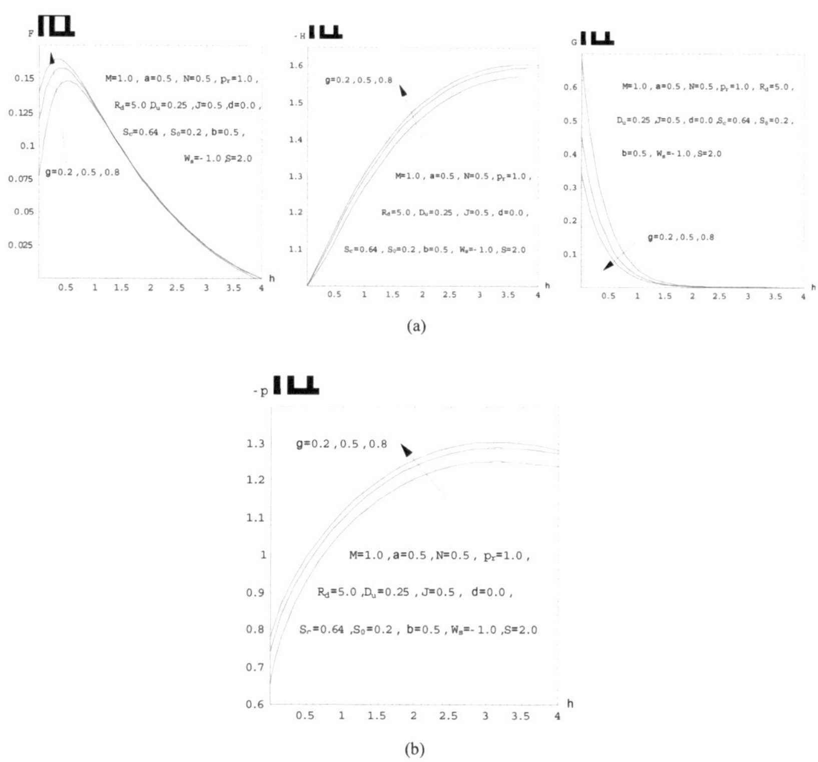

Figure 14.
Effect of slip parameter on (a) the velocity (radial, axial and tangential) profile, and (b) the pressure profile.

M	S	α	N	P_r	R_d	$D_u \& S_0$	J	δ	S_c	β	W_s	γ	$R_e^{-1/2}C_{f1}$	$-R_e^{-1/2}C_{f2}$	$-N_u$	$-S_h$
0.1	2.0	0.5	0.5	1.0	5.0	0.25 & 0.2	0.5	0.0	0.64	0.5	−1.0	0.2	0.3857	1.6257	0.3144	1.0486
2.0													0.3323	1.7275	0.3116	1.0367
3.0													0.3112	1.7732	0.3105	1.0319
0.1	0.2												0.5534	1.3978	0.3220	1.0870
	1.0												0.4629	1.5071	0.3185	1.0663
	2.0												0.3857	1.6257	0.3144	1.0486
0.1	2.0	0.0											0.1970	−1.6005	0.3039	1.0056
		0.5											0.3857	1.6257	0.3144	1.0486
		0.8											0.4953	1.6397	0.3198	1.0725
0.1	2.0	0.5	0.0										0.2510	1.6096	0.3093	1.0247
			0.5										0.3857	1.6257	0.3144	1.0486
			0.9										0.4894	1.6377	0.3180	1.0667
0.1	2.0	0.5	0.5	0.71									0.3879	1.6261	0.2944	1.0522
				1.0									0.3857	1.6257	0.3144	1.0486
				2.3									0.3757	1.6241	0.4155	1.0326
0.1	2.0	0.5	0.5	1.0	1.0								0.3683	1.6229	0.5022	1.0194
					5.0								0.3857	1.6257	0.3144	1.0486
					10.0								0.3892	1.6263	0.2828	1.0541
0.1	2.0	0.5	0.5	1.0	5.0	1.0 & 0.05							0.3910	1.6266	0.2324	1.0591
						0.25 & 0.2							0.3857	1.6257	0.3144	1.0486
						0.2 & 0.25							0.3856	1.6256	0.3195	1.0454
0.1	2.0	0.5	0.5	1.0	5.0	0.25 & 0.2	0.1						0.3858		0.3157	1.0487

M	S	α	N	P_r	R_d	$D_u \& S_0$	J	δ	S_c	β	W_s	γ	$R_e^{-1/2}C_{f1}$	$-R_e^{-1/2}C_{f2}$	$-N_u$	$-S_h$
							0.5						0.3857		0.3144	1.0486
							1.5						0.3854		0.3111	1.0485
0.1	2.0	0.5	0.5	1.0	5.0	0.25 & 0.2	0.5	−0.5					0.3795	1.6247	0.3938	1.0370
								0.0					0.3857	1.6257	0.3144	1.0486
								0.5					0.3931	1.6270	0.2239	1.0625
0.1	2.0	0.5	0.5	1.0	5.0	0.25 & 0.2	0.5	0.0	0.22				0.6083	1.4121	0.3451	0.5590
									0.64				0.3857	1.6257	0.3144	1.0486
									0.78				0.3777	1.6278	0.3084	1.2094
0.1	2.0	0.5	0.5	1.0	5.0	0.25 & 0.2	0.5	0.0	0.64	0.2			0.3927	1.6268	0.3193	0.9130
										0.5			0.3857	1.6257	0.3144	1.0486
										1.0			0.3770	1.6244	0.3075	1.2368
0.1	2.0	0.5	0.5	1.0	5.0	0.25 & 0.2	0.5	0.0	0.64	0.5	−1.0		0.3857	1.6257	0.3144	1.0486
											−1.5		0.3737	1.7684	0.3445	1.2854
											−2.0		0.3590	1.9095	0.3764	1.5404
0.1	2.0	0.5	0.5	1.0	5.0	0.25 & 0.2	0.5	0.0	0.64	0.5	−1.0	0.2	0.3857	1.6257	0.3144	1.0486
												0.5	0.2403	1.0968	0.3156	1.0573
												0.8	0.1757	0.8264	0.3162	1.0613

Table 2.
Numerical of the values skin-friction coefficient $\left(R_e^{-1/2}C_{f1}, -R_e^{-1/2}C_{f2}\right)$*, Nusselt number* N_u *and Sherwood number* S_h *at the surface with* M, S, α, N, P_r, R_d, $D_u \& S_0$, J, δ, S_c, β, W_s *and* γ*.*

parameter. The (radial, axial and tangential) components of the velocity, temperature, concentration and pressure profiles decrease with increase of suction parameter. The effects of γ on the velocity (radial, axial and tangential) and pressure profiles are shown in **Figure 14a** and **b**, respectively. It is observed that the (radial and axial) components of the velocity, and pressure profiles increase with the increasing slip parameter. While the tangential component of the velocity profile decrease.

The radial and tangential skin frictions and the heat and mass transfer coefficients are tabulated in **Table 2** for various values of M, S, α, N, P_r, R_d, $D_u\&S_0$, J, δ, S_c, β, W_s and γ. We observed that increase for all magnetic field parameter M and porosity parameter S leads to an decrease in the all tangential skin friction $\left(-G^{/}(0)\right)$, heat transfer rate $\left(-\theta^{/}(0)\right)$ and mass transfer rate $\left(-\varphi^{/}(0)\right)$, while an increase in the radial skin friction $\left(F^{/}(0)\right)$, the increase for all radial skin friction, tangential skin friction, heat transfer rate and mass transfer rate, with increasing of the temperature buoyancy parameter α and the concentration buoyancy parameter N. We found that the radial skin friction, tangential skin friction, mass transfer rate decreases while heat transfer rate increase with increasing of Prandtl number, Dufour number decreases and Soret number increases. It can that be seen that the radial skin friction, tangential skin friction and mass transfer rate increase while heat transfer rate decrease with increasing values of R_d and δ. It is observed that an increase in the Joule heating parameter, results in a decrease in the tangential Skin-friction coefficient, Nusselt number and Sherwood number. The tangential skin friction and heat transfer rate decrease but the radial skin friction and mass transfer rate increase with increasing the Schmidt number. It also can be seen from this table that increasing the chemical reaction parameter to decrease in the radial skin friction, tangential skin friction and heat transfer rate while increase the mass transfer rate. We found also the tangential skin friction increase but the radial skin friction, Nusselt number and Sherwood number decrease with increasing the suction parameter. Finally, the radial skin friction and the tangential skin friction decrease but Nusselt number and Sherwood number increase with increasing slip parameter.

5. Conclusions

In this work, thermal radiation and thermal diffusion effects over an electrically conducting, Newtonian fluid in a steady laminar magnetohydrodynamic convective flow over a porous rotating infinite disk with the consideration of heat and mass transfer in the presence of Soret and Dufour's diffusion effects have been obtained and studied numerically. Magnetic field parameter, porosity parameter, temperature buoyancy parameter, concentration buoyancy parameter, Prandtl number, radiation parameter, Soret and Dufour's number, Joule heating parameter, heat source parameter, Schmidt number, chemical reaction parameter, suction parameter, slip parameter effects were considered in the separate cases. The subsequent outcome may be drawn as:

1. The components (radial, axial and tangential) of the velocity, temperature, concentration and pressure profiles under the porosity parameter. The (radial, axial and tangential) components of the velocity and pressure profile decrease with increasing porosity parameter, while the temperature and the concentration profiles increase with increasing porosity parameter. And also, that increase porosity parameter S leads to a decrease in the all tangential skin

friction $\left(-G^{/}(0)\right)$, heat transfer rate $\left(-\theta^{/}(0)\right)$ and mass transfer rate $\left(-\varphi^{/}(0)\right)$, while an increase in the radial skin friction $\left(F^{/}(0)\right)$.

2. The (radial and axial) components of the velocity and pressure increase with increasing of temperature buoyancy parameter and concentration buoyancy parameter, while the tangential component of the velocity, the temperature and the concentration profiles decrease with increasing temperature buoyancy parameter and concentration buoyancy parameter. We found that the increase for all radial skin friction, tangential skin friction, heat transfer rate and mass transfer rate, with increasing of the temperature buoyancy parameter α and the concentration buoyancy parameter N.

3. The (radial and axial) components of the velocity, temperature and pressure profiles increase with the increase of heat source parameter. While the concentration profile decrease. And also, the radial skin friction, tangential skin friction and mass transfer rate increase while heat transfer rate decrease with increasing value of δ.

4. The (radial and axial) components of the velocity, concentration and pressure profiles decrease with increase of chemical reaction parameter. And in **Table 2**, an increase the chemical reaction parameter, results in an decrease in the (radial and tangential) Skin-friction coefficient and Nusselt number, while Sherwood number increases.

Nomenclature

B	external uniform magnetic field
B_0	constant magnetic flux density
b	induced magnetic field
C	concentration distribution
C_w	uniform concentration
C_∞	constant concentration
c_p	specific heat at constant pressure
c_s	concentration susceptibility
C_{f1}	tangential skin-friction coefficient
C_{f2}	radial skin-friction coefficient
D	molecular diffusion coefficient
D_u	Dufour number $\left[= Dk_T(C_w - C_\infty)/\nu c_s c_p(T_w - T_\infty)\right]$
g	gravitational acceleration, $\left[\text{m s}^{-2}\right]$
k	thermal conductivity
k_1	rate of chemical reaction
K_T	thermal-diffusion rate
K_1^*	permeability of the porous medium
k^*	Rosseland mean absorption coefficient
F, G, H	radial (F), tangential (G) and axial (H) components of dimensionless velocity
M	magnetic field parameter $\left[= \sigma B_0^2/\rho\Omega\right]$
N	concentration buoyancy parameter $\left[= g\beta_c(C_w - C_\infty)/L\overline{R}\Omega^2\right]$
N_u	Nusselt number
n	normal direction to the wall

$\vec{J}$	electric current density
J	Joule heating parameter $\left[= \sigma B_0^2 \Omega L / \rho c_p (T_w - T_\infty)\right]$
P_r	Prandtl number $\left[= \nu \rho c_p / k\right]$
p	pressure distribution, $[\mathrm{N\ m^{-2}}]$
p_∞	constant pressure
Q	volumetric heat generation/absorption rate, $[\mathrm{W\ m^{-3}}]$
q_r	radiative heat flux, $[\mathrm{W\ m^{-2}}]$
R_d	radiation parameter $\left[= k^* k / 4\sigma^* T_\infty^3\right]$
S	porosity parameter $\left[= \overline{\nu} / \overline{k_1}\right]$
S_0	Soret number $[= D k_T (T_w - T_\infty) / \nu T_m (C_w - C_\infty)]$
S_c	Schmidt number $[= \nu / D]$
S_h	Sherwood number
T	temperature distribution [K]
T_m	mean fluid temperature [K]
T_w	uniform temperature [K]
T_∞	constant temperature [K]
(u, v, w)	components of the flow velocity are in the directions of increasing (r, φ, z), respectively, $[\mathrm{m\,s^{-1}}]$
U_t	tangential velocity, $[\mathrm{m\,s^{-1}}]$
w_0	uniform suction
Greek symbols	
μ_m	magnetic permeability, $[\mathrm{kg\ m^{-1} s^{-1}}]$
ξ	tangent momentum accommodation coefficient
λ	mean free path
ρ	fluid density, $[\mathrm{kg\ m^{-3}}]$
μ	dynamic viscosity, $[\mathrm{kg\ m^{-1} s^{-1}}]$
ν	kinematic viscosity of the ambient fluid, $[\mathrm{m^2\,s^{-1}}]$
δ	heat source parameter $\left[= Q / \Omega \rho c_p\right]$
α	temperature buoyancy parameter $\left[= g \beta_T (T_w - T_\infty) / L \overline{R} \Omega^2\right]$
γ	slip factor
η	similarity variable
θ	dimensionless temperature distribution
φ	dimensionless concentration distribution
σ	electrical conductivity
σ	Stephan-Boltzmann constant
Ω	constant angular velocity
β	chemical reaction parameter $\left[= \overline{k} / \Omega L^2\right]$
β_T	coefficient of temperature
β_C	coefficient of concentration
Subscripts	
w, ∞	conditions at the surface and in the free stream

Author details

Gamal M. Abdel-Rahman[1]* and Faiza M.N. El-fayez[2]

1 Department of Mathematics, Faculty of Science, Benha University, Benha, Egypt

2 Mathematical Science Department, College of Sciences, Princess Nourah Bint Abdulrahman University, Riyadh, KSA

*Address all correspondence to: gamalm60@yahoo.com

IntechOpen

References

[1] Von Karman T. Uber laminare und turbulente Reibung. Zeitschrift für Angewandte Mathematik und Mechanik. 1921;**1**:1233-1255

[2] Cochran WG. The flow due to a rotating disk. Proceedings of the Cambridge Philosophical Society. 1934; **30**:365-375

[3] Benton ER. On the flow due to a rotating disk. Journal of Fluid Mechanics. 1966;**24**:781-800

[4] Nield DA, Bejan A. Convection in Porous Media. 2nd ed. New York: Springer-Verlag; 1999

[5] Eckert ERG, Drake RM. Analysis of Heat and Mass Transfer. New York: McGraw-Hill; 1972

[6] Northrup EF. Some newly observed manifestations of forces in the interior of an electrical conductor. Physics Review. 1907;**24**:474-497

[7] Tillack MS, Morley NB. Magnetohydrodynamics. Standard Handbook for Electrical Engineer. McGraw Hill; 1998

[8] Hayat T, Nadeem S, Siddiqui A, Asghar S. An oscillating hydromagnetic non-Newtonian flow in a rotating system. Applied Mathematics Letters. 2004;**17**:609-614

[9] Ahn CH, Allen MG. Fluid micropumps based on rotary magnetic actuators. Proceedings of MEMS. 1995; **95**:408-412

[10] Mensing G, Pearce T, Bebe DJ. An in-plane active magnetic mixer for microfluidic applications. In: IEEEEMB Conference. 2002. pp. 531-534

[11] Mahmud S, Tasnim SH, Mamun MAH. Thermodynamic analysis of mixed convection in a channel with transverse hydromagnetic effect. International Journal of Thermal Sciences. 2003;**42**:731-740

[12] Miklavcic M. The flow due to a rough rotating disk. Zeitschrift für Angewandte Mathematik und Physik. 2004;**55**:235-246

[13] Mizzi S, Emerson DR, Stefanov SK, Barber RW, Reese JM. Effects of rarefaction on cavity flow in the slip regime. Journal of Computational and Theoretical Nanoscience. 2007;**4**: 817-822

[14] Sparrow EM, Beavers GS, Hung LY. Flow about a porous surface rotating disk. International Journal of Heat and Mass Transfer. 1971;**14**:993-996

[15] Frusteri F, Osalusi E. On MHD and slip flow over a rotating porous disk with variable properties. International Communications in Heat and Mass Transfer. 2007;**14**:34, 492-501

[16] Sparrow EM, Minkowycz WJ, Eckert ERG. Transpiration induced buoyancy and thermal diffusion-diffusion thermo in helium-air free convection boundary layer. ASME Journal of Heat Transfer. 1964;**86**: 508-520

[17] Abdul Maleque K. Magnetohydrodynamic convective heat and mass transfer flow due to a rotating disk with thermal diffusion effect. Journal of Heat Transfer. 2009;**131**:1-8

[18] Afify Ahmed A. Effects of thermal diffusion and diffusion thermo on free convective heat and mass transfer over a stretching surface considering suction or injection. Communications in Nonlinear Science and Numerical Simulation. 2009;**14**:2202-2214

[19] Beg A, Bakier AY, Prasad VR. Numerical study of free convection magnetohydrodynamic heat and mass transfer from a stretching surface to a saturated porous medium with Soret and Dufour effects. Computational Materials Science. 2009;**46**:57-65

[20] Gad-el-Hak M. The fluid mechanics of microdevices. The freeman scholar lecture. Journal of Fluids Engineering-Transactions of the ASME. 1999;**121**: 5-33

[21] Rashed GMA. Chemical entropy generation and MHD effects on the unsteady heat and fluid flow through a porous medium. Journal of Applied Mathematics. 2016;**2016:**9 pages, Article ID 1748312

[22] Cobble MH. Magnetohydrodynamic flow with a pressure gradient and fluid injection. Journal of Engineering Mathematics. 1977;**11**:249-256

[23] Osalusi E, Sibanda P. On variable laminar convective flow properties due to a porous rotating disk in a magnetic field. Romanian Journal of Physics. 2006;**51**(9–10):933-944

[24] Maleque AK, Sattar AM. Steady laminar convective flow with variable properties due to a porous rotating disk. Journal of Heat Transfer. 2005;**127**: 1406-1409

Chapter 8

MHD Flow and Heat Transfer of Casson Nanofluid through a Porous Media over a Stretching Sheet

Ayesha Siddiqui and Bandari Shankar

Abstract

The present chapter aims at investigating the magnetohydrodynamic (MHD) boundary layer flow and heat transfer of a non-Newtonian fluid over a stretching surface through a porous medium. Casson fluid model is utilized to describe the non-Newtonian fluid behavior. Two types of nanofluids, that is, Ag-water and Cu-water, are studied. The governing partial differential equations are transformed into a system of coupled non-linear ordinary differential equations using similarity transformations and then solved numerically using the Keller box method. Numerical results are obtained for the velocity, temperature, skin friction coefficient and Nusselt number. The influence of the various governing parameters viz. Casson parameter, magnetic parameter, porosity parameter and Prandtl number on the flow and heat transfer characteristics of the nanofluids is plotted graphically and discussed in detail. The chapter shows that with an increase in the Casson parameter, the velocity field decreases whereas the temperature profile increases. A decrease in the momentum boundary layer thickness and an increase in the thermal boundary layer thickness are noted with an increase in the magnetic parameter.

Keywords: magnetohydrodynamic, Casson fluid, porous medium, stretching sheet

1. Introduction

The boundary layer flow and heat transfer over a stretching sheet have momentous views not only from theoretical point of view but also one can see their practical applications in the paper production, polymer industry, crystal growing, food processing etc. Crane [1] was the first to study the boundary layer flow yielded by a stretching sheet. He gave an exact solution for the originating problem. Later on, the boundary layer flow over linear and non-linear stretching surfaces have pulled in a great deal of interest of many of the researchers [2–5]. Magnetohydrodynamic (MHD) boundary layer flow due to an exponentially stretching sheet with radiation effect has been examined by Ishak [6]. In fluid dynamics, the influence of external magnetic field on magnetohydrodynamic (MHD) flow over a stretching sheet is very significant due to its applications in many engineering problems such as for purification of crude oil, paper production and glass manufacturing. A physiological process in human body can be deciphered by processes like MRI, NMRI and MRT, in which MHD plays an important role [7, 8]. Pavlov [9] analyzed the effect of external magnetic field on MHD flow over a stretching sheet. Andersson [10] studied the MHD flow of viscous fluid over a stretching sheet. A robust numerical method for solving stagnation point flow over a permeable shrinking sheet under the influence of MHD was considered by Bhatti et al. [11]. They observed that as the Hartman number increases, the fluid velocity also increases. Sheikholeslami et al. [12] employed the control volume-based finite element method (CVFEM) to show the influence of external magnetic source on $Fe_3O_4 - H_2O$ nanofluid behavior in a permeable cavity considering shape effect. They remarked that the nanofluid velocity and heat transfer rate decrease with augment of Hartmann number. The rate of heat transfer between the stretching surface and the fluid flow is crucial to obtain the desired quality of the end product. So, in the boundary layer flow problems dealing with the non-Newtonian fluids, heat transfer analysis plays an important role. Barzegar et al. [13] applied neural network for the estimation of heat transfer treatment of $Al_2O_3 - H_2O$ nanofluid through a channel. Sadoughi et al. [14] investigated CuO-water nanofluid heat transfer enhancement in the presence of melting surface. Sheikholeslami et al. [15] investigated nanofluid heat transfer augmentation and exergy loss inside a pipe equipped with innovative turbulators. Sheikholeslami et al. [16] investigated thermal radiation of ferrofluids in the presence of Lorentz force considering variable viscosity and reported that the Nusselt number increases with an increase of buoyancy forces and radiation parameter, but it is reduced with rise of Hartmann number. For an appraisal of technological applications, knowledge of the rheological characteristics of the non-Newtonian fluids [17] is required. In the study of fluid dynamics and heat transfer, an essential component is the fundamental analysis of the non-Newtonian fluid flow field in a boundary layer adjacent to a stretching sheet or an extended surface [18–20]. The flows of various non-Newtonian fluids over stretching or shrinking sheets were analyzed by Liao [21], Hayat et al. [22] and Ishak et al. [23]. Compared to the viscous fluids, the characteristics of the non-Newtonian fluids are different and the governing equations are also extremely non-linear and complicated. Therefore, no single constitutive equation, displaying all properties of such fluids is available [24, 25]. In literature, several models of non-Newtonian fluids have been proposed but most of the models are related with simple models like "the power-law fluid of grade two or three" [26–30]. Casson fluid model is a non-Newtonian fluid model. A Casson fluid can be defined as a shear thinning liquid which is assumed to have an infinite viscosity at zero rate of shear and a zero viscosity at an infinite rate of shear [31]. Examples of Casson fluid are jelly, tomato sauce, honey, soap and concentrated fruit juices. Human blood is also an example of Casson fluid. Rouleaux is a chain-like structure formed by the human red blood cells, due to the presence of substances like globulin, protein and fibrinogen in an aqueous base plasma. If the rouleaux acts like a plastic solid, then there exists a yield stress that can be identified with the constant yield stress in Casson fluid [32]. Casson fluid model (McDonald 1974) [33] describes the blood flow through small vessels at low shear rates. Mukhopadhyay [34] examined the effects of Casson fluid flow and heat transfer over a non-linearly stretching surface. She concluded that temperature increases with an increase in non-linear stretching parameter and the momentum boundary layer thickness decreases with an increase in Casson parameter. The relationship between the fluxes and the driving potentials becomes complicated whenever heat and mass transfer occur simultaneously in a moving fluid. Apart from the temperature gradients, the concentration gradients are also one of the factors to cause energy flux. The generation of energy flux by concentration gradient is named as diffusion-thermo(Dufour) effect and that of mass flux by temperature gradient is termed as thermal-diffusion(Soret) effect. Hayat et al. [35] and Nawaz et al. [36] studied Soret and Dufour effects on the MHD flow of a Casson fluid on a stretching surface. An analysis on heat and mass transfer in stagnation point flow of a polar fluid toward a stretching surface in porous medium in the presence of Soret, Dufour and chemical reaction effects was carried out by Chamkha and Aly [37]. They observed that as the Soret number increases and the Dufour number decreases, the fluid velocity increases. Casson [38] derived the non-linear Casson constitutive equation and it depicts the properties of many polymers. Mustafa et al. [39] studied the stagnation point flow and heat transfer in a Casson fluid flow over a stretching sheet. An exact solution of the steady boundary layer flow of a Casson fluid over a porous stretching sheet was studied by Shehzad et al. [40].

In a boundary layer flow, the flow field gets significantly affected by the presence of porous media and as a result, the rate of heat transfer at the surface also gets influenced. Practical applications of the flow and heat transfer through a porous media can be seen in geophysical fluid dynamics such as limestone, wood, beach sand, sandstone, the human lungs and in small blood vessels [41]. Sheikholeslami [42] analyzed the exergy and entropy of nanofluids under the impact of Lorentz force through a porous media by incorporating the CVFEM method. He observed that exergy drop diminishes with reduction of magnetic forces. Shehzad et al. [43] simulated nanofluid convective flow inside a porous enclosure by means of a two-temperature model. They remarked that the porosity and temperature gradient are inversely related. Sheikholeslami [44] studied CuO-water nanofluid flow due to magnetic field inside a porous media considering Brownian motion. Shehzad et al. [45] considered the numerical modeling for alumina nanofluid's magnetohydrodynamic convective heat transfer in a permeable medium using Darcy law. They concluded that an increase in radiation parameter makes the thermal boundary layer thinner. Sheikholeslami [46] examined CuO-water nanofluid's free convection in a porous cavity considering the Darcy law. He applied the CVFEM method to interpret his results. Numerical simulation for heat transfer intensification of a nanofluid in a porous curved enclosure considering shape effect of Fe_3O_4 nanoparticles was carried out by Shamlooei et al. [47]. Rokni et al. [48] made a numerical simulation for the impact of Coulomb force on nanofluid heat transfer in a porous enclosure in the presence of thermal radiation. They noticed that the Nusselt number of nanoparticles with platelet shape is the highest. Sibanda et al. [49] considered nanofluid flow over a non-linear stretching sheet in porous media with MHD and viscous dissipation effects. Sheikholeslami et al. [50] examined magnetohydrodynamic nanofluid convection in a porous enclosure considering heat flux boundary condition. Their studies reveal that the temperature gradient shows a reduction with increasing values of Hartman number. The objective of this chapter is to study the MHD flow and heat transfer of a Casson nanofluid through a porous medium over a stretching sheet. The governing partial differential equations of the flow and energy distribution are transformed into a set of non-linear ordinary differential equations by using similarity transformations and are then solved numerically by a finite difference numerical technique called Keller box method [51]. The effects of various governing parameters on the flow and heat transfer characteristics of the nanofluids, that is, Ag-water and Cu-water are analyzed and shown graphically.

2. Equations of motion

Consider the steady two-dimensional MHD flow of an electrically conducting non-Newtonian Casson nanofluid over a stretching sheet situated at $y = 0$. The flow is confined in the region $y>0$. Two equal and opposite forces are applied along the x-axis so that the wall is stretched with the origin fixed. The rheological equation of state for an isotropic and incompressible flow of the Casson nanofluid is

$$\tau_{ij} = \begin{cases} 2\left(\mu_B + \dfrac{P_y}{\sqrt{2\pi}}\right)e_{ij} & \pi > \pi_c \\ 2\left(\mu_B + \dfrac{P_y}{\sqrt{2\pi_c}}\right)e_{ij} & \pi < \pi_c \end{cases} \tag{1}$$

$\pi = e_{ij}e_{ij}$ *and* e_{ij} is the (i,j)th component of deformation rate, n is the product of deformation rate with itself, π_c is a critical value of this product based on the non-Newtonian model, μ_B is the plastic dynamic viscosity of the non-Newtonian fluid and P_y is the yield stress of the fluid. The continuity, momentum and energy equations governing such type of flow are

$$\frac{\partial u}{\partial x} + \frac{\partial v}{\partial y} = 0 \tag{2}$$

$$u\frac{\partial u}{\partial x} + v\frac{\partial u}{\partial y} = \nu_{nf}\left(1 + \frac{1}{\gamma}\right)\frac{\partial^2 u}{\partial y^2} - \left(\frac{\sigma B_0^2}{\rho_{nf}} + \frac{\nu_{nf}}{k_0}\right)u \tag{3}$$

$$u\frac{\partial T}{\partial x} + v\frac{\partial T}{\partial y} = \alpha_{nf}\frac{\partial^2 T}{\partial y^2} \tag{4}$$

where u and v are the velocity components in the x and y directions, respectively. ν_{nf} is the kinematic viscosity, ρ_{nf} is the Casson fluid density, $\gamma = \mu_B\sqrt{\frac{2\pi_c}{P_y}}$ is the parameter of Casson fluid, σ is the electrical conductivity of the fluid, α_{nf} is the thermal diffusivity, T is the temperature and k_0 is the permeability of the porous medium.

The appropriate boundary conditions for the problem are given by

$$\begin{aligned} &u = u_w = bx, \quad v = 0 \\ &T = T_w = T_\infty + A\left(\tfrac{x}{l}\right)^2 \qquad \text{at} \quad y = 0 \\ &u \to 0, \quad T \to T_\infty \qquad \text{as} \quad y \to \infty \end{aligned} \tag{5}$$

where u_w = bx, b>0, is the stretching sheet velocity and A is a constant, l is the characteristic length, T is the temperature of the fluid, T_w is the temperature of the sheet and T_∞ is the free stream temperature.

Introducing the following similarity transformations

$$\begin{aligned} &\eta = y\sqrt{\frac{b}{\nu_f}}, \quad u = bxf'(\eta) \\ &v = -\sqrt{b\nu_f}\,f(\eta), \quad \theta = \frac{T - T_\infty}{T_w - T_\infty} \end{aligned} \tag{6}$$

Making use of Eq. (6), the governing equations (3) and (4) are reduced into the non-dimensional form as follows

$$\left(1 + \frac{1}{\gamma}\right)f''' + \phi_1\left(ff'' - f'^2 - \frac{M}{\phi_2}f'\right) - kf' = 0 \tag{7}$$

$$\theta'' + \left(\frac{k_f}{k_{nf}}\mathrm{Pr}\phi_3\right)f\theta' = 0 \tag{8}$$

The corresponding boundary conditions are

$$\begin{aligned} f = 0, \quad f' = 1, \quad \theta = 1 \qquad & \text{at} \quad \eta = 0 \\ f' \to 0, \quad \theta \to 0 \qquad & \text{as} \quad \eta \to \infty \end{aligned} \tag{9}$$

where prime denotes differentiation with respect to η, M = $\frac{\sigma B_0^2}{b\rho_f}$ is the magnetic parameter, $k = \frac{\nu_f}{bk_0}$ is the porosity parameter and Pr = $\frac{\nu_f}{\alpha_f}$ is the Prandtl number.

The important physical quantities of interest are the skin friction coefficient C_f and the local Nusselt number Nu_x defined as

$$C_f = \frac{\tau_w}{\rho_f u_w^2} \qquad \textit{and} \qquad Nu_x = \frac{x\, q_w}{k\,(T_w - T_\infty)} \tag{10}$$

where τ_w is the skin friction or the shear stress along the stretching surface and q_w is the surface heat flux given by

$$\tau_w = \left(\mu_B + \frac{P_y}{\sqrt{2\,\pi_c}}\right)\left(\frac{\partial u}{\partial y}\right)_{y=0} \qquad \textit{and} \qquad q_w = -k_{nf}\left(\frac{\partial T}{\partial y}\right)_{y=0} \tag{11}$$

Substituting the transformations in (6) in equations (10) and (11), we obtain

$$C_f\ \mathrm{Re}_x^{\frac{1}{2}} = \left(1 + \frac{1}{\gamma}\right) f''(0) \qquad \textit{and} \qquad Nu_x\ (\mathrm{Re}_x)^{\frac{-1}{2}} = -\theta'(0) \tag{12}$$

where $Re_x = \frac{u_w^2\ x}{\nu}$ is the local Reynolds number.

3. Numerical solution

A finite difference scheme known as Keller box method is used to solve numerically the system of non-linear ordinary differential equations (7) and (8) together with the boundary conditions in (9). The method was developed by Keller [51] and is described in Cebeci and Bradshaw [52]. The main steps involved in this method are:

1. Reduce the governing equations of the problem to a system of first-order ordinary differential equations.

2. Convert the resulting system of first-order ordinary differential equations into difference equations by using the central difference scheme.

3. Newton's method is used to linearize the non-linear finite difference equations so obtained and then write them in matrix-vector form.

4. Solve the linearized system of difference equations by using the block tri-diagonal elimination technique.

The method is highly adaptable to solve non-linear problems. In this method, the choice of the initial guess is very important to give the most accurate solution to the problem and it is made based on the convergence criteria along with the boundary conditions of the flow into consideration. In boundary layer flow calculations, the greatest error appears in the wall shear stress, as mentioned in Cebeci and Bradshaw [52]. So, in accordance with it, the values of the wall shear stress, in our case $f''(0)$ is commonly used as a convergence criteria. We used this convergence criterion in the present chapter. A convergence criteria of 10^{-4} is chosen which gives about a four decimal places accuracy for most of the prescribed quantities.

4. Results and discussion

In order to analyze the results, numerical computation has been carried out to calculate the velocity profiles, temperature profiles, skin friction coefficient and local Nusselt number for various values of the parameters that describe the flow characteristics, that is, magnetic parameter (M), Casson parameter (γ), porosity parameter (k) and nanoparticle volume fraction (ϕ). The numerical results are presented graphically in **Figures 1–9**. To know the accuracy of the applied numerical scheme, a comparison of the present results corresponding to the values of the skin friction coefficient $-f''(0)$ for various values of M and ϕ when Prandtl number Pr = 6.2 is made with the available results of Hamad [53] and presented in **Table 1**. It is observed that as M and ϕ increase, the skin friction coefficient also increases. This is due to the fact that an increase in M results in an increase in Lorentz force which opposes the motion of flow. A comparison of the numerical results of the local Nusselt number $-\theta'(0)$ is also done with Vajravelu [54] for various values of the Prandtl number Pr and presented in **Table 2**. It is clear from **Table 2** that the heat transfer rate coefficient increases with an increase in Prandtl number, which is the ratio of momentum diffusivity to thermal diffusivity. So, as Prandtl number increases, the momentum diffusivity increases whereas the thermal diffusivity decreases. Hence, the rate of heat transfer at the surface increases with increasing values of Pr. The results are found to be in excellent agreement. **Figure 1** shows the effects of Casson parameter γ on the velocity profile $f'(\eta)$, that is, $f'(\eta)$ is a decreasing function of γ. The momentum boundary layer thickness decreases with an increase in γ, because as the Casson parameter γ increases, the yield stress decreases and as a result the velocity of the fluid is suppressed and the reverse can be seen in **Figure 2**, which shows the effects of Casson parameter γ on the temperature profile $\theta(\eta)$. It can be seen that the temperature of the nanofluids is enhanced with the increasing values of γ and hence the thermal boundary layer thickness increases as the elasticity stress parameter is increased. **Figure 3** illustrates the effects of

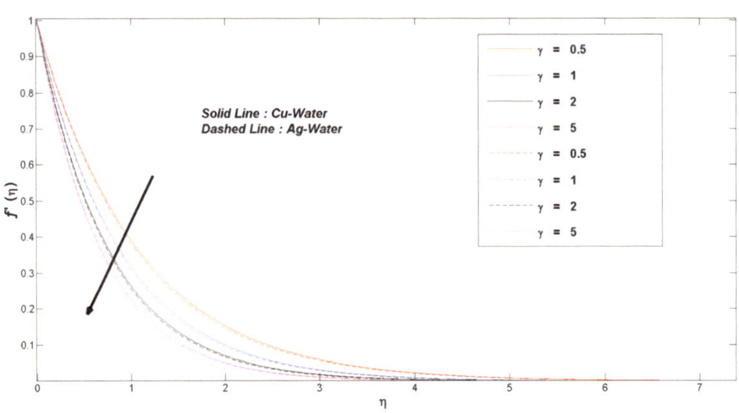

Figure 1.
Velocity profiles $f'(\eta)$ for various values of γ with $M = 1, k = 0.5, \phi = 0.1$ and $\mathrm{Pr} = 6.2$.

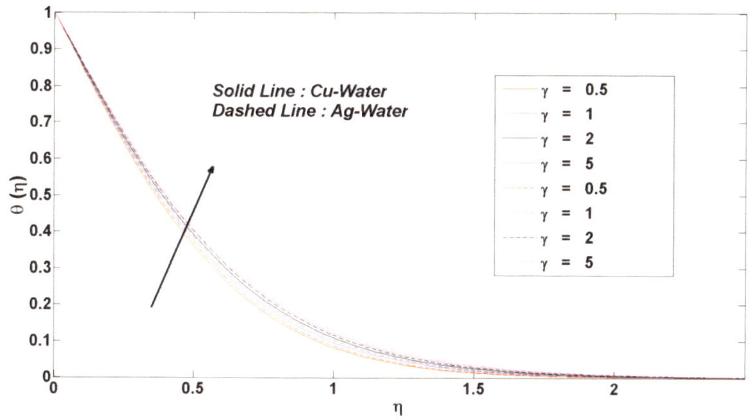

Figure 2.
Temperature profiles $\theta(\eta)$ for various values of γ with $M = 1,\ \ k = 0.5,\ \ \phi = 0.1\ \ and\ \ \Pr = 6.2.$

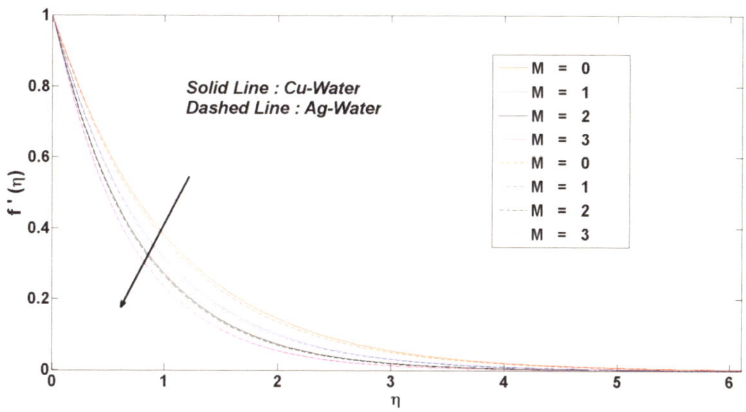

Figure 3.
Velocity profiles $f'(\eta)$ for various values of M with $\gamma = 1,\ \ k = 0.5,\ \ \phi = 0.1\ \ and\ \ \Pr = 6.2.$

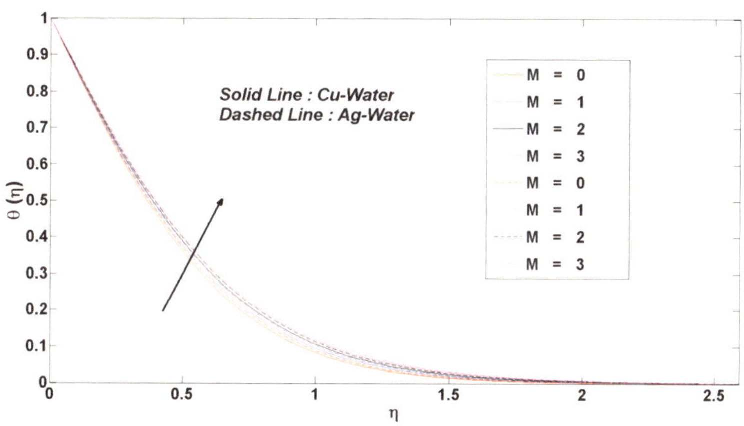

Figure 4.
Temperature profiles $\theta(\eta)$ for various values of M with $\gamma = 1,\ \ k = 0.5,\ \ \phi = 0.1\ \ and\ \ \Pr = 6.2.$

magnetic parameter M on the velocity distribution. From the figure, we can observe that as 'M' increases, the fluid velocity decreases. This is because Lorentz force is induced by the transverse magnetic field and it opposes the motion of the fluid.

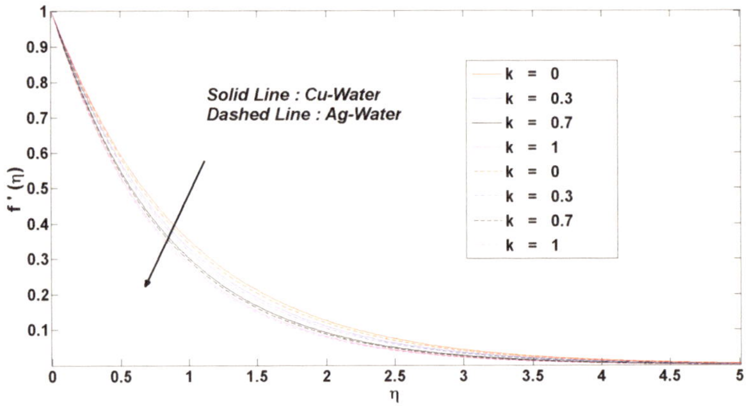

Figure 5.
Velocity profiles $f'(\eta)$ *for various values of k with* $M = 1, \gamma = 1, \ \phi = 0.1$ *and* $\mathrm{Pr} = 6.2$.

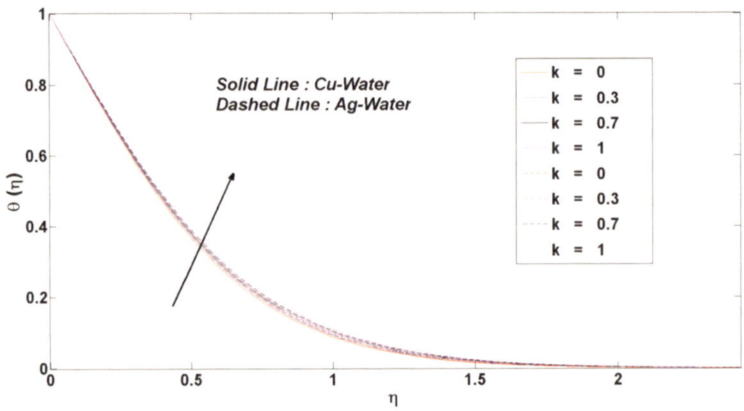

Figure 6.
Temperature profiles $\theta(\eta)$ *for various values of k with* $M = 1, \gamma = 1, \ \phi = 0.1$ *and* $\mathrm{Pr} = 6.2$.

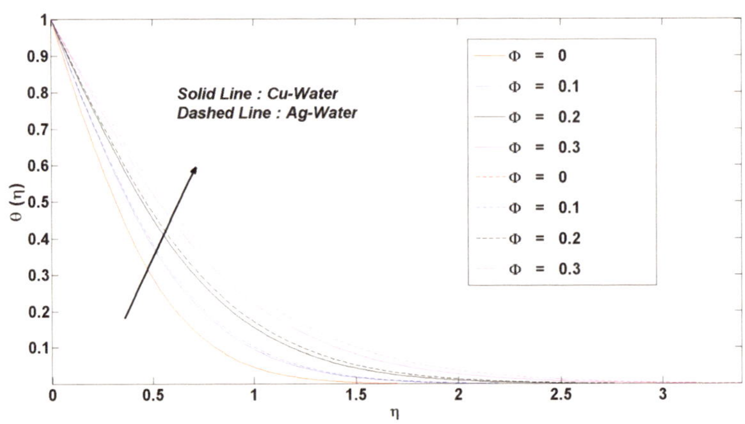

Figure 7.
Temperature profiles $\theta(\eta)$ *for various values of* ϕ *with* $\gamma = 1, \ \mathrm{Pr} = 6.2, \ M = 1$ *and* $k = 0.5$.

As the magnetic parameter M increases, the momentum boundary layer thickness decreases. The velocity distribution in the case of Cu-water nanofliud is higher as compared to Ag-water nanofluid and **Figure 4** shows the effect of magnetic parameter 'M' on the temperature profile $\theta(\eta)$. As 'M' increases, the thermal

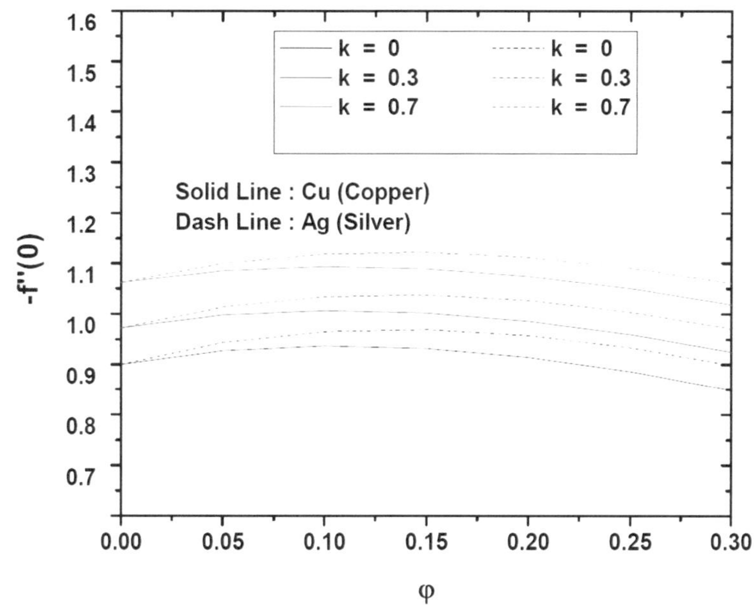

Figure 8.
Variation of skin friction coefficient $-f''(0)$ with nanoparticle volume fraction ϕ for various values of porosity parameter k.

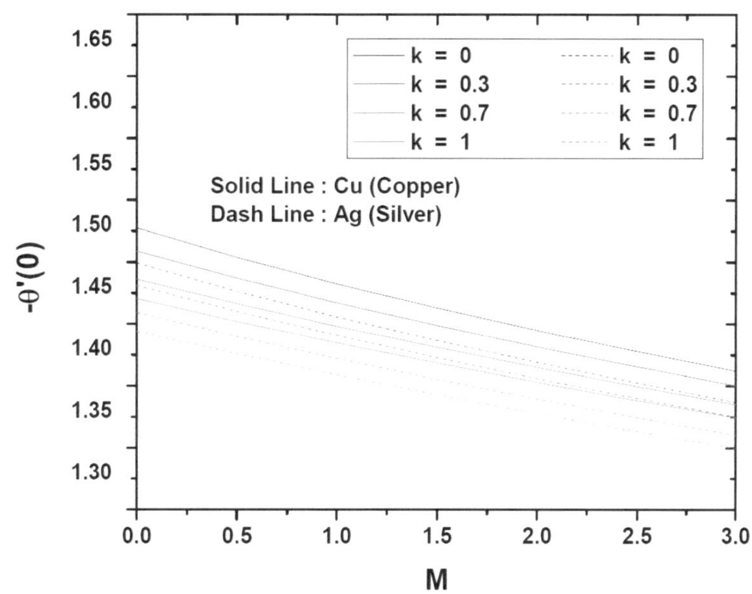

Figure 9.
Variation of heat transfer coefficient $-\theta'(0)$ with magnetic parameter M for various values of porosity parameter k.

boundary layer thickness also increases since the presence of magnetic field enhances the fluid's temperature within the boundary layer. The Ag-water nanofluid has a thicker thermal boundary layer than the Cu-water nanofluid because the thermal conductivity of Ag is more than that of Cu. **Figure 5** shows the effects of porosity parameter k on the velocity distribution. It is observed that with the increasing values of k, the velocity field decreases. The fluid velocity decreases because the presence of a porous medium increases the resistance to flow. In the

		Cu-Water		Ag-Water	
M	Φ	**Hamad [53]**	**Present**	**Hamad [53]**	**Present**
0	0.05	1.10892	1.1089	1.13966	1.1397
	0.1	1.17475	1.1747	1.22507	1.2251
	0.15	1.20886	1.2089	1.27215	1.2722
	0.2	1.21804	1.2180	1.28979	1.2898
0.5	0.05	1.29210	1.2921	1.31858	1.3186
	0.1	1.32825	1.3282	1.37296	1.3730
	0.15	1.33955	1.3396	1.39694	1.3969
	0.2	1.33036	1.3304	1.39634	1.3963
1	0.05	1.45236	1.4524	1.47597	1.4760
	0.1	1.46576	1.4658	1.50640	1.5064
	0.15	1.45858	1.4586	1.51145	1.5115
	0.2	1.43390	1.4339	1.49532	1.4953
2	0.05	1.72887	1.7289	1.74875	1.7487
	0.1	1.70789	1.7079	1.74289	1.7429
	0.15	1.67140	1.6714	1.71773	1.7177
	0.2	1.62126	1.6213	1.67583	1.6758

Table 1.
Comparison of results of the skin friction coefficient $-f''(0)$ *for various values of M and* ϕ.

Pr	**Vajravelu [54]**	**Present**
0.72	0.4590	0.4596
7	1.8953	1.8954

Table 2.
Comparison of values of local Nusselt number $-\theta'(0)$ *for various values of Pr.*

case of Ag-water nanofluid, the velocity is slightly less as compared with Cu-water nanofluid. **Figure 6** illustrates the impact of the porosity parameter k on the temperature profile and it is noted that with the increasing values of the porosity parameter, the temperature of the fluid increases. The temperature of Ag-water nanofluid is more as compared with Cu-water nanofluid. **Figure 7** shows the effect of nanoparticle volume fraction ϕ on the temperature of the nanofluids. From the figure, it is clear that the fluid temperature increases as the nanoparticle's volume fraction ϕ increases. The temperature distribution in Ag-water nanofluid is higher than that of Cu-water nanofluid. It is also observed that as the nanoparticle volume fraction increases, the thermal boundary layer thickness increases because as volume fraction increases, the thermal conductivity of the fluid increases. **Figure 8** shows the effect of the porosity parameter k and the nanoparticle volume fraction ϕ on the wall skin friction. It is observed that the skin friction increases with the increase in the porosity parameter k and the nanoparticle volume fraction ϕ for both the Cu-water and Ag-water nanofluids. The wall skin friction is higher in the case of Ag-water nanofluid than in Cu-water. Hence, the Ag-water nanofluid gives a higher drag in opposition to the flow than the Cu-water nanofluid. **Figure 9** shows the effect of the magnetic parameter M and porosity parameter k on the wall heat transfer rate $-\theta'(0)$. The influence of the magnetic field is to reduce the wall heat transfer rate. The porous media effect reduces the wall heat transfer rate. Moreover, the rate of heat transfer at the wall is less in the case of the Ag-water nanofluid as compared with Cu-water nanofluid.

5. Conclusion

MHD flow and heat transfer of Casson nanofluid through a porous medium over a stretching sheet have been investigated. The governing boundary layer equations are transformed into ordinary differential equations using similarity transformations and are then solved by the Keller box method. The effects of the various governing parameters viz. magnetic parameter M, Casson parameter γ, porosity parameter k and the nanoparticle volume fraction ϕ on the flow and heat transfer characteristics of two types of nanofluids, that is, Ag-water and Cu-water are determined [55, 56]. The present chapter leads to the following observations:

1. An increase in the Casson parameter γ suppresses the velocity field of the nanofluids whereas the temperature is enhanced.

2. With an increase in the magnetic parameter M, the momentum boundary layer thickness decreases while the thermal boundary layer thickness increases.

3. The temperature and the thermal boundary thickness increase as the nanoparticle volume fraction ϕ increases.

4. Ag-water nanofluid has thicker thermal boundary layer than Cu-water nanofluid.

5. The velocity of the nanofluids decreases as the porosity parameter k increases and the reverse is observed in the case of temperature.

6. The skin friction increases with an increase in nanoparticle volume fraction ϕ and the porosity parameter k.

7. The rate of heat transfer at the surface of the sheet decreases with an increase in magnetic parameter M and porosity parameter k.

Nomenclature

C_f	skin friction coefficient
C_p	specific heat capacity at constant pressure
f	dimensionless velocity
k_f	thermal conductivity of the base fluid
k_s	thermal conductivity of the nanoparticle
k_{nf}	thermal conductivity of the nanofluid
k^*	mean absorption coefficient
l	characteristic length
k_0	permeability of the porous medium
M	magnetic field parameter
Nu_x	local Nusselt number

Pr	Prandtl number
P_y	yield stress of the fluid
q_w	heat flux
Re_x	local Reynolds number
T	fluid temperature
T_w	temperature at the stretching surface
T_∞	temperature of the fluid far away from the stretching surface
u, v	velocity components along x-axis and y-axis, respectively
u_w	velocity of the stretching surface
x, y	Cartesian coordinates measured along stretching surface
π	product of the component of deformation rate with itself
π_c	critical value of the product
α_{nf}	thermal diffusivity of the nanofluid
γ	Casson fluid parameter
η	dimensionless similarity variable
σ	electrical conductivity
θ	dimensionless temperature
ϕ	nanoparticle volume fraction
τ_w	shear stress
μ_f	dynamic viscosity of the base fluid
μ_{nf}	dynamic viscosity of the nanofluid
ν_f	kinematic viscosity of the base fluid
ν_{nf}	kinematic viscosity of the nanofluid
ρ_f	density of the base fluid
ρ_s	density of the nanoparticle
ρ_{nf}	density of the nanofluid
$(\rho c_p)_f$	heat capacity of the base fluid
$(\rho c_p)_s$	heat capacity of the nanoparticle
$(\rho c_p)_{nf}$	heat capacity of the nanofluid
μ_B	plastic dynamic viscosity of the fluid
$'$	denotes differentiation with respect to η

Author details

Ayesha Siddiqui[1*] and Bandari Shankar[2]

1 Department of Mathematics, Nizam College, OU, Hyderabad, Telangana, India

2 Department of Mathematics, Osmania University, Hyderabad, Telangana, India

*Address all correspondence to: ayesha_siddiquias@yahoo.com

References

[1] Crane LJ. Flow past a stretching plate. Journal of Applied Mathematics and Physics. 1970;**21**:645-647

[2] Cortell R. Effects of viscous dissipation and work done by deformation on the MHD flow and heat transfer of a viscoelastic fluid over a stretching sheet. Physics Letters A. 2006;**357**:298-305

[3] Bhattacharyya K, Hayat T, Alsaedi A. Analytic solution for magnetohydrodynamic boundary layer flow of Casson fluid over a stretching/ shrinking sheet with wall mass transfer. Chinese Physics B. 2013;**22**

[4] Mukhopadhyay S. Casson fluid flow and heat transfer over a nonlinearly stretching surface. Chinese Physics B. 2013;**22**

[5] Rashidi MM, Mohimanian Pour SA. Analytic approximate solutions for unsteady boundary-layer flow and heat transfer due to a stretching sheet by homotopy analysis method. Nonlinear Analysis: Modelling and Control. 2010; **15**:83-95

[6] Ishak A. MHD boundary layer flow due to an exponentially stretching sheet with radiation effect. Sains Malaysiana. 2011;**40**:391-395

[7] Bachok N, Ishak A, Pop I. Boundary-layer flow of nanofluids over a moving surface in a flowing fluid. International Journal of Thermal Sciences. 2010;**49**: 1663-1668

[8] Mandal IC, Mukhopadhyay S. Heat transfer analysis for fluid flow over an exponentially stretching porous sheet with surface heat flux in porous medium. Ain Shams Engineering Journal. 2013;**4**:103-110

[9] Pavlov KB. Magneto hydrodynamic flow of an incompressible viscous fluid caused by the deformation of a plane surface. Magneto Hydrodynamics. 1974; **10**:146-148

[10] Andersson HI. MHD flow of a viscoelastic fluid past a stretching surface. Acta Mechanica. 1992;**95**: 227-230

[11] Bhatti MM, Abbas MA, Rashidi MM. A robust numerical method for solving stagnation point flow over a permeable shrinking sheet under the influence of MHD. Applied Mathematics and Computation. 2018;**316**:381-389

[12] Sheikholeslami M, Shehzad SA. CVFEM for influence of external magnetic source on $Fe_3O_4 - H_2O$ nanofluid behavior in a permeable cavity considering shape effect. International Journal of Heat and Mass Transfer. 2017;**115**:180-191

[13] Sheikholeslami M, Barzegar Gerdroodbary M, Moradi R, Shafee A, Li Z. Application of neural network for estimation of heat transfer treatment of $Al_2O_3 - H_2O$ nanofluid through a channel. Computer Methods in Applied Mechanics and Engineering. 2019;**344**: 1-12

[14] Sheikholeslami M, Sadoughi MK. Simulation of CuO-water nanofluid heat transfer enhancement in presence of melting surface. International Journal of Heat and Mass Transfer. 2018;**116**:1-12

[15] Sheikholeslami M, Jafaryar M, Saleem S, Li Z, Shafee A, Jiang Y. Nanofluid heat transfer augmentation and exergy loss inside a pipe equipped with innovative turbulators. International Journal of Heat and Mass Transfer. 2018;**126**:156-163

[16] Sheikholeslami M, Shehzad SA. Thermal radiation of ferrofluid in existence of Lorentz forces considering

variable viscosity. International Journal of Heat and Mass Transfer. 2017;**109**: 82-92

[17] Abel MS, Mahesha N. Heat transfer in MHD viscoelastic fluid flow over a stretching sheet with variable thermal conductivity, non-uniform heat source and radiation. Applied Mathematical Modeling. 2008;**32**:1965-1983

[18] Hsiao KL. Conjugate heat transfer of magnetic mixed convection with radiative and viscous dissipation effects for second-grade viscoelastic fluid past a stretching sheet. Applied Thermal Engineering. 2007;**27**:1895-1903

[19] Mukhopadhyay S. Heat transfer analysis of the unsteady flow of a maxwell fluid over a stretching surface in the presence of a heat source/sink. Chinese Physics Letters. 2012;**29**

[20] Mukhopadhyay S, Bhattacharyya K. Unsteady flow of a Maxwell fluid over a stretching surface in presence of chemical reaction. Journal Egyptian Math. 2012;**20**:229-234

[21] Liao SJ. On the analytic solution of magnetohydrodynamic flows of non-Newtonian fluids over a stretching sheet. Journal of Fluid Mechanics. 2003; **488**:189-212

[22] Hayat T, Javed T, Sajid M. Analytic solution for MHD rotating flow of a second grade fluid over a shrinking surface. Physics Letters A. 2008;**372**: 3264-3273

[23] Ishak A, Lok Y, Pop I. Non-Newtonian power-law fluid flow past a shrinking sheet with suction. Chemical Engineering Communications. 2012; **199**:142-150

[24] Mukhopadhyay S. Upper-convected Maxwell fluid flow over an unsteady stretching surface embedded in porous medium subjected to suction/blowing. Zeitschrift fr Naturforschung A A, Journal of Physical Sciences. 2012;**67**: 641-646

[25] Mukhopadhyay S, Vajravelu K. Effects of transpiration and internal heat generation/absorption on the unsteady flow of a Maxwell fluid at a stretching surface. The American Society of Mechanical Engineers Journal of Applied Mechanics. 2012;**79**

[26] Andersson HI, Bech KH, Dandapat BS. Magnetohydrodynamic flow of a power-law fluid over a stretching sheet. International Journal of Non-Linear Mechanics. 1992;7:929-936

[27] Hassanien I. A, mixed convection in micropolar boundary-layer flow over a horizontal semi-infinite plate. The American Society of Mechanical Engineers Journal of Fluids Engineering. 1996;**118**:833-838

[28] Haroun MH. Effect of Deborah number and phase difference on peristaltic transport of a third-order fluid in an asymmetric channel. Communications in Nonlinear Science and Numerical Simulation. 2007;**12**: 1464-1480

[29] Siddiqui AM, Zeb A, Ghori QK, Benharbit AM. Homotopy perturbation method for heat transfer flow of a third grade fluid between parallel plates. Chaos, Solitons & Fractals. 2008;**36**:182-192

[30] Sajid M, Ahmad I, Hayat T, Ayub M. Unsteady flow and heat transfer of a second grade fluid over a stretching sheet. Communications in Nonlinear Science and Numerical Simulation. 2009;**14**:96-108

[31] Dash RK, Mehta KN, Jayaraman G. Casson fluid flow in a pipe filled with a homogeneous porous medium. International Journal of Engineering Science. 1996;**34**:1145-1156

[32] Fung YC. Biodynamics Circulation. Springer Verlag; 1984

[33] McDonald DA. Blood flows in arteries. Edward Arnold; 1974

[34] Casson MS. Fluid flow and heat transfer over a nonlinearly stretching surface. Chinese Physics B. 2013;**22**: 577-585

[35] Hayat T, Shehzad SA, Alsaedi A, Alhothuali MS. Mixed convection stagnation point flow of Casson fluid with convective boundary conditions. Chinese Physics Letters. 2012;**29**

[36] Nawaz M, Hayat T, Alsaed A. Dufour and Soret effects on MHD flow of viscous fluid between radially stretching sheets in porous medium. Applied Mathematics and Mechanics. 2012;**33**:1403-1418

[37] Chamkha AJ, Aly AM. Heat and mass transfer in stagnation point flow of a polar fluid towards a stretching surface in porous media in the presence of Soret, Dufour and chemical reaction effects. Chemical Engineering Communications. 2010;**198**:214-234

[38] Casson N, Flow A. Equation for pigment-oil suspensions of the printing ink type. In: Mill CC, editor. Rheology of Disperse Systems. Oxford: Pergamon Press; 1959. pp. 84-104

[39] Mustafa M, Hayat T, Pop I, Hendi A. A Stagnation-point flow and heat transfer of a Casson fluid towards a stretching sheet. Zeitschrift fr Naturforschung A A Journal of Physical Sciences. 2012;**67**:70-76

[40] Shehzad SA, Hayat T, Qasim M, Asghar S. Efects of mass transfer on MHD flow of Casson fluid with chemical reaction and suction. Brazilian Journal of Chemical Engineering. 2013; **30**:187-195

[41] Khani F, Farmany A, Ahmadzadeh Raji M, Aziz A, Samadi F. Analytic solution for heat transfer of a third grade viscoelastic fluid in non-Darcy porous media with thermophysical effects. Brazilian Journal of Chemical Engineering Communications in Nonlinear Science and Numerical Simulation. 2009;**14**:3867-3878

[42] Sheikholeslami M. New computational approach for exergy and entropy analysis of nanofluid under the impact of Lorentz force through a porous media. Computer Methods in Applied Mechanics and Engineering. 2019;**344**:319-333

[43] Sheikholeslami M, Shehzad SA. Simulation of water based nanofluid convective flow inside a porous enclosure via non-equilibrium model. International Journal of Heat and Mass Transfer. 2018;**120**:1200-1212

[44] Sheikholeslami M. CuO-water nanofluid flow due to magnetic field inside a porous media considering Brownian motion. Journal of Molecular Liquids. 2018;**249**:921-929

[45] Sheikholeslami M, Shehzad SA, Li Z, Shafee A. Numerical modeling for alumina nanofluid magnetohydrodynamic convective heat transfer in a permeable medium using Darcy law. International Journal of Heat and Mass Transfer. 2018;**127**:614-622

[46] Sheikholeslami M. CuO-water nanofluid free convection in a porous cavity considering Darcy law. The European Physical Journal Plus. 2017; **132**:55

[47] Sheikholeslami M, Shamlooei M, Moradi R. Numerical simulation for heat transfer intensification of nanofluid in a porous curved enclosure considering shape effect of Fe_3O_4 nanoparticles. Chemical Engineering and Processing Process Intensification. 2018;**124**:71-82

[48] Sheikholeslami M, Rokni HB. Numerical simulation for impact of Coulomb force on nanofluid heat transfer in a porous enclosure in

presence of thermal radiation. International Journal of Heat and Mass Transfer. 2018;**118**:921-929

[49] Sibanda P, Khidir AA. Nanofluid flow over a nonlinear stretching sheet in porous media with MHD and viscous dissipation effects. Journal of Porous Media. 2014;**17**(5):391-403

[50] Sheikholeslami M, Shehzad SA. Magnetohydrodynamic nanofluid convection in a porous enclosure considering heat flux boundary condition. International Journal of Heat and Mass Transfer. 2017;**106**:1261-1269

[51] Keller HB. A New Difference Scheme for Parabolic Problems. Vol. 2. Academic Press; 1971. pp. 327-350

[52] Cebeci T, Bradshaw P. Physical and Computational Aspects of Convective Heat Transfer. Springer; 1988

[53] Hamad MAA. Analytical solution of natural convection flow of a nanofluid over a linearly stretching sheet in the presence of magnetic field. International Communications in Heat and Mass Transfer. 2011;**38**:487-492

[54] Vajravelu K. Viscous flow over a nonlinearly stretching sheet. Applied Mathematics and Computation. 2001; **124**:281-288

[55] Khan WA, Pop I. Free convection boundary layer flow past a horizontal flat plate embedded in a porous medium filled with a nanofluid. Journal of Heat Transfer. 2011;**133**

[56] Rashad AM, Reddy Gorla RS, Mansour MA, Ahmed SE. Free convection boundary layer flow past a horizontal flat plate embedded in a porous medium filled with a nanofluid. Journal of Heat Transfer. 2011;**133**:9

Chapter 9

Heat Transfer and Thermal Radiation at a General Three-Dimensional in a Nanofluid through a Porous Medium

Gamal M. Abdel-Rahman and Faiza M.N. El-fayez

Abstract

In this chapter, the magnetohydrodynamic effects on heat transfer and thermal radiation at a stagnation point flowing in a nanofluid containing different types of nanoparticles namely, copper (Cu), alumina (Al_2O_3) and titania (TiO_2) through a porous medium have been investigated numerically. By using appropriate transformation for velocity and temperature into a set of non-linear coupled ordinary differential equations which are solved numerically. Numerical results are presented for velocity and temperature profiles for different parameters of the problem. Also, the effects of the pertinent parameters on the skin friction and the heat fluxes are obtained and discussed numerically and illustrated graphically.

Keywords: nanofluid, MHD, porous medium, thermal radiation, heat transfer

1. Introduction

Nanofluids have novel properties that make them potentially useful in many applications. They exhibit enhanced thermal conductivity and the convective heat transfer coefficient compared to the base fluid [1], so the nanofluids transfer heat at a higher rate than ordinary fluids (for example, water) which allows for more efficient heating or cooling while reducing energy consumption. Since nanofluid consists of very small sized solid particles, therefore in low solid concentration it is reasonable to consider nanofluid as a single-phase flow [2].

The term "nanofluid" refers to a liquid containing a suspension of submicronic solid particles (nanoparticles). This interest is generated by a variety of applications, ranging from laser-assisted drug delivery to electronic chip cooling. The term was coined by Choi [3]. The characteristic feature of nanofluids is thermal conductivity enhancement, a phenomenon observed by Masuda et al. [4]. This phenomenon suggests the possibility of using nanofluids in advanced nuclear system [5].

In recent years, numerous investigations have been conducted on the magnetohydrodynamic (MHD) flows and heat transfer because of its important applications in metallurgical industry, such as the cooling of continuous strips and filaments, drawn through a quiescent fluid and the purification of molten metals from non-metallic inclusions. It is known that the properties of the final product depend considerably on the rate of cooling during the manufacturing processes. The rate of cooling can be

IntechOpen

controlled by drawing the strips in an electrically conducting fluid subject to a magnetic field, so that a final product of desired characteristics can be achieved [6]. One can solve this problem by using solid particles as an additive suspended into the base fluid as Choi [3] did, who first utilized and used fluids suspended by nanometer-sized solid particles. The resulting mixture referred to as a nanofluid possesses a substantially larger thermal conductivity compared to that of the traditional fluids [7]. These suspended nanoparticles can change the transport and thermal properties of the base fluid [8]. Therefore, by mixing the nanoparticles in the fluid, thermal conductivity of the fluid increases and the heat transfer capability improves.

The flow and heat transfer characteristics at a stagnation point, for both two-dimensional and axisymmetric, have been studied extensively in the literature [9]. These studies have been motivated by the fundamental nature of the boundary layer flows near such points by the exact applicability of similar solutions and by their relevance to the leading edge and nose regions of bodies in high speed flows. These two cases of two-dimensional and axisymmetric flows can be recognized as special cases of more general stagnation point flows. Both two-dimensional and axisymmetric flows were extended to three-dimensional by Howarth [10]. Bhattacharyya and Gupta [11] studied the flow and heat transfer in an incompressible viscous and electrically conducting fluid near a three-dimensional stagnation point of a body permeated by a uniform magnetic field. Analysis of such flows is very important in both theory and in practice. From a theoretical point of view, flows of this type are fundamental in fluid mechanics and forced convective heat transfer. On the other hand, from a practical point of view, these flows have applications in many manufacturing processes in industry such as the boundary layer along material handling conveyers, the aerodynamic extrusion of plastic sheet and petrochemical industries.

It is worth mentioning that the nanofluid model proposed by Buongiorno [12] was very recently used by Nield and Kuznetsov [13], Kuznetsov and Nield [14] and Bachok et al. [15] in their papers. Different from the above model.

The aim of the present chapter is to study the Magnetohydrodynamic effects on heat transfer and thermal radiation in an incompressible viscous fluid near the three-dimensional stagnation point of a body that is placed in a water based nanofluid through a porous medium containing different types of nanoparticles: copper (Cu), alumina (Al_2O_3) and titania (TiO_2). Numerical results are presented for velocity and temperature profiles for different parameters of the problem.

2. Mathematical analysis

Consider a flow of an electrically conducting fluid with heat transfer flow at a stagnation point of an incompressible viscous fluid past a body that is placed in a nanofluid of ambient uniform temperature T_∞ through a porous medium in the presence of radiation has been considered, where the body surface is kept at a constant temperature T_w. The stagnation point is located at the origin o of the Cartesian coordinate system $oxyz$. It was shown by Howarth [10] that the inviscid irrational flow near o has the velocity components

$$u_e(x) = ax, \quad v_e(y) = by \tag{1}$$

with the constants $a > 0$ and $b \geq 0$ or $b \leq 0$.

The MHD body forces $\bar{J} \times \bar{B}$ the Maxwell's equations:

$$\operatorname{div} \bar{B} = 0, \quad Curl\, \bar{B} = \mu_m \bar{J} \;\; and \quad \operatorname{div} \bar{E} = 0$$

where $\bar{J}$ is the electric current density, $\bar{B} = B + d$ is the total magnetic field, μ_f is the magnetic permeability and d is the induced magnetic field. The magnetic Reynolds number of the flow is taken to be small, so that the flow induction distortion of the applied magnetic field can be neglected as in the case with most of conducting fluids. The magnetic body force $\bar{J} \times \bar{B}$ takes the form $\sigma(\bar{V} \times \bar{B}) \times \bar{B}$, therefore, $\sigma(\bar{V} \times \bar{B}) \times \bar{B} = -\sigma B^2 \bar{V}$, where $\bar{V}$ is velocity vector $\bar{V} = (u, v, w)$ and $\bar{B} = (0, 0, B)$. The Lorentz force (MHD body force) has two components:

$$F_x = -\sigma B^2 u, F_y = -\sigma B^2 v.$$

We now make the standard boundary-layer approximation, based on a scale analysis, and write the governing equations (see [6, 11]):

$$\frac{\partial u}{\partial x} + \frac{\partial v}{\partial y} + \frac{\partial w}{\partial z} = 0, \tag{2}$$

$$u\frac{\partial u}{\partial x} + v\frac{\partial u}{\partial y} + w\frac{\partial u}{\partial z} = a^2 x + \frac{\mu_{nf}}{\rho_{nf}}\frac{\partial^2 u}{\partial z^2} - \frac{\sigma B^2}{\rho_{nf}}u - \frac{\mu_{nf}}{\rho_{nf}\kappa}u, \tag{3}$$

$$u\frac{\partial v}{\partial x} + v\frac{\partial v}{\partial y} + w\frac{\partial v}{\partial z} = b^2 x + \frac{\mu_{nf}}{\rho_{nf}}\frac{\partial^2 v}{\partial z^2} - \frac{\sigma B^2}{\rho_{nf}}v - \frac{\mu_{nf}}{\rho_{nf}\kappa}v, \tag{4}$$

$$u\frac{\partial T}{\partial x} + v\frac{\partial T}{\partial y} + w\frac{\partial T}{\partial z} = \alpha_{nf}\frac{\partial^2 T}{\partial z^2} - \frac{1}{(\rho c)_{nf}}\frac{\partial q_r}{\partial z}, \tag{5}$$

Subject to the boundary conditions

$$\begin{array}{l} u = v = w = 0, \quad T = T_w \quad at \quad z = 0 \\ u \to u_e(x), \quad v \to v_e(y), \quad T \to T_\infty \quad as \quad z \to \infty \end{array}. \tag{6}$$

Here μ_{nf} the viscosity of the nanofluid, α_{nf} the thermal diffusivity of the nanofluid and ρ_{nf} the density of the nanofluid, which are given by

$$\begin{array}{l} \mu_{nf} = \dfrac{\mu_f}{(1-\varphi)^{5/2}}, \quad \alpha_{nf} = \dfrac{k_{nf}}{(\rho C_p)_{nf}}, \quad \rho_{nf} = (1-\varphi)\rho_f + \varphi\rho_s, \\ (\rho C_p)_{nf} = (1-\varphi)(\rho C_p)_f + \varphi(\rho C_p)_s, \\ \dfrac{k_{nf}}{k_f} = \dfrac{(k_s + 2k_f) - 2\varphi(k_f - k_s)}{(k_s + 2k_f) + \varphi(k_f - k_s)} \end{array} \tag{7}$$

where q_r is the energy flux which is relative to a frame moving with the nanofluid velocity. Using the Rosseland approximation (Rashed [16]), the radiative heat flux q_r could be expressed by

$$q_r = -k_1\frac{\partial T^4}{\partial z} \tag{8}$$

where $k_1 = 4\sigma^*/3k^*$ is the nanofluid thermal conductivity.

Assuming that the temperature difference within the flow is sufficiently small such that T^4 could be approached as the linear function of temperature;

$$T^4 \cong 4T_\infty^3 T - 3T_\infty^4 \tag{9}$$

Further, we look for a solution of Eqs. (2)–(5) of the form

$$u = axf^{/}(\eta), \quad v = byg^{/}(\eta), \quad w = -\sqrt{a\upsilon_f}(f + cg),$$
$$\theta(\eta) = \frac{(T - T_\infty)}{(T_w - T_\infty)}, \quad \eta = \sqrt{a/\upsilon_f z} \tag{10}$$

where $c = b/a$ is the ratio of the gradient of velocities in the y- and x-directions, and primes denote differentiation with respect to z. Substituting Eqs. (8)–(10) in Eqs. (2)–(5), we obtain the following nonlinear ordinary differential equations:

$$\left(\frac{1}{(1-\varphi)^{5/2}\left(1-\varphi\left(1-\rho_s/\rho_f\right)\right)}\right)\left[f^{///} - M(1-\varphi)^{5/2}f^{/} - Sf^{/}\right] + (f + cg)f^{//} - f^{/2} + 1 = 0, \tag{11}$$

$$\left(\frac{1}{(1-\varphi)^{5/2}\left(1-\varphi\left(1-\rho_s/\rho_f\right)\right)}\right)\left[g^{///} - M(1-\varphi)^{5/2}g^{/} - Sg^{/}\right] + (f + cg)g^{//} + c\left(1 - g^{/2}\right) = 0, \tag{12}$$

$$\left(\frac{((k_{nf}/k_f) + Q)}{P_r\left(1-\varphi\left(1-\left((\rho C_p)_s/(\rho C_p)_f\right)\right)\right)}\right)\theta^{//} + (f + cg)\theta^{/} = 0, \tag{13}$$

With the appropriate boundary conditions:

$$f^{/}(0) = 0, g^{/}(0) = 0, f(0) = -c\,g(0), \theta(0) = 1,$$
$$f^{/}(\infty) = 1, g^{/}(\infty) = 1, \theta(\infty) = 0. \tag{14}$$

where the prime denotes a partial differentiation with respect to η, $M = \sigma B^2/a\rho_f$ is the Magnetic field parameter, $S = \upsilon_f/a\kappa$ is the Porous medium parameter and $Q = 4k_1 T_\infty^3$ is the Thermal radiation parameter.

There is no loss of generality in the requirement that $|a| \geq |b|$ with $a > 0$. Clearly $b = 0$ corresponds to the plane stagnation point flow case, while $b = a$ is the axisymmetric case. The case $0 < c < 1$ displays the nodal stagnation point of attachment ($b > 0$), while $-1 < c < 0$ displays the saddle points of attachment ($b < 0$) [5].

3. Skin-friction coefficient and Nusselt number

The parameters of engineering interest for the present problem are the local skin-friction coefficients C_{fx} and C_{fy}, along the x- and y-directions, respectively, and the local Nusselt number Nu_x, which are defined as

$$C_{fx} = \frac{\tau_{wx}}{ax\rho_f u_w^2}, \quad C_{fy} = \frac{\tau_{wy}}{by\rho_f u_w^2}, \quad Nu_x = \frac{u_w^2 q_w}{ak_f(T_w - T_\infty)} \tag{15}$$

where τ_{wx} and τ_{wy} are the surface shear stresses along the x- and y-directions, respectively, and q_w is the surface heat flux, which are given by

$$\tau_{wx} = \mu_{nf}\left(\frac{\partial u}{\partial z}\right)_{z=0}, \quad \tau_{wy} = \mu_{nf}\left(\frac{\partial v}{\partial z}\right)_{z=0}, \quad q_w = -k_{nf}\left(\frac{\partial T}{\partial z}\right)_{z=0} \tag{16}$$

Using Eqs. (10), (15), and (16), we obtain

$$\begin{aligned} Re_x^{1/2} C_{fx} &= \frac{1}{(1-\varphi)^{5/2}} f^{//}(0), \quad Re_x^{1/2} C_{fy} = \frac{1}{(1-\varphi)^{5/2}} g^{//}(0) \\ Re_x^{-1/2} Nu_x &= -\frac{k_{nf}}{k_f} \theta^{/}(0) \end{aligned} \tag{17}$$

.

where $Re_x = u_w^4 / a\nu_f$ is the local Reynolds number.

4. Results and discussion

Numerical solutions to the nonlinear ordinary differential Eqs. (11)–(14) were obtained using the modified fourth order Runge-Kutta method along with Nachtsheim-Swigert shooting technique [17]. In order to gain physical insight, the velocity and the temperature profiles have been discussed by assigning numerical values to the parameter, encountered in the problem which the numerical results are tabulated and displayed with the graphical illustrations **Figure 1**. We find the missing slopes $f^{//}(0), g^{//}(0)$ and $-\theta^{/}(0)$ for some values of the governing parameters, namely the nanoparticle volume fraction φ and the ratio of the gradient of velocities in the i- and x-directions c, where $0 < c < 1$ displays the nodal stagnation point of attachment and $-1 < c < 0$ displays the saddle points of attachment. Three types of nanoparticles were considered, namely, copper (Cu), alumina (Al_2O_3) and titania (TiO_2). Following Oztop and Abu-Nada [18], the value of the Prandtl number P_r is taken as 6.2 (for water) and the volume fraction of nanoparticles is from 0 to 0.2 ($0 \leq \varphi \leq 0.2$) in which φ corresponds to the regular Newtonian fluid. The numerical results are summarized in **Figures 2–6**. It is worth mentioning that we have used data related to the thermophysical properties of the fluid and nanoparticles as listed in **Table 1** [18] to compute each case of the nanofluid.

Figures 2, 3 and **5a–c**, display the velocity $f^{/}(\eta), g^{/}(\eta)$ and the temperature $\theta(\eta)$ profiles under the different parameters of the problem as magnetic field parameter, porous medium parameter and nanoparticle parameter. It is observed that the velocity profile decreases, while the temperature profile increases with the increase in each of magnetic field parameter, porous medium parameter and nanoparticle parameter.

The effects of the nanoparticle volume fractions on the velocity and the temperature profiles are shown in **Figure 4a–c**, respectively. It is observed that the

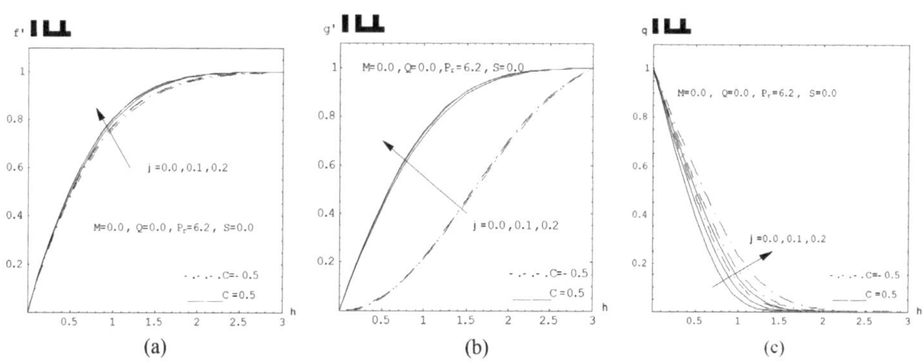

Figure 1.
Effect of nanoparticle volume fractions on (a) the velocity profile $f^{/}(\eta)$, (b) the velocity profile $g^{/}(\eta)$ and (c) the temperature profile $\theta(\eta)$ at c = 0.5 and −0.5 for copper-water nanofluid.

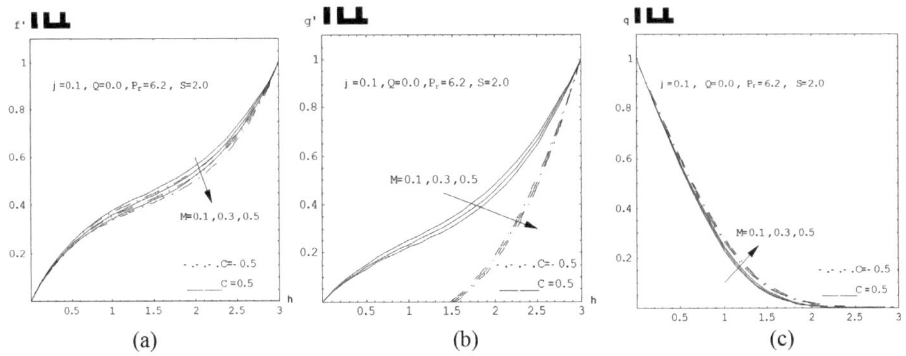

Figure 2.
Effect of magnetic field parameter on (a) the velocity profile $f^{/}(\eta)$, *(b) the velocity profile* $g^{/}(\eta)$ *and (c) the temperature profile* $\theta(\eta)$.

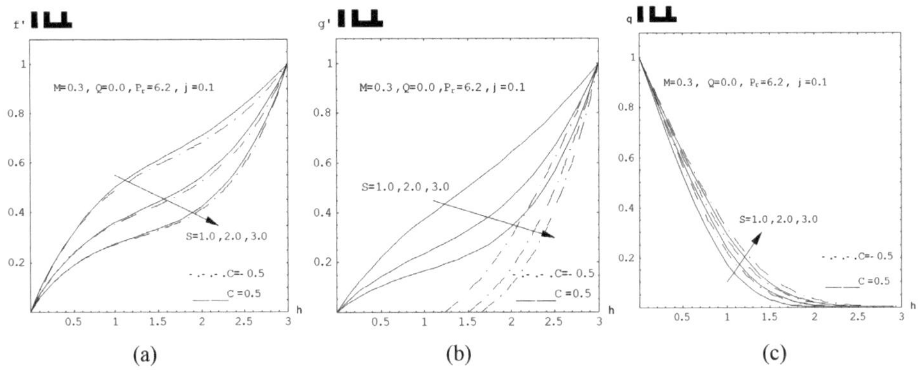

Figure 3.
Effect of porous medium parameter on (a) the velocity profile $f^{/}(\eta)$, *(b) the velocity profile* $g^{/}(\eta)$ *and (c) the temperature profile* $\theta(\eta)$.

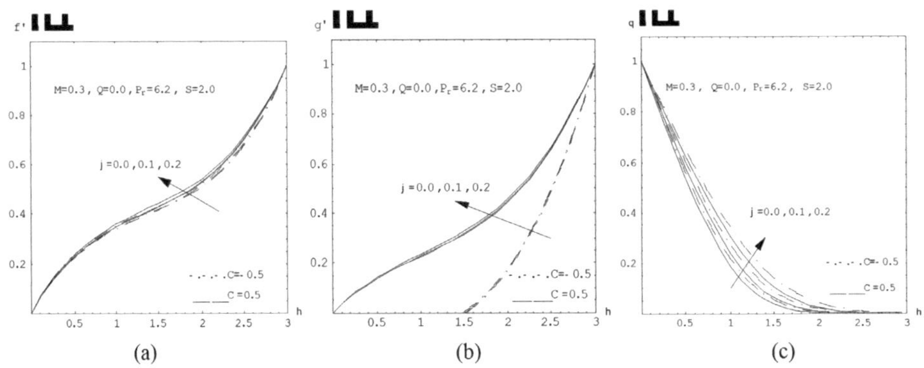

Figure 4.
Effect of nanoparticle volume fractions on (a) the velocity profile $f^{/}(\eta)$, *(b) the velocity profile* $g^{/}(\eta)$ *and (c) the temperature profile* $\theta(\eta)$.

velocity and the temperature profiles increase with the increase of the nanoparticle volume fractions. The effects of thermal radiation parameter on the temperature profile are shown in **Figure 6**; also, we found that the temperature profiles increase with the increase of thermal radiation parameters. i.e. as expected, since the effect of thermal radiation is to decrease the rate of energy transport to the fluid.

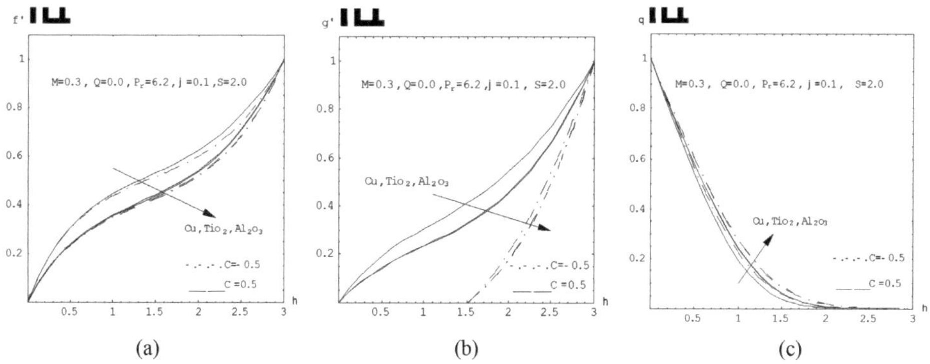

Figure 5.
Effect of nanoparticle on (a) the velocity profile $f'(\eta)$, (b) the velocity profile $g'(\eta)$ and (c) the temperature profile $\theta(\eta)$.

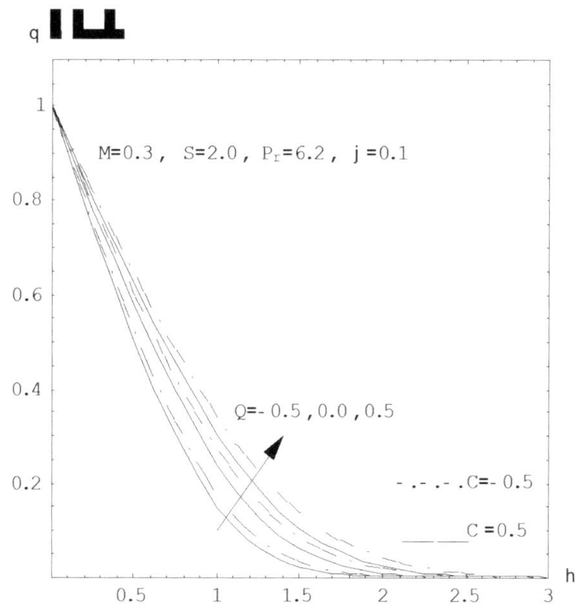

Figure 6.
Effect of the thermal radiation parameter on the temperature profile $\theta(\eta)$.

Physical properties	Fluid phase (water)	Cu	Al_2o_3	Tio_2
$C_p\,(J/kg\,K)$	4179	385	765	686.2
$\rho(kg/m^3)$	997.1	8933	3970	4250
$k\,(W/mK)$	0.613	400	40	8.9538
$\alpha\times10^7\,(m^2/s)$	1.47	1163.1	131.7	30.7

Table 1.
Thermophysical properties of fluid and nanoparticles [18].

M	S	φ	Q	$\text{Re}_x^{1/2} C_{fx}$		$\text{Re}_x^{1/2} C_{fy}$		$\text{Re}_x^{-1/2} Nu_x$	
				$c = -0.5$	$c = 0.5$	$c = -0.5$	$c = 0.5$	$c = -0.5$	$c = 0.5$
0.1	2.0	0.1	0.0	0.905057	0.912069	-0.359099	0.524052	1.08178	1.14672
0.3				0.87735	0.883391	-0.356488	0.501349	1.06709	1.12757
0.5				0.85163	0.856843	-0.353334	0.480884	1.05291	1.10938
0.3	1.0	0.1	0.0	1.10357	1.11976	-0.352305	0.706599	1.17147	1.27281
	2.0			0.87735	0.88339	-0.356488	0.501349	1.06709	1.12757
	3.0			0.73820	0.74059	-0.358357	0.397432	0.98337	1.02393
0.3	2.0	0.0	0.0	0.654066	0.658315	-0.267656	0.371336	1.18524	1.2446
		0.1		0.87735	0.88339	-0.356488	0.501349	1.06709	1.26757
		0.2		1.11575	1.12365	-0.444151	0.638333	1.01473	1.29428
0.3	2.0	0.1	-0.5					1.28272	1.34486
			0.0					1.06709	1.12757
			0.5					0.94130	1.00087

Table 2.
Numerical of the values skin-friction coefficient $(\text{Re}_x^{1/2}C_{fx})$ *and* $(\text{Re}_x^{1/2}C_{fy})$ *along x- and y-directions and the local Nusselt number* $(\text{Re}_x^{-1/2}Nu_x)$ *with* M, S, φ *and* Q, *when* $P_r = 6.2$.

From **Table 2**, the numerical values of the skin-friction and the local Nusselt number are given in **Table 2**. For an increase in the magnetic field parameter M, we observe that the skin-friction coefficient along x- and y-directions and the local Nusselt number decrease when $c = 0.5$, but at $c = -0.5$, the skin-friction coefficient along x-direction and the local Nusselt number decrease, while the skin-friction coefficient along y-direction increases. While, with an increase in the porous medium parameter S, we observe that the skin-friction coefficient and the local Nusselt number decrease when $c = -0.5$ and $c = -0.5$.

With an increase in nanoparticle volume fractions φ, we observe that the skin-friction coefficient increases and the local Nusselt number increases when $c = 0.5$, but at $c = -0.5$, the skin-friction coefficient along y-direction and the local Nusselt number decrease, while the skin-friction coefficient along x-direction increases. While, with an increase in the thermal radiation parameter Q, we observe that the local Nusselt number decreases when $c = -0.5$ and $c = -0.5$.

5. Conclusions

In the present chapter, the magnetohydrodynamic effects on heat transfer and thermal radiation at a stagnation point flowing in a nanofluid containing different types of nanoparticles namely, copper (Cu), alumina (Al_2O_3) and titania (TiO_2) through a porous medium have been investigated numerically. By using appropriate transformation for velocity and temperature into a set of non-linear coupled ordinary differential equations which are solved numerically, the governing equations were:

1. With the increasing values of the magnetic field parameter, porous medium parameter and nanoparticle parameter, it is observed that the velocity profile decreases, while the temperature profile increases.

2. The velocity and the temperature profiles increase with the increase of the nanoparticle volume fractions.

3. For the effects of thermal radiation parameter on the temperature profile, we found that the temperature profiles increase with the increase of thermal radiation parameters.

4. For an increase in the magnetic field parameter M, we observe that the skin-friction coefficient along x- and y-directions and the local Nusselt number decrease when $c = 0.5$, but at $c = -0.5$, the skin-friction coefficient along x-direction and the local Nusselt number decrease, while the skin-friction coefficient along y-direction increases.

5. About the effects of the porous medium parameter S, we observe that the skin-friction coefficient and the local Nusselt number decrease when $c = -0.5$ and $c = -0.5$.

6. The local Nusselt number decreases when $c = -0.5$ and $c = -0.5$, with the increase of the thermal radiation parameter Q.

Nomenclature

a	constant
B	magnetic field
b	constant
c	the ratio of the gradient of velocities in the y- and x-directions
C_{fx}, C_{fy}	skin friction in the y- and x-directions
d	the induced magnetic field
k_{nf}	the thermal conductivity of the nanofluid
k_f	the thermal conductivities of the fluid
k_s	the thermal conductivities of the solid fractions
k_1	the nanofluid thermal conductivity
k^*	mean absorption coefficient
Nu_x	Nusselt number
M	magnetic field parameter
Q	thermal radiation parameter
q_r	the energy flux which is relative to a frame moving with the nanofluid velocity
q_w	the surface heat flux
T	temperature distribution
T_w	constant temperature
T_∞	ambient uniform temperature
S	porous medium parameter
Re_x	the local Reynolds number
u	velocity in the x-direction
v	velocity in the y-direction
w	velocity in the z-direction
x	horizontal distance
y	vertical distance
z	normal distance
Greek symbols	
α_{nf}	the thermal diffusivity of the nanofluid
η	similarity variable

θ	dimensionless temperature distribution
μ_f	the viscosity of the fluid
μ_{nf}	the viscosity of the nanofluid
ρ_f	the densities of the fluid
ρ_s	density of the solid fractions
ρ_{nf}	density of the nanofluid
ν_f	kinematic viscosity of the fluid
$(\rho C_p)_{nf}$	the heat capacity of the nanofluid
$(\rho C_p)_f$	the heat capacity of the fluid
$(\rho C_p)_s$	the heat capacity of the solid fractions
κ	the permeability of the porous medium
σ	electrical conductivity
σ^*	Stephan-Boltzmann constant
τ_{wx}, τ_{wy}	the surface shear stresses along the x- and y-directions
φ	nanoparticle volume fraction
Subscripts	
w, ∞	conditions at the surface and in the free stream

Author details

Gamal M. Abdel-Rahman[1]* and Faiza M.N. El-fayez[2]

1 Department of Mathematics, Faculty of Science, Benha University, Benha, Egypt

2 Mathematical Science Department, College of Sciences, Princess Nourah Bint Abdulrahman University, KSA

*Address all correspondence to: gamalm60@yahoo.com

References

[1] Kakaç S, Pramuanjaroenkij A. Review of convective heat transfer enhancement with nanofluids. International Journal of Heat and Mass Transfer. 2009;**52**:3187-3196

[2] Xuan YM, Li Q. Heat transfer enhancement of nanofluid. International Journal of Heat and Fluid Flow. 2000;**21**:58-64

[3] Choi S. Enhancing thermal conductivity of fluids with nanoparticle. In: Siginer DA, Wang HP, editors. Developments and Applications of Non-Newtonian Flows. ASME MD vol. 231 and FED vol. 66; 1995. pp. 99-105

[4] Masuda H, Ebata A, Teramae K, Hishinuma N. Alteration of thermal conductivity and viscosity of liquid by dispersing ultra-fine particles. Netsu Bussei. 1993;**7**:227-233

[5] Buongiorno J, Hu W. Nanofluid coolants for advanced nuclear power plants. In: Paper no. 5705, Proceedings of ICAPP'05; Seoul. May 2005. pp. 15-19

[6] Vajravelu K, Hadjinicolaou A. Convective heat transfer in an electrically conducting fluid at a stretching surface with uniform free stream. International Journal of Engineering Science. 1997;**35**:1237-1244

[7] Eastman JA, Choi SUS, Li S, Yu W, Thompson J. Anomalously increased effective thermal conductivities of ethylene glycol-based nanofluids containing copper nanoparticles. Applied Physics Letters. 2001;**78**(6): 718-720

[8] Lotfi R, Saboohi Y, Rashidi AM. Numerical study of forced convective heat transfer of nanofluids: Comparison of different approaches. International Communications in Heat and Mass Transfer. 2010;**74**:37

[9] Hiemenz K. Die Grenzschicht an einem in den gleichförmigen Flüessigkeitsstrom eingetauchten geraden Kreiszylinder. Dingler's Polytechnisches Journal. 1911;**326**:321-410

[10] Howarth L. The boundary layer in three dimensional flow—Part II. The flow near a stagnation point stagnation point. Philosophical Magazine Series. 1951;**7**(42):1433-1440

[11] Bhattacharyya S, Gupta AS. MHD flow and heat transfer at a general three-dimensional at a stagnation point. International Journal of Non-Linear Mechanics. 1998;**33**:125-134

[12] Buongiorno J. Convective transport in nanofluids. ASME Journal of Heat Transfer. 2006;**128**:240-250

[13] Nield DA, Kuznetsov AV. Thermal instability in a porous medium layer saturated by a nanofluid. International Journal of Heat and Mass Transfer. 2009;**52**:5796-5801

[14] Kuznetsov AV, Nield DA. Natural convective boundary-layer flow of a nanofluid past a vertical plate. International Journal of Thermal Sciences. 2010;**49**:243-247

[15] Bachok N, Ishak A, Pop I. Boundary-layer flow of nanofluids over a moving surface in a flowing fluid. International Journal of Thermal Sciences. 2010; **49**(9):1663-1668

[16] Rashed GMA. Chemical entropy generation and MHD effects on the unsteady heat and fluid flow through a porous medium. Journal of Applied Mathematics. 2016;**2016**:9. Article ID 1748312

[17] Adams JA, Rogers DF. Computer-Aided Heat Transfer Analysis. McGraw-Hill; 1973

[18] Oztop HF, Abu-Nada E. Numerical study of natural convection in partially heated rectangular enclosures filled with nanofluids. International Journal of Heat and Fluid Flow. 2008;**29**:1326-1336

Chapter 10

Effects of MHD on Modified Nanofluid Model with Variable Viscosity in a Porous Medium

Sohail Nadeem and Nadeem Abbas

Abstract

A computational simulation for two-dimensional steady flow of modified nanofluid over an exponential stretching surface in a porous medium with magnet hydrodynamics and variable viscosity is presented in this study. Modified nanofluids are generalization of both hybrid nanofluids and simple nanofluids. Here, we consider three nanoparticles which drastically enhance the thermal conductivity of nanofluid. The viscous model associated with variable viscosity and MHD flow is employed. Well-known similarity transformations are utilized to convert the partial differential equations to system of ordinary differential equations. These converted equations are solved by utilizing the numerical technique Runge–Kutta-Fehlberg method. The impacts of variable viscosity, porosity parameter, Nusselt number, thermal and velocity slip, skin friction coefficient, solid nanoparticle, and magnetic field are observed. The computational results accomplished in the present investigation are validated and felt to be a good agreement with decayed results. It is highlighted that modified nanofluid model enhances the heat transfer rate much higher than the case of hybrid nanofluid and simple nanofluid model.

Keywords: variable viscosity, exponential stretching, modified nanofluid, MHD, porous medium, shooting method

1. Introduction

Porous medium is one of the most useful studies due to its applications in the industry and medical sciences. In the medical sciences, it is used in the transport process in the human lungs and kidneys, gall bladder in the presence of stone, clogging in arteries, and also little blood vessels which cannot be opposed. There are several examples of the naturally porous medium such as limestone, wood, seepage of water in river beds, etc. Many researchers are interested to discuss the porous medium due to scientific and technically importance such as earth's science and metallurgy. Such kinds of the flow are analyzed at low Reynolds number in the presence of porous space theoretically. Few researchers were analyzed analytically and experimentally on the porous medium with respect to different aspects (see [1–3]). Recently, the Carreau fluid flow over porous medium in the presence of pressure-dependent viscosity has been discussed by Malik et al. [4]. Some

IntechOpen

significant results are analyzed on the porous medium for Newtonian fluids and non-Newtonian fluids with respect to different aspects (see [5–15]) (**Figure 1**).

In the depth study, flow phenomenon focusing on the variable viscosity and exponentially stretching surface is an important rule in the study of fluid mechanics and has attracted the investigators after its valuable applications in the industry as well as flows detected over the tip of submarine and aircrafts. Numerous methods have been established in recent past years to enhance the fluid thermal conductivity which is suspended with micro−/nano-sized particle mix with base fluid. The nanoparticle possesses chemical and physical properties uniquely because it has been used widely in nanotechnology. The nano-sized particle which is suspended with fluid is called nanofluid. Many investigators investigate about the enhancement of thermal conductivity [16–20] by using the nano-sized particles.

Several experiments have been done in two types of the particles suspended in the base fluid, namely, "hybrid nanofluid." Basically, such type of fluids is enhances thermal conductivity which was proven through experimental research. Suresh et al. [21, 22] were the first to discuss the idea of hybrid nanofluid through their experimental and numerical results. According to their views, the hybrid nanofluid boosts the heat transfer rate at the surface as compared to nanofluid and simple fluid. These results open a new horizon to the researchers to do a work in the field of hybrid nanofluid. Baghbanzadeh et al. [23] also discussed about the mixture of multiwall/spherical silica nanotube hybrid nanostructures and analysis of thermal conductivity of associated nanofluid. The analysis of Al_2O_3−MWCNTs with base fluid water and their thermal properties are discussed by Nine et al. [24]. According to them [24], spherical particles with hybrid nanofluid reveal a bit increment in thermal conductivity as related to cylindrical-shaped particle. The hybrid nanofluids are considered experimentally and theoretically by a number of the researchers [25–28].

The physical characteristics of hybrid nanofluid and nanofluid are usually considered constant. It is prominent that the significant physical characteristics of nanofluid and hybrid nanofluid can vary with temperature. For lubricating fluids,

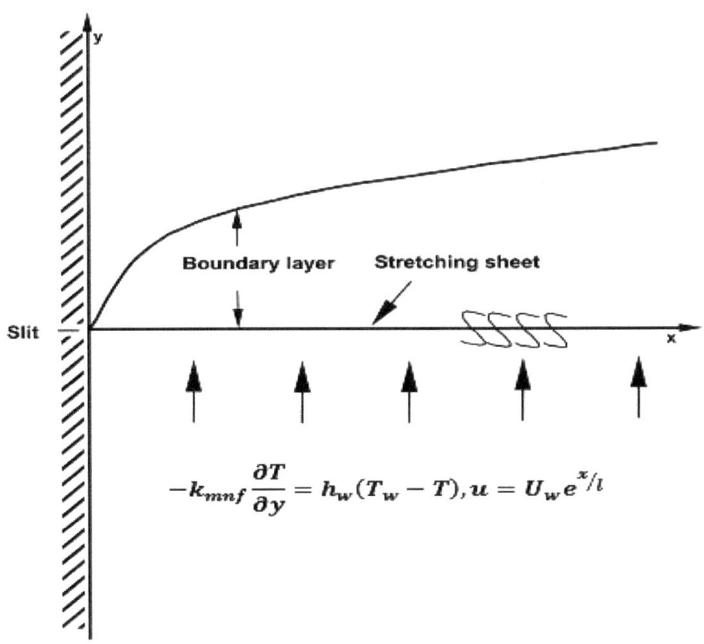

Figure 1.
Flow pattern of modified nanofluid.

heat generated by the internal friction and the corresponding rise in temperature affects the viscosity of the fluid and so the fluid viscosity can no longer be assumed constant. The increase of temperature leads to a local increase in the transport phenomena by reducing the viscosity across the momentum boundary layer and so the heat transfer rate at the wall is also affected. The impact of thermal radiation and dependent viscosity of fluid on free convective and heat transfer past a porous stretching surface were discussed by Mukhopadhyay and Layek [29]. They gain some significant results for the variable viscosity on the temperature profile and velocity profile. The velocity profile increases and temperature profile decreases for large values of the variable viscosity parameter. The existing literature survey on the variable fluid characteristics and hybrid nanofluid [30–33] reveals that the work is not carried out for hybrid nanofluid over an exponentially stretching surface.

The investigation about the stretching surface has attracted the interest of scientists because of its several applications in the fields of engineering including glass blowing, cooling of microelectronics, quenching in metal foundries, wire drawing, polymer extrusion, rapid spray, etc. Crane [34] discussed about the theoretical boundary layer flow over stretching surface. Various researchers analyzed the exponentially stretching surface [35–38], major applications in the industry and technology.

Here, we study the temperature-dependent viscosity effects on the modified nanofluid flow over an exponentially stretching porous medium in the presence of MHD and Biot number. It is highlighted here that the idea of modified nanofluid has been proposed by us from whom the hybrid nanofluid and simple nanofluid cases can be recovered as a special case. The temperature depends on the Biot number, nanoparticle, and variable viscosity. The system of the flow is illustrated in the form of partial differential equations (PDEs). The system of PDEs is converted into the form of ordinary differential equations (ODEs) by utilizing acceptable similarity transformations. These nonlinear ODEs are solved "numerically" through MATLAB built-in technique. The outcomes are represented through table and graphs.

2. Flow formulation

Investigation of steady laminar flow of two-dimensional electrically conducting modified nanofluid over exponentially stretching surface in the presence of variable viscosity has been taken into consideration which is revealed in **Figure 1**.

The fluid flows in the x-direction and is maintains at a constant wall temperature T_w. The working fluid is water-based modified nanofluid involving with different types of solid particles (Al_2O_2, *Cu and Ni*) while these particles having nanosized. These three particles are suspended with base fluid water. Some assumptions of these solid particles are following in this study like as negligible internal heat generation, incompressible flow, negligible radiative heat transfer and no chemical reaction. The thermophysical characteristics of modified nanofluid are represented in **Table 1**.

Under these assumptions with the usual boundary layer approximation, the governing differential equations of mass, momentum, and energy for the problem under consideration are defined as follows:

$$\frac{\partial u}{\partial x}+\frac{\partial v}{\partial y}=0, \tag{1}$$

$$u\frac{\partial u}{\partial x}+v\frac{\partial u}{\partial y}=\frac{1}{\rho_{mnf}}\frac{\partial}{\partial y}\left(\mu_{mnf}\frac{\partial u}{\partial y}\right)-\frac{\sigma B_\circ^2}{\rho_{mnf}}u-\frac{\nu_{mnf}}{R}u, \tag{2}$$

Thermophysical properties	**Fluidphase (water)**	Al_2O_3	Cu	**Ni**
C_p(j/kg)K	4179	765	385	444
ρ(kg/m^3)	997.1	3970	8933	8900
k(W/mK)	0.613	40	400	90.7

Table 1.
Numerical values of nanoparticles and water.

$$u\frac{\partial T}{\partial x}+v\frac{\partial T}{\partial y}=\alpha_{mnf}\frac{\partial^2 T}{\partial y^2}, \tag{3}$$

The appropriated boundary conditions are stated as

$$u\rightarrow U_w,\ -k_{mnf}\frac{\partial T}{\partial y}=h(T_w-T),\ \ as\ \ y\rightarrow 0, \tag{4}$$
$$u\rightarrow 0,\quad T\rightarrow T_\infty, as\ \ y\rightarrow\infty,$$

where u and v are the fluid velocity components in the x and y directions, respectively, T is the fluid temperature, U_0 is denoted as the stream velocity, and T_∞ represents as the temperature of the fluid far away from the surface. The thermophysical properties of nanofluid, hybrid nanofluid, and modified nanofluid are represented in **Tables 2** and **3**.

An extraordinary type of physical characteristics is acquainted in the present examination to investigate the boundary layer equations for modified nanofluid. Modified nanofluid is deliberated through taking the combination of Al_2O_3 and Cu with base fluid water. The nanoparticles Al_2O_3 and Cu ($\Phi_1=0.05\ vol$ and $\Phi_2=0.05\ vol$, respectively) are fixed throughout this problem. To make it ideal, the final type of the powerful thermophysical characteristics of $(Al_2O_3/water)$ nanofluid, $(Al_2O_3-Cu/water)$ hybrid nanofluid and $(Al_2O_3-Cu-Ni/water)$ modified nanofluid, is assumed in **Tables 2** and **3**, while n = 3 is for spherical nanoparticles. Some subscripts are defined as following, solid nanoparticles of Al_2O_3, s_2 solid nanoparticles of the Cu, s_3 solid nanoparticles of Ni, f for base fluid (water), nf for nanofluid, hnf for Hybrid nanofluid and mnf for modified nanofluid. The thermophysical characteristics of fluid are represented in **Table 1** at 25°C where $U_w=U_0e^{\frac{x}{l}}$, $T_w=T_\infty+T_0e^{\frac{x}{2l}}$. The μ_f is the coefficient of the viscosity which is assumed to vary as an inverse function of temperature [23] as

Properties	**Nanofluid**	**Hybrid nano-fluid**
Density	$\rho_{nf}=(1-\Phi)\rho_f+\Phi\rho_s$	$\rho_{nf}=\{[(1-\Phi_2)(1-\Phi_1)\rho_f\}+\Phi_1\rho_{s_1}]+\Phi_2\rho_{s_2}$
Heat capacity	$(\rho C_p)_{nf}=(1-\Phi)(\rho C_p)_f+\Phi(\rho C_p)_s$	$(\rho C_p)_{hnf}=[\{(1-\Phi_2)(1-\Phi_1)(\rho C_p)_f\}$ $+\Phi_1(\rho C_p)_{s_1}]+\Phi_2\rho(\rho C_p)_{s_2}$
Viscosity	$\mu_{nf}=\frac{\mu_f}{(1-\Phi)^{2.5}}$	$\mu_{hnf}=\frac{\mu_f}{(1-\Phi_1)^{2.5}(1-\Phi_2)^{2.5}}$
Thermal conductivity	$\frac{\kappa_{nf}}{\kappa_f}=\frac{\kappa_s+(n-1)\kappa_f-(n-1)\Phi(\kappa_f-\kappa_s)}{\kappa_s+(n-1)\kappa_f+\Phi(\kappa_f-\kappa_s)}$	$\frac{\kappa_{hnf}}{\kappa_{bf}}=\frac{\kappa_{s_2}+(n-1)\kappa_{bf}-(n-1)\Phi_2(\kappa_{bf}-\kappa_{s_2})}{\kappa_{s_2}+(n-1)\kappa_{bf}+\Phi_2(\kappa_{bf}-\kappa_{s_2})}$ where $\frac{\kappa_{bf}}{\kappa_f}=\frac{\kappa_{s_1}+(n-1)\kappa_f-(n-1)\Phi_1(\kappa_f-\kappa_{s_1})}{\kappa_{s_1}+(n-1)\kappa_f+\Phi_1(\kappa_f-\kappa_{s_1})}$

Table 2.
Physical Properties of Nanofluid and Hybrid Nanofluid.

Properties	Modified nanofluid
Density	$\rho_{mnf} = (1-\Phi_3)([\{(1-\Phi_2)(1-\Phi_1)\rho_f\} + \Phi_1\rho_{s_1}] + \Phi_2\rho_{s_2}) + \Phi_3\rho_{s_3}$
Heat capacity	$(\rho C_p)_{mnf} = (1-\Phi_3)\left(\left[\{(1-\Phi_2)(1-\Phi_1)(\rho C_p)_f\} + \Phi_1(\rho C_p)_{s_1}\right] + \Phi_2(\rho C_p)_{s_2}\right) + \Phi_3(\rho C_p)_{s_3}$
Viscosity	$\mu_{mnf} = \dfrac{\mu_f}{(1-\Phi_3)^{2.5}(1-\Phi_1)^{2.5}(1-\Phi_2)^{2.5}}$
Thermal conductivity	$\dfrac{\kappa_{nf}}{\kappa_f} = \dfrac{\kappa_{s_1} + (n-1)\kappa_f - (n-1)\Phi_1(\kappa_f - \kappa_{s_1})}{\kappa_{s_1} + (n-1)\kappa_f + \Phi_1(\kappa_f - \kappa_{s_1})}$ $\dfrac{\kappa_{hnf}}{\kappa_{nf}} = \dfrac{\kappa_{s_2} + (n-1)\kappa_{nf} - (n-1)\Phi_2(\kappa_{nf} - \kappa_{s_2})}{\kappa_{s_2} + (n-1)\kappa_{nf} + \Phi_2(\kappa_{nf} - \kappa_{s_2})}$ $\dfrac{\kappa_{mnf}}{\kappa_{hnf}} = \dfrac{\kappa_{s_3} + (n-1)\kappa_{hnf} - (n-1)\Phi_3(\kappa_{hnf} - \kappa_{s_3})}{\kappa_{s_3} + (n-1)\kappa_{hnf} + \Phi_3(\kappa_{hnf} - \kappa_{s_3})}$

Table 3.
Physical Properties Modified Nanofluid.

$$\frac{1}{\mu} = \frac{1}{\mu_f}[1 + \delta(T - T_r)], \tag{5}$$

i.e., $\frac{1}{\mu_f} = a(T - T_\infty)$ where $a = \frac{\delta}{\mu_f}$ and $T_r = T_\infty - \frac{1}{\delta}$.

$$u = U_0 e^{x/l} f'(\zeta),\ \ v = -\frac{\nu}{l}\sqrt{\frac{Re}{2}} e^{x/2l}\ (f(\zeta) + \zeta f'(\zeta)), \tag{6}$$

$$T = T_\infty + T_w e^{x/2l}\ \theta(\zeta), \qquad \zeta = \frac{y}{l}\sqrt{\frac{Re}{2}} e^{x/2l}, \tag{7}$$

The mathematical model over exponentially stretching surface is chosen to allow the coupled non-linear partial differential equations are converted into coupled non-linear ordinary differential equations by using the suitable similarity transformation which is given above. Where ζ is the similarity variable and θ and f are the dimensionless temperature and velocity, respectively. Eq. (1) is directly satisfied by using the similarities which is called continuity equation. The momentum and energy equation are written as

$$\frac{\left(\dfrac{f'''}{1-\dfrac{\theta}{\theta_e}} + \dfrac{f''\theta'}{\theta_e\left(1-\frac{\theta}{\theta_e}\right)^2}\right)}{(1-\varphi_3)^{2.5}(1-\varphi_2)^{2.5}(1-\varphi_1)^{2.5}\left[(1-\varphi_3)\left\{(1-\varphi_2)\left(1-\varphi_1+\varphi_1\dfrac{\rho_{s_1}}{\rho_f}\right)+\varphi_2\dfrac{\rho_{s_2}}{\rho_f}\right\}+\varphi_3\dfrac{\rho_{s_3}}{\rho_f}\right]} \\ -\beta f'f' + ff'' - M^2 f' - \beta f' = 0,\ \ \beta = 2; \tag{8}$$

$$\frac{\dfrac{\kappa_{mnf}}{\kappa_f}\theta''}{Pr\left[(1-\varphi_3)\left(\left\{(1-\varphi_2)\left(1-\varphi_1+\varphi_1\dfrac{(\rho C_p)_{s_1}}{(\rho C_p)_f}\right)\right\}+\varphi_2\dfrac{(\rho C_p)_{s_2}}{(\rho C_p)_f}\right)+\varphi_3\dfrac{(\rho C_p)_{s_3}}{(\rho C_p)_f}\right]} \\ -\theta f' + \theta' f = 0 \tag{9}$$

with boundary conditions

$$f'(0) = 1, \quad f(0) = 0, \quad f'(\infty) = 0 \tag{10}$$

$$\theta'(0) = -\gamma\left(\frac{k_f}{k_{mnf}}\right)(1 - \theta(0)), \quad \theta(\infty) = 0 \tag{11}$$

3. Numerical solution method

Boundary layer heat transfer and modified nanofluid flow of an exponentially stretching surface with ($Al_2O_3 - Cu - Ni$) under the assumption of dependent fluid viscosity, magnetic field, and thermal slip effects are computed here. The results of boundary layer problem are obtained numerically through bvp4c method. The notable highlights of the flow and heat transfer characteristics are achieved utilizing the modified nanofluid. Keeping in mind the end goal to get clear knowledge of the physical problem, the results are given through the physical parameter, namely, magnetic field (M), solid nanoparticle (Φ_3) and thermal slip effects (Bi) and (θ_e). The numerical results are represented in the form of tables.

$$y(1) = f(\zeta),$$

$$y(2) = f'(\zeta),$$

$$y(3) = f''(\zeta),$$

$$f''(\zeta) = \left(1 - \frac{y(4)}{\theta_e}\right)\left\{-\frac{y(3)y(5)}{\theta_e\left(1 - \frac{y(4)}{\theta_e}\right)^2}\right.$$
$$+ (1-\varphi_3)^{2.5}(1-\varphi_2)^{2.5}(1-\varphi_1)^{2.5}\left[(1-\varphi_3)\left\{(1-\varphi_2)\left(1-\varphi_1+\varphi_1\frac{\rho_{s_1}}{\rho_f}\right)+\varphi_2\frac{\rho_{s_2}}{\rho_f}\right\}\right.$$
$$\left.\left.+\varphi_3\frac{\rho_{s_3}}{\rho_f}\right] + 2y(2)y(2) - y(1)y(3) + M^2y(2)\right\},$$

$$y(4) = \theta(\zeta),$$

$$y(5) = \theta'(\zeta),$$

$$\theta'' = \frac{\kappa_f Pr\left[(1-\varphi_3)\left(\left\{(1-\varphi_2)\left(1-\varphi_1+\varphi_1\frac{(\rho C_p)_{s_1}}{(\rho C_p)_f}\right)\right\}+\varphi_2\frac{(\rho C_p)_{s_2}}{(\rho C_p)_f}\right)+\varphi_3\frac{(\rho C_p)_{s_3}}{(\rho C_p)_f}\right]}{\kappa_{mnf}}(y(5)y(1) - y(4)y(2)),$$

subject to the boundary conditions

$$y0(2) = 1, \quad y0(1) = 0, \quad yinf(2) = 0,$$

$$y0(5) = -\gamma\left(\frac{k_f}{k_{mnf}}\right)(1 - y0(4)), \quad yinf(4) = 0.$$

For brevity, the points of interest of the solution strategy are not performed here. The heat transfer and modified nanofluid are effected by dependent viscosity parameter and MHD; the fundamental focus of our investigation is to bring out the impacts of these parameters by the numerical solution. It is worth specifying that we have utilized the information displayed in **Tables 1–3** for the thermophysical properties of the fluid, nanofluid, hybrid nanofluid, modified nanofluid, and nanoparticles. Three types of the nanoparticles are used, namely, Al_2O_3, Cu, and Ni. The Nussle number and skin friction coefficient are the most important features of this study. For practical purposes, the functions $\theta(\zeta)$ and $f(\zeta)$ allow to determine the Nusselt number:

$$\mathrm{Nu_x} = -\frac{\mathrm{xk_{mnf}}}{\mathrm{k_f(T_w - T_\infty)}}\left(\frac{\partial T(\mathrm{x,y})}{\partial y}\right)_{\mathrm{y=0}}, \tag{12}$$

$$\frac{\mathrm{Nu_x}}{\sqrt{\mathrm{Re_x}}} = -\frac{\mathrm{k_f}}{\mathrm{k_{mnf}}}\theta'(0) \tag{13}$$

and skin friction coefficient

$$\mathrm{C_f} = \frac{\mu_{\mathrm{mnf}}}{\rho_{\mathrm{f}}\mathrm{u_w^2}}\left(\frac{\partial u(\mathrm{x,y})}{\partial y}\right)_{\mathrm{y=0}}, \tag{14}$$

$$\frac{\mathrm{C_f}}{\sqrt{\mathrm{Re_x}}} = -\frac{1}{(1-\varphi_3)^{2.5}(1-\varphi_2)^{2.5}(1-\varphi_1)^{2.5}}\left(1-\frac{1}{\theta_e}\right)^{-1}\mathrm{f}''(0). \tag{15}$$

Here, the local Reynolds number is $\mathrm{Re}_x = \frac{xu_w}{v_f}$.

4. Numerical results

The impact of dependent viscosity parameter θ_e on the coefficient of skin friction and Nusselt number for negative values of θ_e (for liquids) and for positive values of θ_e (for gases) which reveals in **Table 4**. The variation of $f''(0)$ and $\theta'(0)$ reveals that the same behavior to be noted for large values of θ_e.

The computational results are shown in **Table 5**. The velocity of the flow decreases due to increase in the solid nanoparticle of Ni (Φ_3), as well as skin fraction is decreased. This may be due to more collision between the suspended nanoparticles. The nanoparticles release the energy in the form of heat by physically. Adding more particles may exert more energy which increases the temperature while also thickness of the thermal boundary layer. The increment of solid nanoparticle accelerates the flow velocity which obviously declines the skin fraction which is shown in **Table 5**. It is also seen that $\theta'(0)$ decreases due to increasing the solid nanoparticles (Ni (Φ_3)). It is noted in **Table 5** that the magnetic field parameter increases due to decrease in the velocity flow of the modified nanofluid. For the large values of the magnetic field, the dimensionless rate of heat transfer gains

	$Al_2O_3 - Cu - Ni/water$		$Al_2O_3 - Ni/water$	
θ_e	$f''(0)$	$\theta'(0)$	$f''(0)$	$\theta'(0)$
−10	−1.98532	−1.7779	−1.90976	−1.98809
−5	−2.07138	−1.77154	−1.99315	−1.98175
−1	−2.65068	−1.72622	−2.55803	−1.93639
−0.1	−5.21601	−1.49115	−5.08404	−1.69636
1	−0.768213	−1.85951	−0.731972	−2.06964
5	−1.70487	−1.79801	−1.63866	−2.00813
10	−1.80225	−1.79113	−1.7327	−2.00127

Table 4.
Computational results of $Al_2O_3 - Cu - Ni/water$ and $Al_2O_3 - Ni/water$.

				$Al_2O_3-Cu-Ni/water$		$Al_2O_3-Ni/water$	
γ	*Bi*	*M*	Φ_3	$f''(0)$	$\theta'(0)$	$f''(0)$	$\theta'(0)$
0.0				−1.50718	−1.86066	−1.44127	−2.07585
0.5				−1.6881	−1.83996	−1.61326	−2.05562
1.0				−1.8533	−1.82077	−1.77026	−2.03689
0.5	0.0			−1.58519	−2.79411	−1.52063	−3.06462
	0.2			−1.6881	−1.83996	−1.61326	−2.05562
	0.4			−1.73712	−1.37364	−1.65867	−1.54835
	0.2	0.0		−1.50718	−1.86066	−1.44127	−2.07585
		0.5		−1.6881	−1.83996	−1.61326	−2.05562
		1.0		−1.8533	−1.82077	−1.77026	−2.03689
		0.5	0.005	−1.6158	−1.99761	−1.5041	−2.23451
			0.04	−1.6881	−1.83996	−1.61326	−2.05562
			0.08	−1.74614	−1.67941	−1.70227	−1.87409

Table 5.
Computational results of $Al_2O_3-Cu-Ni/water$ and $Al_2O_3-Ni/water$ fixed at θ_e=0.5.

enhanced in the Modified nanofluid. A dimensionless quantity is the Biot number which compares the relative transport of internal and external resistances. The dimensionless $f''(0)$ and $\theta'(0)$ increase for large values of the Biot numbers. The dimensionless of $\theta'(0)$ gets decreases for the increment of Biot number while the dimensionless of $f''(0)$ gets increases for the increment of Biot number as shown in **Table 5** for the modified nanofluid. The Biot number is directly

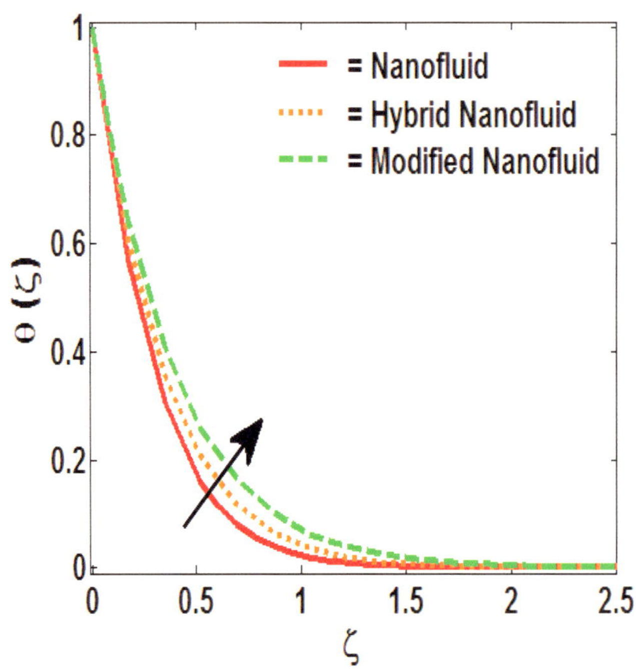

Figure 2.
Comparative results of Nanofluid, Hybrid nanofluid and Modified nanofluid on $\theta(\zeta)$.

proportional to the heat transfer coefficient. It is also seen that the variable viscosity parameter declines for enhancing the dimensionless $f''(0)$ of the modified nanofluid. It is also noted that the heat transfer rate declines due to increase in the variable viscosity parameter as shown in **Table 5**. Effects of porosity parameter on the $f''(0)$ and $\theta'(0)$ are presented in **Table 5**. It is highlighted that $f''(0)$ increased for the higher values of the porosity parameter but had an opposite behavior to be highlighted for $\theta'(0)$. **Figure 2** shows the impacts of comparative study of modified nanofluid, hybrid nanaofluid and simple nanofluid.

5. Graphical results

The temperature profile shows the variation of solid nanoparticle in **Figure 3**. The nanoparticle dissipates energy in the form of heat. So, the mixture of more nanoparticles may exert more energy which increases the thickness of the boundary layer and temperature.

Figure 4 reveals the impacts of solid particle on velocity profiles. The velocity profile gets decelerated due to increase in solid nanoparticle for modified nanofluid. This phenomenon exists due to more collision with suspended nanoparticles.

Figure 5 reveals the effects of magnetic field on the velocity profile. Being there, the transverse magnetic field creates Lorentz force which arises from the attraction of electric field and magnetic field during the motion of an electrically conducting fluid. The velocity profile decreases for the positive values of magnetic field parameter. Because the resisting force increases and consequently velocity declines in the $x-$direction with boundary layer thickness as the magnetic field parameter enhances for modified nanofluid.

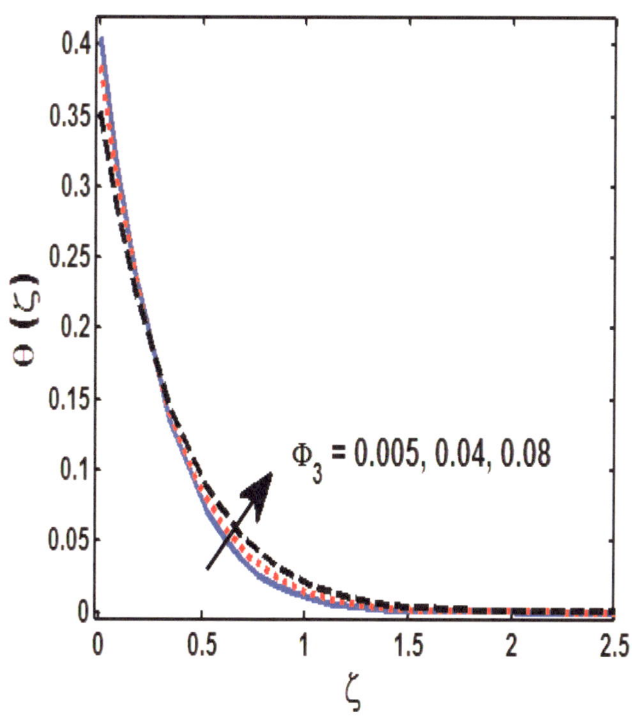

Figure 3.
Impacts of Φ_3 on temperature profile.

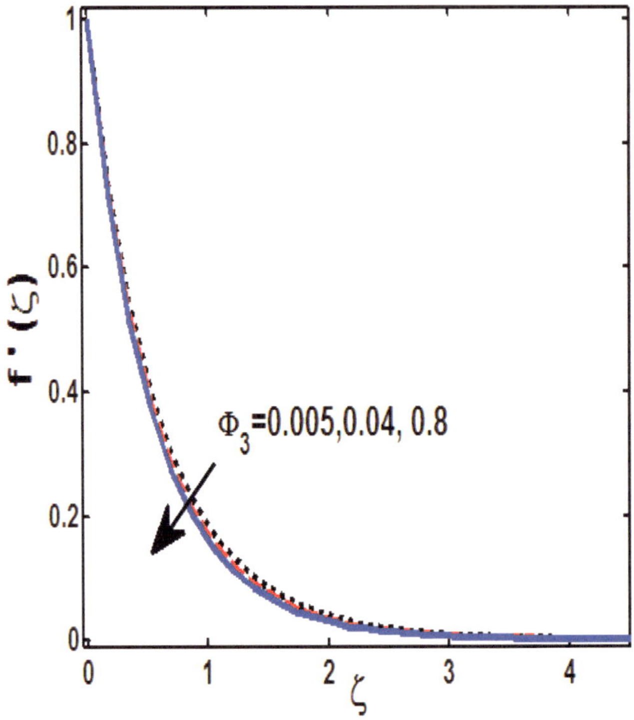

Figure 4.
Impacts of Φ_3 on velocity profile.

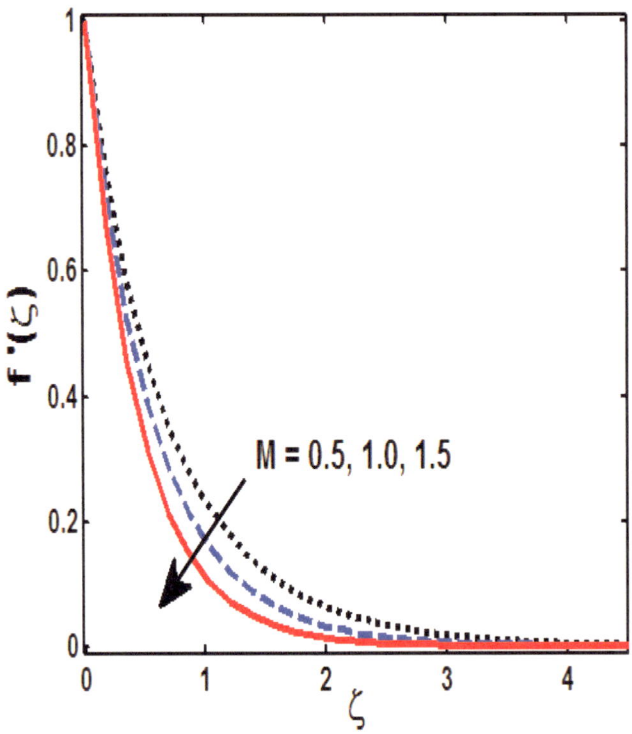

Figure 5.
Impacts of M *on the velocity profile.*

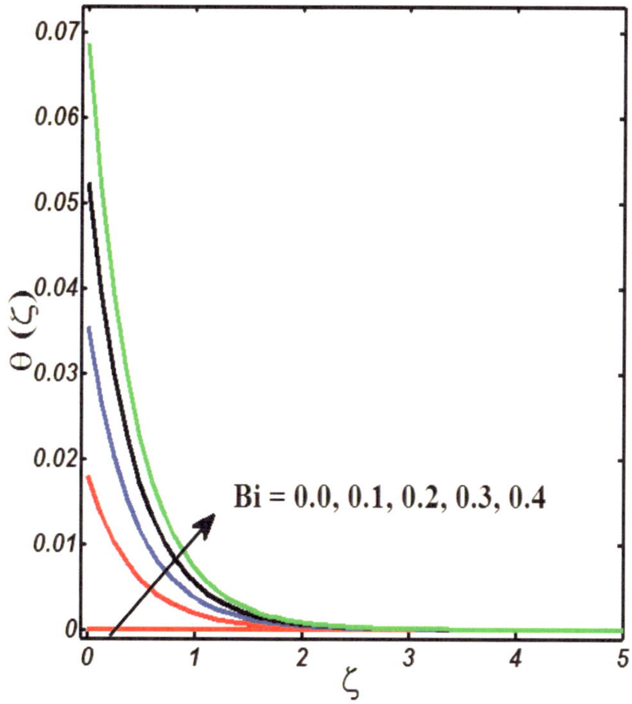

Figure 6.
Impacts of Bi *on the temperature profile.*

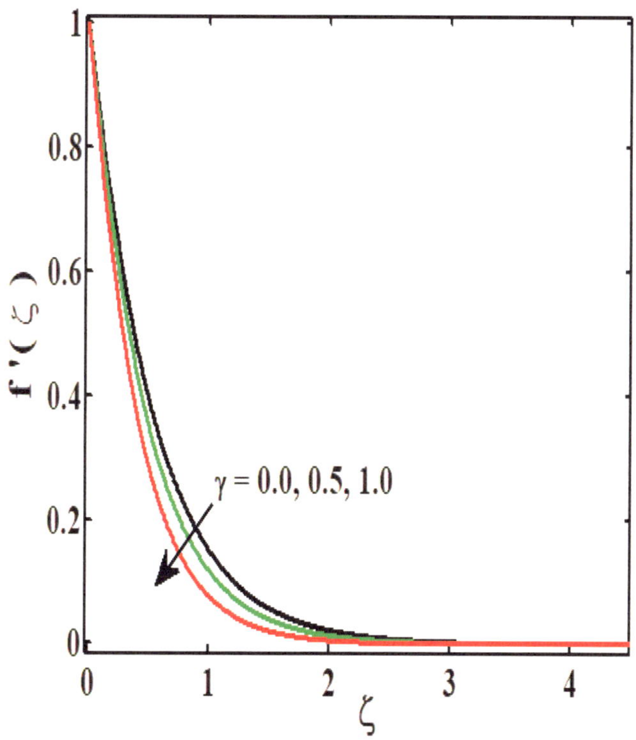

Figure 7.
Impacts of γ on velocity profile.

Figure 6 reveals the variation of dimensionless quantity of Biot number on the temperature profile. The relative transport of internal and external resistances is called the Biot number. The thermal boundary layer increases with increasing in the biot number.

Figure 7 shows the impact of the porosity parameter on the velocity profile. It is noted that velocity profiles decreases for the higher values of the porosity parameter. The boundary layer thickness decreases for large values of porousity parameter.

6. Closing remarks

In this paper, the impacts of dependent viscosity parameter, magnetic field, and solid nanoparticle flow and the heat transfer of modified nanofluid flow at the exponential stretching surface have been analyzed numerically. The governing coupled partial differential equations are converted into ordinary coupled differential equations which are solved numerically by bvp4c method. The parametric analysis is executed to investigate the impacts of the governing physical parameters (magnetic field, variable viscosity (for both cases $\theta_e < 0$ and $\theta_e > 0$), Biot number, and solid nanoparticle) on the flow and heat transfer properties. In particular, we focus on the effect of dependent viscosity when $\theta_e < 0$ and $\theta_e > 0$ of the $Al_2O_3 - Cu - Ni/water$ and $Al_2O_3 - Ni/water$. It is noted that the fluid viscosity and temperature are inverse function. The computational results are presented through graph and tables. The results of modified nanofluid flow and heat transfer properties show many exciting behaviors which deserve further study of modified nanofluid.

Nomenclature

Pr	Prandtl number
Φ_1	nanoparticle volume fraction of Al_2O_3
Φ_2	nanoparticle volume fraction of Cu
Φ_3	nanoparticle volume fraction of Ni
Bi	Biot number
θ	temperature profile
R	permeability
f	velocity profile along x-direction
ρ	density
f	fluid
T_w	wall temperature
T_∞	ambient temperature
ν_f	fluid kinematic viscosity
ν_{nf}	nanofluid kinematic viscosity
ν_{hnf}	hybrid nanofluid kinematic viscosity
ν_{mnf}	modified nanofluid kinematic viscosity
$\left(\rho C_p\right) h_{nf}$	heat capacity of hybrid nanofluid
$\left(\rho C_p\right) m_{nf}$	heat capacity of modified nanofluid
κ_f	thermal conductivity of fluid
κ_{nf}	thermal conductivity of nanofluid
κ_{hnf}	thermal conductivity of hybrid nanofluid

κ_{mnf}	thermal conductivity of modified nanofluid
μ_{hnf}	viscosity of hybrid nanofluid
μ_{mnf}	viscosity of modified nanofluid
μ_{nf}	viscosity of nanofluid
$\left(\rho C_p\right)_{nf}$	heat capacity of nanofluid
α_{hnf}	thermal diffusivity of hybrid nanofluid
α_{mnf}	thermal diffusivity of modified nanofluid
α_{nf}	thermal diffusivity of nanofluid
U, V	velocity components
X, Y	direction components
θ_e	variable viscosity parameter
γ	porosity parameter

Author details

Sohail Nadeem and Nadeem Abbas*
Department of Mathematics, Quaid-I-Azam University, Islamabad, Pakistan

*Address all correspondence to: nabbas@math.qau.edu.pk

References

[1] Ping C, Chang ID. Buoyancy induced flows in a saturated porous medium adjacent to impermeable horizontal surfaces. International Journal of Heat and Mass Transfer. 1976;**19**(11): 1267-1272

[2] Caltagirone JP. Thermoconvective instabilities in a porous medium bounded by two concentric horizontal cylinders. Journal of Fluid Mechanics. 1976;**76**(2):337-362

[3] Yamamoto K, Iwamura N. Flow with convective acceleration through a porous medium. Journal of Engineering Mathematics. 1976;**10**(1):41-54

[4] Malik MY, Zehra I, Nadeem S. Flows of Carreau fluid with pressure dependent viscosity in a variable porous medium: Application of polymer melt. Alexandria Engineering Journal. 2014; **53**(2):427-435

[5] Subhani M, Nadeem S. Numerical analysis of 3D micropolar nanofluid flow induced by an exponentially stretching surface embedded in a porous medium. The European Physical Journal Plus. 2017;**132**(10):441

[6] Sheikholeslami M, Jafaryar M, Saleem S, Li Z, Shafee A, Jiang Y. Nanofluid heat transfer augmentation and exergy loss inside a pipe equipped with innovative turbulators. International Journal of Heat and Mass Transfer. 2018;**126**:156-163

[7] Sheikholeslami M. Finite element method for PCM solidification in existence of CuO nanoparticles. Journal of Molecular Liquids. 2018;**265**:347-355

[8] Sheikholeslami M, Li Z, Shafee A. Lorentz forces effect on NEPCM heat transfer during solidification in a porous energy storage system. International Journal of Heat and Mass Transfer. 2018;**127**:665-674

[9] Sheikholeslami M. Application of Darcy law for nanofluid flow in a porous cavity under the impact of Lorentz forces. Journal of Molecular Liquids. 2018;**266**:495-503

[10] Ahmad S, Farooq M, Anjum A, Javed M, Malik MY, Alshomrani AS. Diffusive species in MHD squeezed fluid flow through non-Darcy porous medium with viscous dissipation and joule heating. Journal of Magnetics. 2018;**23**(2):323-332

[11] Sheikholeslami M, Shehzad SA, Li Z, Shafee A. Numerical modeling for alumina nanofluid magnetohydrodynamic convective heat transfer in a permeable medium using Darcy law. International Journal of Heat and Mass Transfer. 2018;**127**:614-622

[12] Sheikholeslami M, Jafaryar M, Li Z. Second law analysis for nanofluid turbulent flow inside a circular duct in presence of twisted tape turbulators. Journal of Molecular Liquids. 2018;**263**: 489-500

[13] Sheikholeslami M. Influence of magnetic field on Al_2O_3-H_2O nanofluid forced convection heat transfer in a porous lid driven cavity with hot sphere obstacle by means of LBM. Journal of Molecular Liquids. 2018;**263**: 472-488

[14] Sheikholeslami M. Numerical approach for MHD Al_2O_3-water nanofluid transportation inside a permeable medium using innovative computer method. Computer Methods in Applied Mechanics and Engineering. 2019;**344**:306-318

[15] Sheikholeslami M. New computational approach for exergy and entropy analysis of nanofluid under the impact of Lorentz force through a porous media. Computer Methods in

Applied Mechanics and Engineering. 2019;**344**:319-333

[16] Choi S, Zhang ZG, Yu W, et al. Anomalously thermal conductivity enhancement in nanotube suspensions. Applied Physics Letters. 2001;**79**(14): 2252-2254

[17] Khan I, Fatima S, Malik MY, Salahuddin T. Exponentially Varying Viscosity of Magnetohydrodynamic Mixed Convection Eyring-Powell Nanofluid Flow Over An Inclined Surface. Results in Physics; 2018;**8**: 1194-1203

[18] Nadeem S, Abbas N. On Both MHD and Slip Effect in Micropolar Hybrid Nanofluid Past a Circular Cylinder Under Stagnation Point Region. Canadian Journal of Physics, (ja)

[19] Sadaf H, Akbar MU, Nadeem S. Induced magnetic field analysis for the peristaltic transport of non-Newtonian nanofluid in an annulus. Mathematics and Computers in Simulation. 2018;**148**: 16-36

[20] Sheikholeslami M, Gerdroodbary MB, Moradi R, Shafee A, Li Z. Application of Neural Network for estimation of heat transfer treatment of Al_2O_3-H_2O nanofluid through a channel. Computer Methods in Applied Mechanics and Engineering. 2019;**344**: 1-12

[21] Suresh S, Venkitaraj KP, Selvakumar P, Chandrasekar M. Synthesis of Al_2O_3–Cu/water hybrid nanofluids using two step method and its thermo physical properties. Colloids and Surfaces A: Physicochemical and Engineering Aspects. 2011;**388**(1–3): 41-48

[22] Suresh S, Venkitaraj KP, Selvakumar P, Chandrasekar M. Effect of Al_2O_3–Cu/water hybrid nanofluid in heat transfer. Experimental Thermal and Fluid Science. 2012;**38**:54-60

[23] Baghbanzadeh M, Rashidi A, Rashtchian D, Lotfi R, Amrollahi A. Synthesis of spherical silica/multiwall carbon nanotubes hybrid nanostructures and investigation of thermal conductivity of related nanofluids. Thermochimica Acta. 2012; **549**:87-94

[24] Esfe MH, Arani AAA, Rezaie M, Yan WM, Karimipour A. Experimental determination of thermal conductivity and dynamic viscosity of Ag–MgO/ water hybrid nanofluid. International Communications in Heat and Mass Transfer. 2015;**66**:189-195

[25] Hayat T, Nadeem S. Heat transfer enhancement with Ag–CuO/water hybrid nanofluid. Results in Physics. 2017;7:2317-2324

[26] Muhammad N, Nadeem S. Ferrite nanoparticles Ni-$ZnFe_2O_4$, Mn-$ZnFe_2O_4$ and Fe_2O_4 in the flow of ferromagnetic nanofluid. The European Physical Journal Plus. 2017;**132**(9):377

[27] Ijaz S, Nadeem S. Biomedical theoretical investigation of blood mediated nanoparticles (Ag-Al_2O_3/ blood) impact on hemodynamics of overlapped stenotic artery. Journal of Molecular Liquids. 2017;**248**:809-821

[28] Nadeem S, Abbas N, Khan AU. Characteristics of Three Dimensional Stagnation Point Flow of Hybrid Nanofluid Past A Circular Cylinder. Results in Physics; 2018

[29] Mukhopadhyay S, Layek GC. Effects of thermal radiation and variable fluid viscosity on free convective flow and heat transfer past a porous stretching surface. International Journal of Heat and Mass Transfer. 2008; **51**(9–10):2167-2178

[30] Vajravelu K, Prasad KV, Chiu-On NG. The effect of variable viscosity on the flow and heat transfer of a viscous

Ag-water and Cu-water nanofluids. Journal of Hydrodynamics, Ser. B. 2013; **25**(1):1-9

[31] Sheikholeslami M, Gorji-Bandpy M, Vajravelu K. Lattice Boltzmann simulation of magnetohydrodynamic natural convection heat transfer of Al_2O_3–water nanofluid in a horizontal cylindrical enclosure with an inner triangular cylinder. International Journal of Heat and Mass Transfer. 2015; **80**:16-25

[32] Khan WA, Makinde OD, Khan ZH. Non-aligned MHD stagnation point flow of variable viscosity nanofluids past a stretching sheet with radiative heat. International Journal of Heat and Mass Transfer. 2016;**96**:525-534

[33] Konch J, Hazarika G. Effects of variable viscosity and variable thermal conductivity on hydromagnetic dusty fluid flow due to a rotating disk. Frontiers in Heat and Mass Transfer (FHMT). 2017;**8**

[34] Crane LJ. Flow past a stretching plate. Zeitschrift für angewandte Mathematik und Physik ZAMP. 1970; **21**(4):645-647

[35] Sandeep N, Sulochana C, Kumar BR. Unsteady MHD radiative flow and heat transfer of a dusty nanofluid over an exponentially stretching surface. Engineering Science and Technology, an International Journal. 2016;**19**(1): 227-240

[36] Nayak MK, Akbar NS, Tripathi D, Khan ZH, Pandey VS. MHD 3D free convective flow of nanofluid over an exponentially stretching sheet with chemical reaction. Advanced Powder Technology. 2017;**28**(9):2159-2166

[37] Hayat T, Nadeem S. Flow of 3D Eyring-Powell Fluid by Utilizing Cattaneo-Christov Heat Flux Model and Chemical Processes Over An Exponentially Stretching Surface. Results in Physics; 2017

[38] Rehman FU, Nadeem S, Haq RU. Heat transfer analysis for three-dimensional stagnation-point flow over an exponentially stretching surface. Chinese Journal of Physics. 2017;**55**(4): 1552-1560

Chapter 11

Porous Low-Dielectric-Constant Material for Semiconductor Microelectronics

Yi-Lung Cheng and Chih-Yen Lee

Abstract

To provide high speed, low dynamic power dissipation, and low cross-talk noise for microelectronic circuits, low-dielectric-constant (low-*k*) materials are required as the inter- and intra-level dielectric (ILD) insulator of the back-end-of-line interconnects. Porous low-*k* materials have low-polarizability chemical compositions and the introducing porosity in the film. Integration of porous low-*k* materials into microelectronic circuits, however, poses a number of challenges because the composition and porosity affected the resistance to damage during integration processing and reduced the mechanical strength, thereby degrading the properties and reliability. These issues arising from porous low-*k* materials are the subject of the present chapter.

Keywords: porous low-*k*, porosity, Cu interconnects, BEOL, integration, plasma damage, Cu drift, TDDB, reliability

1. Introduction

To obtain a high operation performance and to pack more chips in microelectronics, the semiconductor industry spent a lot of efforts to accomplish successful integration of the integrated circuits (ICs). As the dimensions of the device are continuously shrinking with the advance of technology node, the carrier's transit time across the length of a transistor channel (called gate delay) decreases, while the signal propagation through the interconnects [called resistance-capacitance (RC) delay] increases, as shown in **Figure 1**. As a result, the effective speed of the device is limited by the RC delay since 0.25 μm technology node [2–4]. The RC delay can be reduced by using metals with low resistivity and dielectric materials with low dielectric constant (*k*). Therefore, copper (Cu) and low-dielectric-constant (low-*k*) materials have been introduced in back-end-of-line (BEOL) interconnects of ICs to replace the conventional Al/SiO_2 interconnects [4–7]. Cu with a resistivity of 1.7 μΩ-cm (2.7 μΩ-cm for Al) is becoming the common metallization material. Low-*k* materials with *k* values lower than 4.0 (*k* value of SiO_2) provide lower capacitance between wires. To effectively reduce the *k* value of a dielectric film, low-polar bonds and porosity are introduced into the film. The produced dielectric materials are called porous low-*k* materials [8–10]. To provide a further low-*k* value, more porosity is introduced into the low-*k* material; however, more integration challenges arise.

IntechOpen

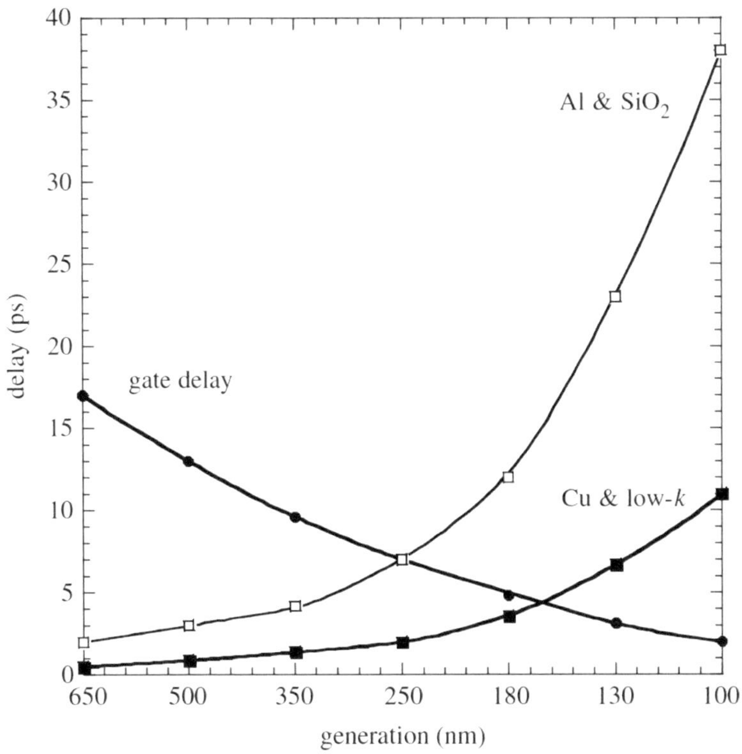

Figure 1.
Gate and interconnect delay with technological generation (International Technology Roadmap for Semiconductors [1]).

This chapter is an attempt to provide an overview of porous low-*k* materials. The resulting issues and reliability during the integration of porous low-*k* material in Cu interconnects are discussed.

2. Low-*k* dielectric materials and deposition method

2.1 Low-*k* dielectric materials

The dielectric constant (k) of a dielectric material is generally described by Clausius-Mossotti Eq. (1):

$$\frac{k-1}{k+2} = \frac{4\pi N}{3}\alpha \tag{1}$$

where $k = \varepsilon/\varepsilon_0$, ε, and ε_0 are the permittivity of the material and vacuum, N is the number of molecules per unit volume (density), and α is the total polarizability, including electronic (α_e), distortion (α_d), and orientation (α_o) polarizabilities. According to Eq. (1), decreasing the total polarizability (α) and/or density (N) is the feasible method to effectively reduce the k value of a dielectric material. Reducing the polarizability can be achieved by the use of low-polar bonds (like C-C, C-H, Si-F, $Si\text{-}CH_3$, etc.). Based on the used type of the low-polar bond, the produced low-*k* dielectric material can be divided into two types: One type is organic polymer that contains saturated and unsaturated and conjugated and aromatic hydrocarbons [11]. However, this type low-*k* dielectric material is thermally unstable and has poor mechanical strength and relatively high coefficient of thermal expansion (CTE). As a result, the successful integration into the BEOL interconnects is still not achieved.

The other type is hybrid silica-based low-*k* dielectric material, which is the mainstream inter-layer-dielectric (ILD) insulator used in BEOL interconnects. This type of low-*k* dielectric material can be produced by doping fluorine or/and carbon into the traditional SiO_2 film. The formation of low-*k* dielectric materials are fluorinated silicon glass (FSG) [11, 12] or carbon-doped silicon glass [SiCOH or called organosilicate glass (OSG)] [11, 13]. Fluorine or carbon substitution lowers the *k* value by decreasing the polarizability and increasing the free volume.

The minimum *k* value of the hybrid silica-based low-*k* dielectric material is limited to be 2.6–2.7. To prevent a huge increase in the parasitic capacitance of BEOL interconnects in the 45 nm or below technology nodes, a new low-*k* dielectric material with *k* value less than 2.6 is required. The air has a minimum *k* value of ~1.0 in the world; as a result, the introduction of air pores in the existing low-*k* dielectric film is the possible strategy to further reduce the *k* value. The produced low-*k* dielectrics are porous, which are called "porous low-*k* dielectrics" [14, 15]. The *k* value of porous low-*k* dielectrics depends on the porosity and dielectric constant of the film skeleton (k_2) [16]:

$$\frac{k-1}{k+2} = V\frac{k_1-1}{k_1+2} + (1-V)\frac{k_2-1}{k_2+2} \tag{2}$$

where k_1 is the dielectric constant of the material inside the pores and *V* is the average pore volume. The first term in the right side of Eq. (2) equals to zero if the air is inside the pore ($k_1 \sim 1.0$). As a result, porous low-*k* dielectrics with relatively small k_2 value and higher porosity can provide much lower *k* value. Currently, porous low-*k* dielectrics have been successfully integrated into Cu interconnects since 45 nm technology node. The widely used method to produce the porous low-*k* dielectrics is co-deposition of a silica-like matrix together with a sacrificial organic polymer (porogen) using plasma-enhanced chemical vapor deposition (PECVD). Following, the sacrificial organic polymer in the deposited low-*k* dielectric material is removed by ultraviolet (UV)-assisted thermal curing at a temperature range of 300–450°C in order to form the pores in the film. The precise composition and porosity depend on the type of precursor molecules, the matrix/porogen ratio used during deposition, and the curing conditions [17, 18].

2.2 Deposition method for porous low-*k* materials

Porous low-*k* dielectric materials can be produced by either spin-on technology or chemical vapor deposition (CVD) method [14, 15, 17–20]. In the CVD method, the deposition rate of CVD method is strongly dependent of the deposition temperature. To obtain a suitable deposition rate, increasing the deposition temperature is required to deposit the porous low-*k* dielectric material. However, the temperature of BEOL interconnects is limited to be less than 450°C because of melting concern for metal conductors. With an assistant of plasma technology, the deposition precursors are dissociated to form the active radicals under the electron collision in the cold plasma. The generated active radicals with high reactivity accelerate the deposition process, thus reducing the deposition temperature.

2.2.1 Spin-on technology

Spin-on technology has been used in semiconductor processing for photoresist coating. It can also use to deposit the low-*k* dielectric material. The used dispensing liquid contains the deposition precursors for low-*k* materials, which is dropping into the center of the substrate. The created centrifugal forces by rotating of the substrate help to distribute the material on the surface. After the spinning step, a heating (or bake) is required to remove solvent. The temperature is typically below 250°C. Finally, a curing at temperatures varying from 350 to 600°C is required to obtain a stable film.

There are two methods to introduce the porosity into the film to produce porous low-*k* dielectric materials by spin-on technology. One is through sol–gel process, and the other is formed through the use of sacrificial particles (porogens) that are desorbed during the curing process. In the sol–gel process, the formation of subtractive porosity can be achieved by two approaches: the aging process and the hierarchical organization of the primary particles in the sol (self-assembly) [21, 22]. The other method is the use of sacrificial porogens, in which molecular or supramolecular particles are added in the low-*k* dielectric precursor with the purpose of tailoring the thermal stability. In the final curing process, these added molecular particles are removed by pyrolysis effect. The detailed description about spin-on technology to form porous low-*k* materials can be found elsewhere [23].

2.2.2 PECVD technology

PECVD is a complex process, involving a wide variety of scientific and technical principles, including gas-phase reaction chemistry, thermodynamics, heat and material transfer, fluid mechanics, surface and plasma reactions, thin film growth mechanism, and reactors engineering. During the deposition process, the active intermediates and structural units are formed in the gas phase and then absorbed in the solid substrate. Finally, they migrate and react to form the matrix of the growing layer [11].

In the current semiconductor industry, the production of the porous low-*k* dielectric material is relied on PECVD technology because the formation material is more thermally stable and the *k* value can be lower than 2.0. The subtractive porosity approach is the widely accepted method. In this method, a low-*k* (generally is SiCOH) skeleton precursor mixed with a porogen precursor is introduced into the reactor during the deposition. After the deposition, a dual-phase SiCOH-CH_x material is formed after the deposition. Tetramethylcyclotetrasiloxane (TMCTS), octamethylcyclotetrasiloxane (OMCTS), decamethylcyclopentasiloxane (DMCPS), and diethoxymethylsilane (DEMS) are the widely used skeleton precursors [24–27]. These skeleton precursors have a common property with a sufficiently low dissociation level under rf power in order to keep the sufficient hardness for the produced porous low-*k* dielectric material. The porogen precursor is organic molecule with sufficient volatility. Unsaturated cyclic hydrocarbons like terpinenes or norbornenes, linear alkenes, or molecules with strained rings like cycloalkene oxides or butadiene monoxide are the commonly used porogen precursors [11, 28].

Following, it is necessary to remove the labile organic fraction C_xH_y from the as-deposited SiCOH-C_xH_y film to form pores in the film. Thermal annealing, electron beam, or ultraviolet (UV) irradiation methods are provided to remove the labile organic fraction C_xH_y [29–31]. To reach better removal efficiency, it can be done by UV-assisted curing. However, the temperature of the curing has to be limited at ~400°C. The mechanical strength (elastic modulus and hardness) of the porous low-*k* dielectric material can also be improved by UV-assisted curing because the UV curing can rearrange and enhance the cross-linking of the skeleton of the low-*k* material by breaking a fraction of mainly the Si-CH_3 (Si-Me) bonds. The improvement effect is associated to the used wavelength, temperature, and time of the UV curing [32, 33].

Currently, a promising method to deposition of the porous low-*k* dielectric film is using a single precursor molecule consisting of skeleton with embedded (or grafted) porogen precursor. An example of such a porous SiCOH material is Applied Materials' Black Diamond 3 (BD3) dielectric film. The UV curing is also modified to create more uniform porosity and improve the mechanical properties [8, 33].

2.3 Characterizations of porous low-*k* dielectric materials

In order to successfully integrate the porous low-*k* dielectric material into Cu interconnects, their physical, chemical, mechanical, and electric properties are important consideration factors. **Table 1** lists the main characterization techniques for porous low-*k* dielectric materials. Detailed principles and operation procedures can be found elsewhere [34, 35].

Table 2 lists the main properties of porous low-*k* dielectric materials and compares to other generations of ILD materials (including SiO_2, FSG, and OSG) [36–38]. In addition to providing a lower *k* value, porous low-*k* dielectric materials possess the degrading material properties. The degradation is more pronounced with increasing porosity (for the reduction of *k* value) for porous low-*k* dielectric materials. Therefore, the use of porous low-*k* dielectric materials in the ICs is becoming more challenging.

Characteristics	Characterization techniques	Properties
Physical	Spectroscopic ellipsometry (SE)	Thickness, Refractive index
	X-ray reflectivity (XRR)	Density, Thickness
	transmission electron microscopy (TEM)	Microstrurure,Thickness
	Atomic force microscopy (AFM)	Surface roughness
Chemical	X-Ray Photoelectron Spectroscopy (XPS)	Composition, Chemical bonding
	Rutherford backscattering spectroscopy (RBS)	Composition
	Fourier transform infrared spectroscopy (FTIR)	Composition,Chemical bonding
	electron energy loss spectroscopy (EELS)	composition, Chemical bonding
	Nuclear magnetic resonance (NMR)	Composition, Bonding structure
Mechnaical	Nanoindentation.	Hardness, Elastic modulus
	sandwich structure four-point bending (SS4PB)	Adnesion
	modified-edge liftoff test (m-ELT)	Adnesion
	Ccross-section nanoindentation (CSN)	Adnesion
Pore structure	Positron annihilation lifetime spectroscopy (PALS,PAS)	Pore size
	Scattering techniques (SANS,SAXS)	Pore size, Porosity
	Adsorption techniques (BJH, EP)	Pore size, Porosity
Electrical	Mercury probe measurements	Dielectric constant
	Metal/dielectric/metal (MIM) or metal/dielectric/Si (MIS)	Dielectric constant, Leakage, TDDB
	Comb-comb or Comb-serpentine structures	Leakage, TDDB

Table 1.
*Characterization techniques for porous low-*k *dielectric materials.*

Properties	SiO_2	FSG	Dense low-*k* (OSG)	Porous low-*k*
Density (g/cm^3)	2.2	2.2	1.8~1.2	1.2~1.0
Dielectric constant (k)	4	3.5~3.8	2.8~3.2	1.9~2.7
Modulus (Gpa)	55~70	~50	10~20	3~10
Hardness (GPa)	3.5	3.36	2.5~1.2	0.3~1.0
CTE (ppm/K)	0.6	~0.6	1~5	10~18
Thermal Conductivity (W/mK)	1.0	1.0	~0.8	0.26
Porosity (%)	NA	NA	<10	25~50
Average Pore Size (nm)	NA	NA	<1.0	2.0~10
Breakdown Filed (MV/cm)	>10	>10	8~10	<8

Table 2.
Properties of various dielectric materials.

3. Integration of porous low-k dielectric materials in Cu interconnects

As Cu metallization replaced Al metallization in BEOL interconnects, the fabrication process was also switched to damascene approach from metal etching approach because the Cu etching formation compounds are hardly volatile at low temperature or the etch rate is relatively slow [39]. In the damascene pattering process, a dielectric is firstly etched, and then a Cu metallization is filled and polished. To prevent Cu diffusion and improve the adhesion with the dielectric layer, a barrier is required to surround the Cu wire [40, 41].

Dual-damascene patterning process is widely used to fabricate BEOL interconnects. In this method, both trench and via are patterned in a dielectric film simultaneously, and Cu metallization is filled into both trench and via. Compared to single-damascene patterning process, this method can reduce the processing step of Cu metallization. According to the order of via and trench pattering, dual-damascene patterning process has two types: "Via first" and "Trench first" processes [42, 43]. Generally, "Via first" dual-damascene process is widely used, plotted in **Figure 2**.

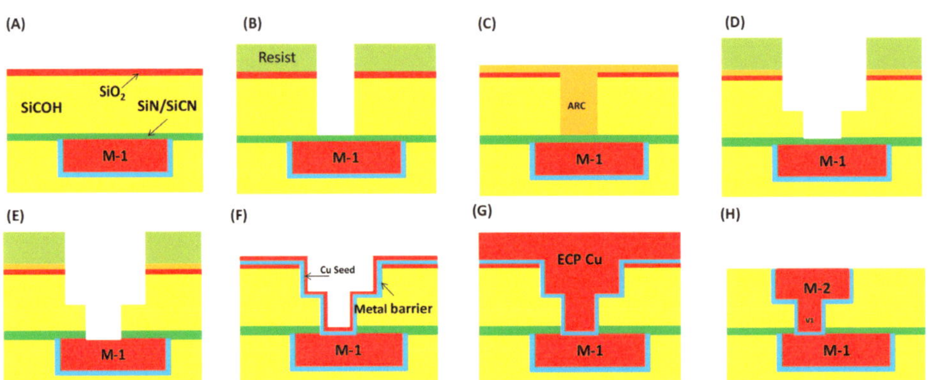

Figure 2.
Via first dual-damascene patterning process: (A) Dielectrics (SiN/SiCN, SiCOH, SiO_2) deposition. (B) Via-1 lithography and RIE. (C) ARC plug. (D) M-2 trench lithography and RIE. (E) Etching stop layer opening. (F) Metal barrier and Cu seed deposition. (G) Electroplating Cu deposition. (H) Cu CMP.

During the fabrication of BEOL interconnects, the used porous low-*k* dielectric material as an interconnecting insulator undergoes dielectric deposition, photoresist, etching, stripping, Cu metallization deposition, and chemical mechanical polishing (CMP) processes. Plasma damage, moisture/chemicals adsorption, Cu diffusion, and mechanical stress occurred on the porous low-*k* dielectric materials. These issues would reduce the electrical characteristics and reliability of the porous low-*k* dielectric materials. The mechanism and the resulting effect will be discussed in the following section.

In order to reduce the plasma-induced damage and pattern small features, the metal hardmask method and the multilayer resist method, as plotted in **Figures 3** and **4**, respectively, are proposed since 32 nm technology node [44–46]. In the metal hardmask process, the resist is stripped prior to the trench and via etching into the porous low-*k* ILD; therefore, resist-stripping process-induced damage can be minimal. However, the polymer may remain on the sidewalls of the trenches during the trench etching step. The remaining polymer must be removed without damaging the porous low-*k* dielectric material. Additionally, the stress in the metal layer must be minimized to avoid pattern deformation after the etching process. Metal residues can form on the etched surfaces and block etching of the porous low-*k* dielectric material.

In the advanced technology nodes, the multilayer resist method is preferred because it has an advantage to pattern small features. However, the porous low-*k* dielectric material is fully exposed to the resist strips. In order to avoid

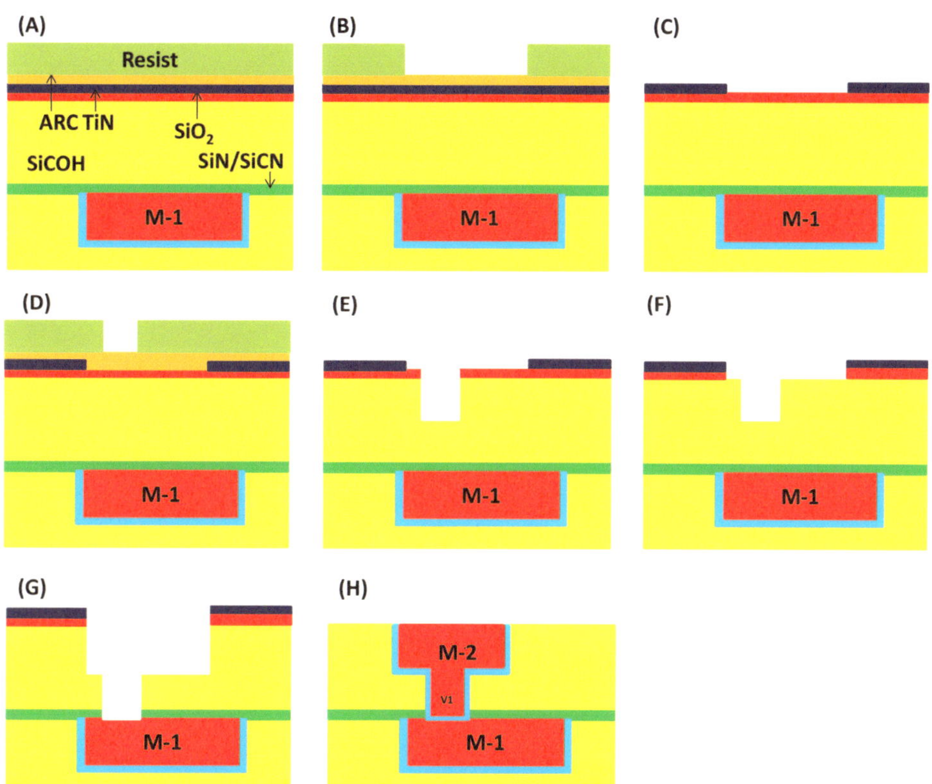

Figure 3.
Metal hardmask dual-damascene patterning process: (A) TiN, ARC, and resist deposition. (B) M-2 metal hardmask RIE. (C) M-2 trench lithography. (D) Via-1 lithography. (E) Via-1 RIE. (F) M-2 oxide hardmask RIE. (G) M-2/Via-1 RIE and M-1 capping layer RIE. (H) M-2/Via-1 Cu metallization.

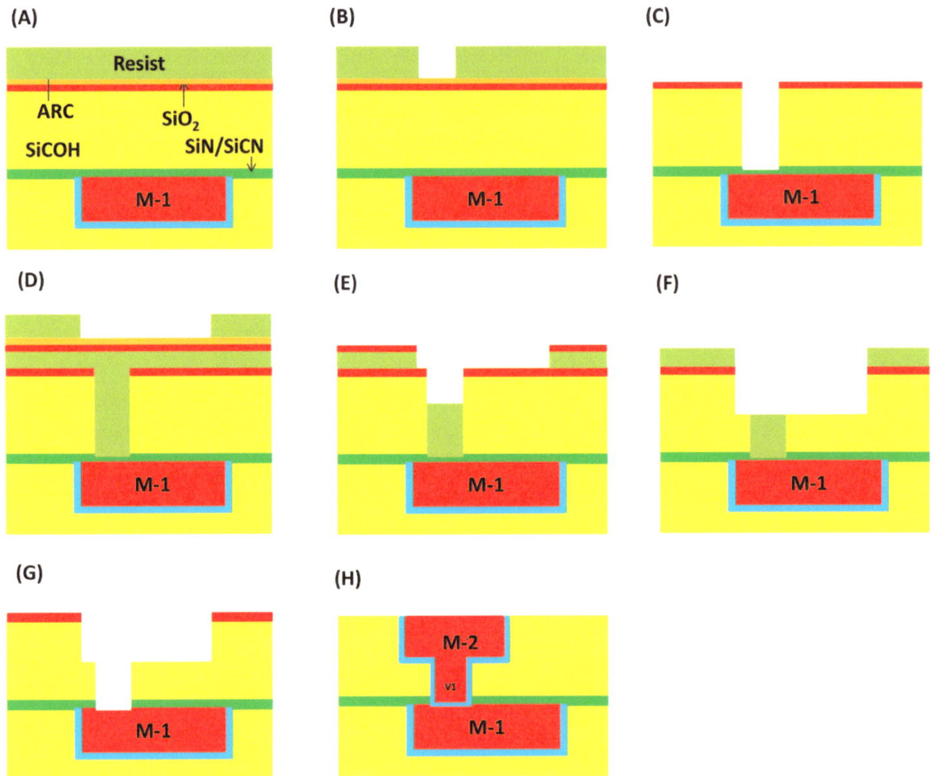

Figure 4.
Multilayer resist dual-damascene process: (A) ARC and resist coating. (B) Via-1 lithography. (C) Via-1 RIE. (D) Multilayer resist coating and M-2 trench lithography. (E) LTO and OPL RIE. (F) M-2 trench RIE. (G) OPL strip and M-1 capping layer RIE. (H) M-2/Via-1 Cu metallization.

plasma-induced damage on the porous low-*k* dielectric material, low-plasma-damage resist-stripping process is required for the multilayer resist method.

4. Electrical and reliability characteristics of porous low-*k* dielectric materials

As porous low-*k* dielectric materials are used in the BEOL interconnects, the change in the *k* value during the integration must be minimal. Additionally, the electrical properties and reliability are the most important concerns. As a result, the leakage current of the porous low-*k* dielectric between metal lines should be maintained low. The time-dependent dielectric breakdown (TDDB) failure time of the integrated BEOL structure at operating conditions should meet the specifications.

4.1 Conduction mechanisms in porous low-*k* dielectrics

In a crystalline solid, as the electrons overcome the bandgap (or called energy gap), the resulting current is detected. The bandgap is defined as the difference between the energy of the lowest conduction band and that of the highest valence band. For thermally deposited SiO_2 dielectric film, the bandgap is around 8.9 eV [47]. As carbon is doped into SiO_2 dielectric film to form SiOCH low-*k* dielectric material, the bandgap was determined to be between 8.0 and 10.0 eV, depending on the low-*k* dielectric types and the characterization techniques [48–50]. If the carbon content in the low-*k* dielectric film is not incorporated in the matrix network but primarily exists as terminal methyl groups, its bandgap is similar to that of SiO_2 film. However, if the carbon content is present in the network bonds by forming Si-C-Si bridging structure, the bandgap value would drop dramatically. As porosity is introduced into the SiOCH low-*k* dielectric material, the bandgap of porous SiOCH low-*k* dielectrics (k = 2.0–3.3) is in the range between 7.5 and 10 eV [51]. The effect of porosity on the bandgap of porous SiOCH low-*k* dielectrics is not pronounced. More investigation about bandgap determination for porous low-*k* dielectric materials is required.

The conduction mechanisms of low-*k* dielectric materials are commonly described by Schottky emission (SE), Poole-Frenkel (PF) emission, and Fowler-Nordheim (FN) tunneling [52–54], as shown in the following Eqs. (3)–(5):

- Schottky emission (SE)

$$\mathrm{J_{SE}} = \mathrm{A^*T^2} \exp\left[-q\left(\frac{\phi_{\mathrm{SE}} - \sqrt{qE/4\pi\varepsilon_0\varepsilon_r}}{kT}\right]\right. \tag{3}$$

- Poole-Frenkel (PF) emission

$$\mathrm{J_{SE}} \sim \mathrm{E} \exp\left[-q\left(\frac{\phi_{\mathrm{PF}} - \sqrt{qE/4\pi\varepsilon_0\varepsilon_r}}{kT}\right]\right. \tag{4}$$

- Fowler-Nordheim (FN) tunneling

$$\mathrm{J_{FN}} \sim \mathrm{E^2} \exp\left[\frac{-8\pi\sqrt{2m^*}(q\phi_{FN})^{3/2}}{3qhE}\right] \tag{5}$$

where J is current density, A^* is Richardson constant, T is temperature, q is the elementary charge, φ is barrier height, E is electric field, ε_o is permittivity of free space, ε_r is dielectric constant, m^* is effective electron mass, and h is Planck's constant.

SE and PF emissions are field-enhanced thermal excitation conduction models. The excited electrons enter the conduction band from the low-*k* interface and the trap states with coulomb potentials for SE and PF emissions, respectively. FN tunneling conduction is caused by electrons tunneling from the metal Fermi energy or trapping sites in the material itself into the low-*k* dielectric conduction band. SE and PF emission currents are associated with the field and temperature. The former exhibits a strong temperature dependency. However, FN tunneling current exhibits a strong field dependency and is independent of temperature. Generally, PF emission is more likely the dominant conduction mechanism in low-*k* dielectric materials, especially at low fields. At high field, the dominant conduction mechanism transfers to FN tunneling [55, 56].

In the integrated interconnects, the barrier height at both the low-*k*/metal and the low-*k*/Si interfaces is around 4 eV, and the barrier height at the etching-stop layer/metal interface is less than 2.0 eV [57]. Therefore, the interface-controlled SE emission occurs.

4.2 Reliability of porous low-*k* dielectric materials

The breakdown field and TDDB failure time are the main reliability items for a dielectric material [58, 59]. **Figure 5** plots the relatively breakdown field of various dielectric materials used as BEOL ILDs. Compared to other dielectric materials, the porous low-*k* dielectrics have relatively weak breakdown field, and the decreasing magnitude is amplified with increasing the porosity [60]. The pores in the porous low-*k* dielectrics are treated as defective cells, shortening the percolation path. Additionally, porous low-*k* dielectrics have weaker bonds, higher trap densities, or lower barrier heights at the metal–insulator interface.

TDDB testing is performed by applying an electric stress on a tested dielectric material for a period of time. The stressing field is lower than the breakdown field of the tested dielectric material. The leakage current is monitored with the stressing time. During the electric stress, electric damage occurs in a dielectric material, converting the resistance state of a dielectric material from high to low. This leads to the loss of the insulating properties for a dielectric material. As a conducting path between a dielectric is formed, the leakage sharply increases. Therefore, the dielectric breakdown occurs. This stressing time is defined as the breakdown time of a dielectric material.

TDDB is strongly related to the property of a tested dielectric film and the applied electric field. As a result, as the technology node advances to 45 nm or below technology nodes, TDBB is becoming a critical reliability issue. In addition to using porous low-*k* dielectrics with a lower breakdown field, the interconnect dimensions are reduced which increases the lateral electric field across the BEOL dielectric. However, in real Cu damascene interconnects, the integration performance strongly dominates TDDB results. The interface of Cu/capping layer, line-edge-roughness line-to-line overlay errors, and via-to-line misalignment are the dominated TDDB failure mechanisms [61–65].

Typically, TDDB testing is done at high fields (voltages) to accelerate the test. To predict lifetime from high voltage/field conditions to operating conditions,

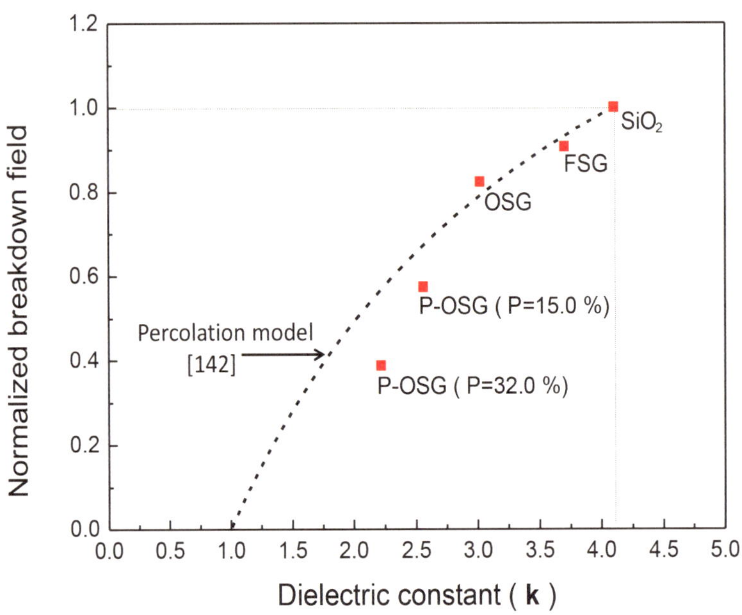

Figure 5.
Relative breakdown field of various dielectric films.

Model	Life-time	Mechanism
E	$A\exp(-\gamma E)$	Field-assisted bond breakage
1/E	$A\exp(\gamma/E)$	Fowler-Nordheim injection
$E^{1/2}$	$A\exp(-\gamma E^{1/2})$	Poole-Frenkel or Schottky
Power-law	AV^{-n}	Very thin dielectric (<3 nm)

A: Pre-exponential factor.
E: Applied electric field.
γ: Field acceleration factor.
n: Power law exponent.

Table 3.
TDDB lifetime models for dielectric materials.

TDDB lifetime model is required and critical for prediction. The commonly used TDDB lifetime models are summarized in **Table 3** [66–68]. Each TDDB lifetime model has its theoretical fundamentals, but cannot explain all observed TDDB phenomenon. Moreover, for the choice of TDDB lifetime model, it is necessary to consider that the breakdown mechanism under testing conditions is also the dominant mechanism under operating conditions.

In these used TDDB lifetime models, E, 1/E, and power-law models are field-driven models, while $E^{1/2}$ model is a current-driven model. Moreover, E model is the most conservative model because it gives the shortest dielectric lifetime in the lower-field conditions, and 1/E model is the optimistic model providing the longest predicted lifetime. The $E^{1/2}$ mode is widely accepted TDDB lifetime model for porous low-*k* dielectrics.

5. Integration issues of porous low-*k* dielectric materials

During the integration of porous low-*k* dielectrics into Cu interconnects, the fabricating processes can seriously degrade material properties, electrical characteristics, and reliability. Moreover, the porosity can act as a fast penetration media for reactive species or contamination during the integration, accelerating degradations.

The main key issues associated with porous low-*k* dielectrics are schematically shown in **Figure 6**. The key issues will be discussed and the improvement actions will be provided in this section.

5.1 Plasma-induced damage

Plasma is an aggressive medium which produces vacuum ultraviolet (VUV) and ultraviolet (UV) photons, energetic ions, electrons, and highly reactive radicals [69]. Exposure to plasma causes physical damage and chemical modifications on porous low-*k* dielectric materials [70, 71]. Under plasma irradiation, Si-CH_3 and Si-H groups in the porous SiCOH low-*k* dielectric material are extracted from the network and then converted into the Si-O or Si-OH groups, leading to densification

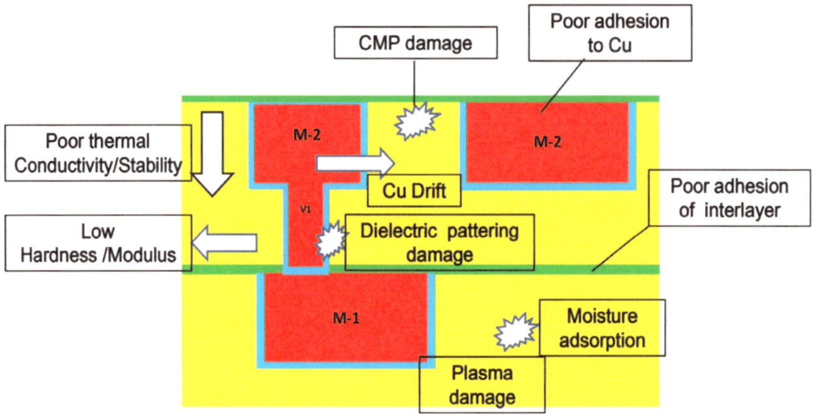

Figure 6.
*Main integration issues of porous low-*k *dielectrics in BEOL interconnects.*

and *k*-value increase. Moreover, plasma-induced damage makes porous low-*k* dielectric materials hydrophilic from hydrophobic, facilitating moisture uptake.

Plasma-induced damage on the porous low-*k* dielectric materials depends on the porosity, the used plasma reactors, power, and gas [72–76]. Therefore, for porous low-*k* dielectric materials that are irradiated under a plasma with higher density, inductively coupling plasma (ICP) reactor, or O_2 plasma, more damage on low-*k* dielectrics is expected.

To minimize the plasma-induced damage on the porous low-*k* dielectric materials, H_2-based plasma in remote-plasma (RP) system is an alternative for resist-stripping process. [77–81]. **Figure** 7(**a**) and (**b**) exhibits the breakdown field and TDDB failure time (TTF) of the porous low-*k* dielectric film after H_2/He plasma treatment [80, 81]. For porous low-*k* dielectric films operated in RP system, a higher breakdown field and a longer TTF were observed as compared to those operated in capacitance coupling plasma (CCP) system. In the RP system, neither deep UV light radiation nor ion bombardment is acted on the porous low-*k* dielectric film, mitigating plasma-induced damage. Additionally, the trends of temperature dependence of reliability characteristics are different for H_2/He plasma treatments in the CCP and RP systems. The breakdown field and TTF of H_2/He plasma-treated porous low-*k* dielectric film in CCP system decrease, while those in CCP system improve with increasing of the operation temperature. Moreover, as

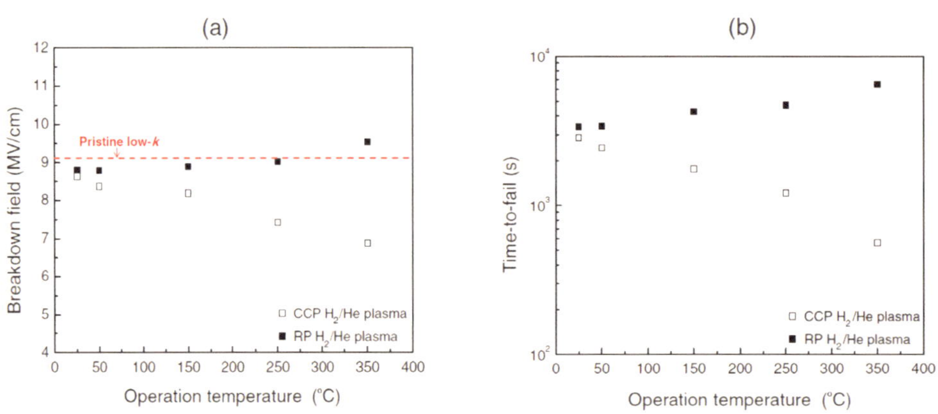

Figure 7.
*(a) Breakdown field. (b) Time-to-fail of H_2/He plasma-treated porous low-*k *dielectric films operated in CCP and RP systems as a function of operation temperature [81].*

the operation temperature of H_2/He plasma treatment in RP system is increased to 350°C, the plasma-treated porous low-*k* dielectric films have better reliability than the pristine samples. The improvement mechanism is attributed to the removal of carbon-based porogen residues from the porous low-*k* dielectric film by H_2/He plasma treatment at 350°C [82].

The dielectric property of the plasma-damaged low-*k* dielectrics can be recovered by applying silylation agents such as hexamethyldisilazane (HMDS), trimethylchlorosilane (TMCS), and dichlorodimethylsilane (DMDCS), depositing hydrophobic agents from hydrocarbon plasma and using a thermal treatment to eliminate the adsorbed hydroxyl (OH) groups and the physisorbed water [83–86].

5.2 Moisture uptake

During the integration processing, the porous low-*k* dielectric films are damaged and are transferred to be hydrophilic. The hydrophilic surface tends to uptake moisture in subsequent process steps. Due to a high *k* value of water (~80), only a small amount of moisture adsorption in the low-*k* dielectric film increases the effective *k* value significantly [87]. As the porosity increases in the porous low-*k* dielectric film, the pores connect each other to form "open pores," which serve as the fast diffusion path for moisture. The adsorbed moisture degrades reliability performance of porous low-*k* dielectric films, as shown in **Figure 8** [88]. The TDDB failure time is reduced by a factor of approximately 10 for the moisture-uptake low-*k* dielectric film and slightly decreases as the moisture immersion time increases. An annealing step is demonstrated to remove moisture and improve the film reliability, as also presented in **Figure 8**. However, even with thermal annealing at 400 C for 1 h, TDDB performance was only partially restored, being poorer than that of the fresh sample.

As the moisture is adsorbed in the low-*k* dielectric film, there are two types: physisorbed and chemisorbed moisture [89]. The physisorbed moisture starts to be

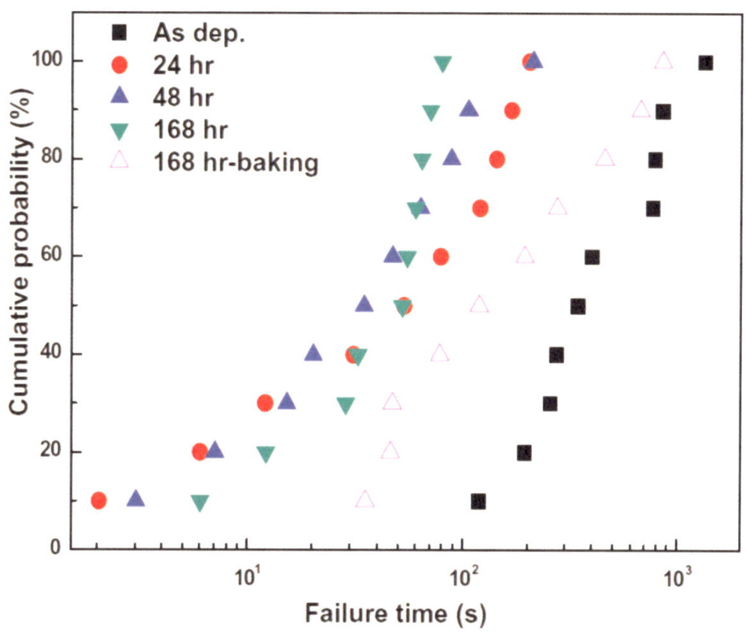

Figure 8.
Cumulative probability of TDDB failure times of porous low-k dielectric films as functions of the moisture immersion time [88].

desorbed at 190°C. After the 400°C annealing, most physically adsorbed moisture is desorbed. The chemisorbed moisture has the higher bonding energy; thus, it can be desorbed by a thermal annealing with the temperature above 600°C. As a result, the temperature of annealing is required to be elevated to 600–1000°C in order to remove the adsorbed water from porous low-*k* dielectric films. However, this temperature is not suitable to use in the BEOL interconnects because porous low-*k* dielectric films become unstable at temperature above 600°C.

To reach a better recovery for moisturized low-*k* dielectric films, a combination of UV curing and silylation process has been provided. UV curing and silylation processes can be done in the same chamber to save the processing step. The UV-assisted restoration is performed at elevated temperatures using a gaseous hydrocarbon in the curing ambient. The efficiency of recovery can be optimized with the process parameters, including UV wavelength and intensity, substrate temperature, UV curing time, chamber pressure, and reactant gas mixture [90, 91].

5.3 Cu drift

Due to a high diffusivity, Cu is easily oxidized to Cu mobile ion and then diffuses into ILDs under thermal and/or electrical bias [92, 93]. The diffused Cu ions could generate shallow energy levels in the bandgap of the porous low-*k* dielectric film [94]. These generated states act as defect centers, facilitating PF type conduction. Additionally, the penetration of Cu atoms or ions contributes to field enhancement locally inside the dielectric or at the electrode of electron injection [95]. These effects result in the significant degradation in the electric characteristics and reliability for the porous low-*k* dielectric films.

To prevent or minimize the diffusion of Cu ions and Cu barriers, including metal and dielectric barriers, are required for Cu metallization. **Figure 9** plots the Cu ion

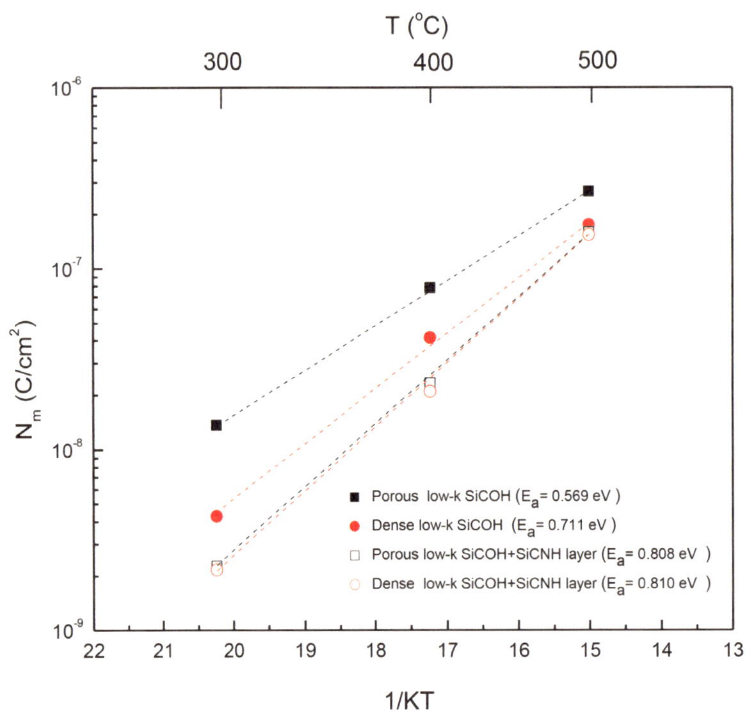

Figure 9.
*Cu ion concentration in dense and porous low-*k *SiOCH films with and without capping SiCNH layer after annealing as function of temperature [96].*

concentration $N_m(T)$ in the various low-*k* dielectric films after thermal stress as a function of annealing temperature [96]. The Cu penetration is enhanced at increased temperatures. The larger Cu ion concentration in the porous low-*k* dielectric film after annealing indicates that the pores in the low-*k* dielectric film induced the rapid migration of Cu ions. Additionally, the porous low-*k* dielectric film had the lowest activation energy (0.57 eV) with a value close to those reported elsewhere (0.42–0.60 eV) [97, 98]. The SiCNH capping layers on the low-*k* dielectric films increased the activation energy to ~0.81 eV for both dense and porous low-*k* films, suggesting that the SiCNH capping layer acts as a Cu barrier and prevents possible Cu migration. The use of SiCNH capping layer as a Cu barrier increases the effective *k* value of BEOL ILD, being a main concern.

The deposition of metal barrier can also prevent Cu migration. However, due to a high resistivity of metal barrier, the overall resistivity of the metal line significantly increases in the scaling interconnect pitch. Additionally, barrier metals like tantalum (Ta) deposited by physical vapor deposition penetrate into low-*k* dielectric in a way similar to Cu, causing low-*k* dielectric degradation. Moreover, the metal barrier-induced damage increases as the porosity of the low-*k* dielectric increases [99, 100].

Currently, self-forming barrier [101], atomic layer deposition (ALD) barrier [102], and self-assembled monolayer (SAM) [103, 104] processes are promising methods to prevent metal penetration. However, the integration with the porous low-*k* dielectric must be controlled precisely to meet all requirements.

5.4 CMP-induced damage

The purpose of chemical mechanical polishing (CMP) is to produce planarization topography by means of both mechanical polishing and chemical reaction. A simultaneous interaction between polishing slurry, a semiconductor wafer, and a polyurethane pad occurred. Thus, the chemical, mechanical, and material properties of the pad, wafer surface, and slurry determine the controllability and quality of CMP process.

In Cu metallization, CMP process is used to remove the excess Cu film and the barrier metal. There are three main steps in Cu CMP process. Firstly, the excess Cu film is polished. Then, as reaching the interface, both metal barrier and Cu film are polished. Finally, to ensure that all metals are removed from the field regions in all parts of the wafer, over-polishing in the last step is necessary. Thus, the used dielectric insulator is polished simultaneously. To reach high degree of planarization and avoid Cu dishing, dielectric erosion, and interface quality degradation (dangling bonds, generation, metal contaminants, and moisture presence), precise control CMP process is required [105, 106].

As the porous low-*k* dielectric film is used as an interconnecting insulator, peeling, delamination, and cracking may occur under CMP process because it has not enough mechanical strength to survive the large mechanical stress process. Therefore, improving the elastic modulus or hardness of the porous low-*k* dielectric film is required. **Figure 10** shows the change in the hardness of porous low-*k* dielectric materials as a function of UV curing time [107, 108]. By increasing UV curing time after the porous low-*k* dielectric film deposition, the hardness (H) can be improved. Moreover, CMP-induced peeling was checked to determine the minimum hardness for integration of the porous low-*k* dielectric film into BEOL interconnects. At a UV curing time of less than 300 s for the porous low- *k* dielectric films, peeling was observed. Peeling was worse at shorter UV curing times. As UV curing time is greater than 300 s, the wafer exhibited peeling-free for the porous

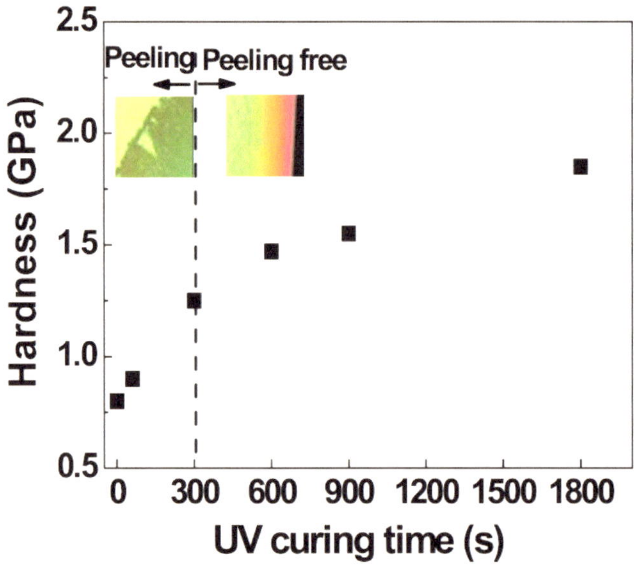

Figure 10.
*Hardness of porous low-*k *dielectric materials as a function of UV curing time [107].*

low-*k* dielectric films, indicating that the minimum hardness for integration of the porous low-*k* dielectric film into BEOL interconnects is 1.2 GPa.

The other problem of Cu CMP problem is that the V-shape corners in the porous low-*k* trenches are formed due to the higher mechanical force. This would become a potential critical path for porous low-*k* dielectric breakdown owing to field enhancement along the CMP interface.

6. Conclusions

To improve the performance of ICs, porous low-*k* dielectric materials have been used as an interconnecting insulator for providing lower parasitic capacitance between the wires to reduce RC time delay. Porous low-*k* dielectric materials can be achieved by introducing low-polarizability chemical bonds and porosity into the film. During the integration, the semiconductor processing induces damage on the porous low-*k* dielectric material, making the dielectric material densification hydrophilic, facilitating moisture uptake, and inducing Cu and barrier metal penetration. These lead to *k* value increase and reliability degradation for the porous low-*k* dielectric material. Moreover, high porosity and large pore size in the porous low-*k* dielectric materials make them sensitive to integration-induced damages. Moreover, porosity in the low-*k* dielectric material weakens the hardness and enhances the local field of the film, resulting in CMP damage and reliability challenges. Therefore, in order to achieve a successful implementation of advanced porous low-*k* dielectric films in the future BEOL interconnects, optimization and innovation of material science and integration processing are needed.

DOI: http://dx.doi.org/10.5772/intechopen.81577

Author details

Yi-Lung Cheng* and Chih-Yen Lee
Department of Electrical Engineering, National Chi-Nan University, Taiwan, ROC

*Address all correspondence to: yjcheng@ncnu.edu.tw

References

[1] The International Technology Roadmap for Semiconductor; ITRS 1997

[2] Thompson SE, Parthasarathy S. Moore's law: The future of Si microelectronics. Materials Today. 2006;**9**:20-25

[3] Isaac RD. The future of CMOS technology. IBM Journal of Research and Development. 2000;**44**:369-378

[4] Bohr M. MOS transistors: Scaling and performance trends. Semiconductor International. 1995;**18**(6):75-80

[5] Dixit GA, Havemann RH. Overview of Interconnect—Copper and Low-*k* Integration, Handbook of Semiconductor Manufacturing Technology. 2nd ed. NY: CRC Press; 2008. (Chapter 2)

[6] Jeffery G. Process Technology for Copper Interconnects, Handbook of Thin Film Deposition. 3rd ed. William Andrew; 2012. pp. 221-269

[7] Havemann RH, Hutchby JA. High-performance interconnects: An integration overview. Proceedings of the IEEE. 2001;**89**:586-601

[8] Grill A, Gates SM, Ryan TE, Nguyen SV, Priyadarshini D. Progress in the development and understanding of advanced low *k* and ultralow *k* dielectrics for very large-scale integrated interconnects–State of the art. Applied Physics Reviews. 2014;**1**: 011306-011312

[9] Grill A. Porous pSiCOH ultralow-*k* dielectrics for chip interconnects prepared by PECVD. Annual Review of Materials Research. 2009;**39**:49-69

[10] Grill A. Plasma enhanced chemical vapor deposited SiCOH dielectrics: From low-*k* to extreme low-*k* interconnect materials. Journal of Applied Physics. 2003;**93**:1785-1790

[11] Maex K, Baklanov MR, Shamiryan D, Iacopi F, Brongersma SH, Yanovitskaya ZS. Low dielectric constant materials for microelectronics. Journal of Applied Physics. 2003;**93**: 8793-8841

[12] Cheng YL, Wang YL, Liu CW, Wu YL, Lo KY, Liu CP, Lan JK. Characterization and reliability of low dielectric constant fluorosilicate glass and silicon rich oxide process for deep sub-micro device application. Thin Solid Films. 2001;**398–399**:544-548

[13] Shapiro MJ, Nguyen SV, Matsuda T, Dobuzinsky D. CVD of fluorosilicate glass for ULSI applications. Thin Solid Films. 1995;**270**:503-507

[14] Chang YM, Chang WY, Leu J, Cheng YL. Effect of thermal treatment on physical, electrical properties and reliability of porogen-containing and porogen-free ultralow-*k* dielectrics. Thin Solid Films. 2013;**528**:67-71

[15] Broussous L, Berthout G, Rebiscoul D, Rouessac V, Ayral A. Mechanical properties of a plasma-modified porous low-*k* material. Microelectronic Engineering. 2010;**87**:466-469

[16] Talebian E, Talebian M. A general review on the derivation of Clausius–Mossotti relation. Optik. 2013;**124**: 2324-2326

[17] Kim CY, Navamathavan R, Lee HS, Woo JK, Hyun MT, Lee KM, et al. Ultraviolet irradiation effect on the properties of leakage current and dielectric break-down plasma damage on low-*k* dielectric materials 23 of low-dielectric-constant SiOC(H) films using comb capacitor structure. Thin Solid Films. 2011;**519**:6732-6736

[18] Chapelon LL, Arnal V, Broekaart M, Gosset LG, Vitiello J, Torres J. Porous ultra low *k* deposited by PECVD: From deposition to material properties. Microelectronic Engineering. 2004;**76**:1-4

[19] Vanstreels K, Ciofi I, Barbarin Y, Baklanov M. Influence of porosity on dielectric breakdown of ultralow-*k* dielectrics. Journal of Vacuum Science and Technology B. 2013;**31**:050604-1-050604-5

[20] Jousseaume V, Zenasni A, Gourhant O, Favennec L, Baklanov MR. In: Baklanov MR, Ho P, Zschech E, editors. Ultra-low-*k* by CVD: Deposition and Curing. Advanced Interconnects for ULSI Technology. Wiley; 2012. pp. 35-77. (Chapter 2)

[21] Sun Y, Negreira AR, Meersschaut J, Hoflijk I, Vaesen I, Conard T, et al. Optimization and upscaling of spin coating with organosilane monolayer for low-k pore sealing. Microelectronic Engineering. 2017;**167**:32-36

[22] Aw KC, Salim NT, Gao W, Li Z. Characterization of spin-on-glass very-low-k polymethylsiloxane with copper metallization. Thin Solid Films 2006; **504**:243-247

[23] Shamiryan D, Abell T, Iacopi F, Maex K, Low-k dielectric materials. Materials Today. 2004;**7**:34-39

[24] Lubguban J Jr, Rajagopalan T, Mehta N, Lahlouh B, Simon SL, Gangopadhyaya S. Low-*k* organosilicate films prepared by tetravinyltetramethylcyclotetrasiloxane. Journal of Applied Physics. 2002;**92**: 1033-1038

[25] Gates SM, Neumayer DA, Sherwood MH, Grill A, Wang X, Sankarapandian M. Preparation and structure of porous dielectrics by plasma enhanced chemical vapor deposition. Journal of Applied Physics. 2007;**101**:094103-1-094103-8

[26] Smirnov E, Ferchichi AK, Huffman C, Baklanov MR. Impact of plasma exposure on organic low-*k* materials. Proceedings of SPIE. 2010;**75217**: 52107-1-52107-8

[27] Uchida Y, Katoh T, Oikawa M. Characterization of low-k porous silica films incorporated with alkylene groups. Materials Science in Semiconductor Processing. 2002;**5**:259-264

[28] Ming Z, Beichao Z. Preparation of porous ultra low *k* films using different sacrificial porogen precursors for 28 nM technological node. Materials Science in Semiconductor Processing. 2015;**36**: 170-178

[29] Chang TC, Tsai TM, Liu PT, Chen CW, Tseng TY. Study on the effect of electron beam curing on low-K porous organosilicate glass (OSG) material. Thin Solid Films. 2004;**469–470**: 383-387

[30] Kemeling N, Matsushita K, Tsuji N, Kagami KI, Kato M, Kaneko S, Sprey H, Roest DD, Kobayashi N. A robust *k* ~2.3 SiCOH low-*k* film formed by porogen removal with UV-cure. Microelectronic Engineering. 2007;**84**: 2575-2581

[31] Dultsev FN, Nekrasov DV. Transformation of porous structure under vacuum ultraviolet irradiation of the films based on silicon dioxide. Thin Solid Films. 2016;**603**:249-254

[32] Prager L, Marsik P, Wennrich L, Baklanov MR, Naumov S, Pistol L, et al. Effect of pressure on efficiency of UV curing of CVD-derived low-k material at different wavelengths. Microelectronic Engineering. 2008;**85**: 2094-2097

[33] Producer® Black Diamond® PECVD-Applied Materials: http://www.appliedmaterials.com/zh-hant/products/producer-black-diamond-pecvd

[34] Marsik P, Urbanowicz AM, Verdonck P, Roest DD, Sprey H, Baklanov MR. Effect of ultraviolet curing wavelength on low-k dielectric material properties and plasma damage resistance. Thin Solid Films 2011;**519**: 3619-3626

[35] Pantouvaki M, Huffman C, Zhao L, Heylen N, Ono Y, Nakajima M, Nakatani K, Beyer GP, Baklanov MR. Advanced organic polymer for the aggressive scaling of low-*k* materials. Japanese Journal of Applied Physics. 2011;**50**:04DB01-1-04DB01-5

[36] Baklanov MR, Vanhaelemeersch S, Bender H, Maex K. Effects of oxygen and fluorine on the dry etch characteristics of organic low-*k* dielectrics. Journal of Vacuum Science and Technology B. 1999;**17**:372-379

[37] Grill A, Neumayer DA. Structure of low dielectric constant to extreme low dielectric constant SiCOH films: Fourier transform infrared spectroscopy characterization. Journal of Applied Physics. 2003;**94**(10):6697-6707

[38] Chang HL, Kuo CT, Liang MS, Mechanical properties and fracture mechanism of porous SiOCH low-k dielectrics. Microelectron. Eng. 2011; **88**(7):1623-1627

[39] Havemann RH, Antonelli GA, Arendt GK, Danek M, McKerrow AJ, Weinberg RS. Copper BEOL solutions for advanced memory. Solid-State Technology. 2009;**52**:10-13

[40] Kim SW. Dielectric barrier, etch stop, and metal capping materials for state of the art and beyond metal interconnects. Journal of Solid State Science and Technology. 2015;**4**(1): N3029-N3047

[41] Cheng YL, Chiu TJ, Wei BJ, Wang HJ, Wu J, Wang YL. Effect of copper barrier dielectric deposition process on characterization of copper interconnects. Journal of Vacuum Science & Technology B. 2010;**28**(3): 567-572

[42] Kriz J, Angelkort C, Czekalla M, Huth S, Meinhold D, Pohl A, et al. Overview of dual damascene integration schemes in Cu BEOL integration. Microelectronic Engineering. 2008;**85**: 2128-2132

[43] Takei S. Resist poisoning studies of gap fill materials for patterning metal trenches in via-first dual damascene process. Japanese Journal of Applied Physics. 2008;**47**:8766-8770

[44] Jeffery G. Process technology for copper interconnects. In: Handbook of Thin Film Deposition; 2012. pp. 221-269

[45] Hamioud K, Arnal V, Farcy A, Jousseaume V, Zenasni A, Icard B, et al. 32 nm node BEOL integration with an extreme low-k porous SiOCH dielectric k=2.3. Microelectronic Engineering. 2010;**87**:316-320

[46] Gambino J, Chen F, He J. Copper interconnect technology for the 32 nm node and beyond. In: IEEE Custom Integrated Circuits Conference Proceedings. PA: Warrendale; 2009. pp. 141-148

[47] Sze SM. Physics of Semiconductor Devices. 2nd ed. New York: Wiley; 1981

[48] King S, French B, Mays E. Detection of defect states in low-k dielectrics using reflection electron energy loss spectroscopy. Journal of Applied Physics. 2013;**113**:044109

[49] Nichols M, Li W, Pei D, Antonelli G, Lin Q, Banna S, Nishi Y, Shohet J. Measurement of bandgap energies in low-k organosilicates. Journal of Applied Physics. 2014;**115**:094105-1-094105-4

[50] Wu C, Li Y, Baklanov MR, Croes K. Electrical reliability challenges of

advanced low-k dielectrics. ECS Journal of Solid State Science and Technology. 2015;**4**(1):N3065-N3070

[51] Zheng H, King S, Ryan V, Nishi Y, Shohet J. Bandgap measurements of low-k porous organosilicate dielectrics using vacuum ultraviolet irradiation. Applied Physics Letters. 2014;**104**: 062904

[52] Vilmay M, Roy D, Volpi F, Chaix JM. Characterization of low-*k* SiOCH dielectric for 45 nm technology and link between the dominant leakage path and the breakdown localization. Microelectronic Engineering. 2008;**85**: 2075-2078

[53] Chen F, Shinosky M. Addressing Cu/low-k dielectric TDDB-reliability challenges for advanced CMOS technologies. IEEE Transactions on Electron Devices. 2009;**56**(1): 2-12

[54] Lloyd JR, Liniger E, Shaw TM. Charge transport model to predict intrinsic reliability for dielectric materials. Journal of Applied Physics. 2005;**98**:084109

[55] Wu C, Li Y, Barbarin Y, Ciofi I, Croes K, Bömmels J, Wolf ID, Tökei Zs. Correlation between field dependent electrical conduction and dielectric breakdown in a SiCOH based low-k (k=2.0) dielectric. Applied Physics Letters. 2013;**103**:032904

[56] Ogawa ET, Kim J, Haase GS, Mogul HC, McPherson IW. Leakage, breakdown, and TDDB characteristics of porous low-k silica-based interconnect dielectrics. In: IEEE International Reliability Physics Symposium Proceedings; 2003. pp. 166-172

[57] Chen F, Shinosky M. Electrical reliability challenges of advanced low-k dielectrics. Microelectronics Reliability. 2014;**54**:529

[58] Nichols MT, Sinha H, Wiltbank CA, Antonelli GA, Nishi Y, Shohet JL. Time-dependent dielectric breakdown of plasma-exposed porous organosilicate glass. Applied Physics Letters. 2012;**100**: 112905-1-112905-4

[59] Chen F, Bravo O, Harmon D, Shinosky M, Aitken J. Cu/low-k dielectric TDDB reliability issues for advanced CMOS technologies. Microelectronics and Reliability. 2008; **48**:1375-1383

[60] Liv J, Gan D, Hu C, Kiene M, Hu PS, Volksen W, et al. Porosity effect on the dielectric constant and thermomechanical properties of organosilicate films. Applied Physics Letters. 2002;**81**:4180-4182

[61] Noguchi J. Dominant factors in TDDB degradation of Cu interconnects. IEEE Transactions on Electron Devices. 2005;**52**(8):1743-1750

[62] Ueno K, Kameyama A, Matsumoto A, Iguchi M, Takewaki T, Oshida D, et al. Timedependent dielectric breakdown characterization of 90-and 65-nm-node Cu/SiOC interconnects with via plugs. Japanese Journal of Applied Physics. 2007;**46**:1444-1451

[63] Tőkei Z, Croes K, Beyer GP. Reliability of copper low-k interconnects. Microelectronic Engineering. 2010;**87**(3):348-354

[64] Lloyd JR, Murray CE, Ponoth S, Cohen S, Liniger E. The effect of Cu diffusion on the TDDB behavior in a low-k interlevel dielectrics. Microelectronics and Reliability. 2006; **46**:1643-1647

[65] Yamada Y, Konishi N, Noguchi J, Jimbo T. Influence of CMP slurries and post-CMP cleaning solutions on Cu interconnects and TDDB reliability. Journal of the Electrochemical Society. 2008;**155**(7):H485-H490

[66] Haase GS, McPherson JW. Modeling of interconnect dielectric lifetime under stress conditions and new extrapolation methodologies for time-dependent dielectric breakdown. In: IEEE International Reliability Physics Symposium Proceedings. 2007. pp. 390-398

[67] Lloyd JR, Liniger E, Shaw TM. Simple model for time-dependent dielectric breakdown in inter- and intralevel low-k dielectrics. Journal of Applied Physics. 2005;**98**:084109

[68] McPherson JW. Time dependent dielectric breakdown physics–models revisited. Microelectronics and Reliability. 2012;**52**:1753-1760

[69] Baklanov MR, de Marneffe J-F, Shamiryan D, Urbanowicz AM, Shi H, Rakhimova TV, Huang H, Ho PS. Plasma processing of low-*k* dielectrics. Journal of Applied Physics. 2013;**113**: 041101-1-041101-35

[70] Kazi H, Kelber JA. Plasma damage mechanisms in low *k* organosilicate glass and their inhibition by Ar ion bombardment. Journal of Vacuum Science and Technology B. 2010;**32**(2): 021302-1-021302-7

[71] Kunnen E, Barkema GT, Maes C, Shamiryan D, Urbanowicz A, Struyf H, et al. Integrated diffusion–recombination model for describing the logarithmic time dependence of plasma damage in porous low-*k* materials. Microelectronic Engineering. 2011;**88**: 631-634

[72] Perret A, Chabert P, Jolly J, Booth J-P. Ion energy uniformity in high-frequency capacitive discharges. Applied Physics Letters. 2005;**86**: 021501-1-021501-3

[73] Rakhimova TV, Braginsky OV, Ivanov VV, Kim TK, Kong JT, Kovalev AS, et al. Experimental and theoretical study of rf plasma at low and high frequency. IEEE Transactions on Plasma Science. 2006;**34**:867-877

[74] Cheng YL, Lin BH, Huang SW. Effect of O2 plasma treatment on physical, electrical, and reliability characteristics of low dielectric constant materials. Thin Solid Films. 2014;**572**: 44-50

[75] Worsley MA, Bent SF, Gates SM, Fuller NCM, Volksen W, Steen M, et al. Effect of plasma interactions with low-*k* films as a function of porosity, plasma, chemistry, and temperature. Journal of Vacuum Science and Technology B. 2005;**23**(2):395-405

[76] Lionti K, Volksen W, Magbitang T, Darnon M, Dubois G. Toward successful integration of porous low-*k* materials: Strategies addressing plasma damage. ECS Journal of Solid State Science and Technology. 2015;**4**(1):N3071-N3083

[77] Amanatides E, Mataras D. Frequency variation under constant power conditions in hydrogen radio frequency discharges. Journal of Applied Physics. 2001;**89**:1556-1566

[78] Han QY, White B, Berry IL, Waldfried C, Escorcia O. Activated He: H2 strip of photoresist over porous low-*k* materials. Solid State Phenomena. 2005;**103–104**:341-345

[79] Urbanowicz AM, Shamiryan D, Zaka A, Verdonck P, De Gendt S, Baklanov MR. Effects of He plasma pretreatment on low-*k* damage during Cu surface cleaning with NH_3 plasma. Journal of the Electrochemical Society. 2010;**157**(5):H565-H573

[80] Cheng YL, Huang JF, Chang YM, Leu J. Impact of plasma treatment on structure and electrical properties of porous low dielectric constant SiCOH material. Thin Solid Films. 2013;**544**:537-540

[81] Cheng YL, Tsai YS, Hung WJ, Sun CR, Lee WH. Effect of H_2/He plasma on

porous low dielectric constant materials. Surface and Coatings Technology. 2016; **38**:182-188

[82] Urbanowicz AM, Vanstreels K, Shamiryan D, De Gendt S, Baklanov MR. Effect of porogen residue on chemical, optical, and mechanical properties of CVD SiCOH low-*k* materials. Electrochemical and Solid-State Letters. 2009;**12**:H292-H295

[83] Urbanowicz AM, Baklanov MR, Heijlen J, Travaly Y, Cockburn A. Damage reduction and sealing of low-*k* films by combined he and NH3 plasma treatment. ECS Journal of Solid State Science and Technology. 2007;**10**(10): G76-G79

[84] Impens NREN, Voort PVD, Vansant EF. Silylation of micro-, meso- and non-porous oxides: Review. Microporous and Mesoporous Materials. 1999;**28**:217-232

[85] Forster A, Wagner C, Schuster J, Gemming S. Theoretical investigation of in situ k-restore processes for damaged ultra-low-k dielectrics. Microelectronic Engineering. 2016;**156**:121-125

[86] Koehler N, Fisher T, Zimmermann S, Schulz SE. A plasma assisted in situ restoration processes for sidewall damaged ULK dielectrics. Microelectronic Engineering. 2016;**156**:116-120

[87] Proost J, Baklanov M, Maex K, Delaey I. Compensation effect during water desorption from siloxane-based spin-on dielectric thin films. Journal of Vacuum Science and Technology B. 2000;**18**(1):303

[88] Cheng YL, Leon KW, Huang JF, Chang WY, Chang YM, Leu J. Effect of moisture on electrical and reliability properties for low dielectric constant materials. Microelectronic Engineering. 2014;**114**:12-16

[89] Raja A, Laibowitz R, Liniger EG, Shaw TM, Heinz TF. Impedance spectroscopy studies of moisture uptake in low-k dielectrics and its relation to reliability. Microelectronic Engineering. 2015;**147**:100-103

[90] Gandhi DD, Singh B, Singh AP, Moore R, Simonyi E, Lane MW, et al. Effects of silylation on fracture and mechanical properties of mesoporous silica films interfaced with copper. Journal of Applied Physics. 2009;**106**: 054502

[91] Kimura Y, Ishikawa D, Nakano A, Kobayashi A, Matsushita K, Roest DD, Kobayashi N. Impact of hydrocarbon control in ultraviolet assisted restoration process for extremely porous plasma enhanced chemical vapor deposition SiOCH films with k =2.0. Japanese Journal of Applied Physics. 2012;**51**: 05EC04

[92] Cui H, Bhat IB, Murarka SP, Lu H, Hsia WJ, Catabay W. Copper drift in methyl-doped silicon oxide film. Journal of Vacuum Science and Technology B. 2002;**20**(5):1987-1993

[93] Cheng YL, Lee CY, Huang YL, Sun CR, Lee WH, Chen GS, et al. Cu-induced dielectric breakdown of porous low dielectric constant film. Journal of Electronic Materials. 2017;**46**(6): 3627-3633

[94] Lee S, Oates A, Chang K. Fundamental understanding of porous low-k dielectric breakdown. IEEE International Reliability Physics Symposium (IRPS);2009. pp. 481-484

[95] Chen F, Shinosky MA. Electron fluence driven, Cu catalyzed, interface breakdown mechanism for BEOL low-k time dependent dielectric breakdown. Microelectronics and Reliability. 2014; **54**(3):529-540

[96] Cheng YL, Lee CY, Hung WJ, Chen GS, Fang JS. Electrical and reliability characteristics of dielectric stack with low dielectric constant SiCOH and

capping SiCNH films. Surface & Coatings Technology. 2018;**350**:57-63

[97] Fisher I, Eizenberg M. Copper ion diffusion in porous and nonporous SiO_2-based dielectrics using bias thermal stress and thermal stress tests. Thin Solid Films. 2008;**516**:4111-4121

[98] Mukaigawa S, Aoki T, Shimizu Y, Kikkawa T. Measurement of copper drift in methylsilsesquiazane-methylsilsesquioxane dielectric films. Japanese Journal of Applied Physics. 2000;**39**:2189-2193

[99] Tan T, Gan C, Du A, Cheng C. Effect of Ta migration from sidewall barrier on leakage current in Cu/SiOCH low-*k* dielectrics. Journal of Applied Physics. 2009;**106**:043517

[100] Zhao L, Volders H, Baklanov M, Tokei Z, Pantouvaki M, Wilson CJ, et al. Study of metal barrier deposition-induced damage to porous low-k materials. Microelectronics Engineering. 2011;**88**:3030-3033

[101] Franz M, Ecke R, Kaufmann C, Kriz J, Schulz SE. Characterisation of the barrier formation process of self-forming barriers with CuMn, CuTi and CuZr alloys. Microelectronics Engineering. 2016;**156**:65-69

[102] Yang CC, Cohen, Shaw ST, Wang PC, Nogami T, Edelstein D. Characterization of ultrathin-Cu/Ru (Ta)/TaN liner stack for copper interconnects, IEEE Electron Device Letter. 2010;**31**:722-724

[103] Uedono A, Armini S, Zhang Y, Kakizaki T, Rehberg RK, Anwand W, et al. Surface sealing using self-assembled monolayers and its effect on metal diffusion in porous low-k dielectric studied using monoenergetic positron beams. Applied Surface Science. 2016;**368**:272-276

[104] Boyd AB, O'Connor R, Armini S, Selvaraju V, Hughes G, Bogan J. On the use of (3-trimethoxysilylpropyl) diethylenetriamine self-assembled monolayers as seed layers for the growth of Mn based copper diffusion barrier layers. Applied Surface Science. 2018; **427**:260-266

[105] Chapelon LL, Chaabouni H, Imbert G, Brun P, Mellier M, Hamioud K, et al. Dense SiOC cap for damage-less ultra low k integration with direct CMP in C45 architecture and beyond. Microelectronic Engineering. 2008;**85**: 2098-2101

[106] Ueki M, Onodera T, Ishikawa A, Hoshino S, Hayashi Y. Defectless monolithic low-k/Cu interconnects produced by chemically controlled chemical mechanical polishing process with in situ end-point-detection technique. Japanese Journal of Applied Physics. 2010;**49**:04C029-1-04C029-6

[107] Kao KC, Chang WY, Chang YM, Leu J, Cheng YL. Effect of UV curing time on physical and electrical properties and reliability of low dielectric constant materials. Journal of Vacuum Science and Technology A. 2014;**32**(3):061514-1-061514-7

[108] Cheng YL, Wang YL, Lan JK, Chen HC, Lin JH, Wu YL, Liu PT, Wu YC, Feng MS. Effect of carrier gas on the structure and electrical properties of low dielectric constant SiCOH film using trimethylsilane prepared by plasma enhanced chemical vapor deposition. Thin Solid Films. 2004;**469**:178-183

Chapter 12

Variation of Coronal Magnetic Field and Solar Flare Eruption

Han He

Abstract

Solar flares are prominent eruptive phenomenon happening in the solar atmosphere. Major flares usually come from solar active regions (ARs) where strong and concentrated bipolar magnetic field exists and manifests as dark sunspots in photosphere. The photospheric magnetic field acts as the bottom boundary of corona system and confines the magnetic structure of the corona. For complex ARs, the coronal magnetic field generally contains electric current around the magnetic polarity inversion lines, which corresponds to the nonpotential magnetic field and manifests as twisted field lines. The coronal magnetic field structure evolves as the response to the variations of the photospheric magnetic field. This coronal evolution can be quasi-steady and approximated by the force-free condition. In some situations, the variations of photospheric magnetic field may cause sudden changes of topological structure of coronal magnetic field at certain sites in the corona. The plasmas at these sites lost equilibrium and are ejected from their original positions. This process is accompanied with magnetic reconnection and leads to the release of magnetic energy in the corona. Part of the released magnetic energy is converted to the electromagnetic emission which manifests as sudden brightening across a broad range of electromagnetic wave spectrum, and hence the flare phenomenon is initiated.

Keywords: solar flare, magnetic field, plasma, corona

1. Introduction

Solar flares are prominent eruptive phenomenon happening in the solar atmosphere [1–3], where the matters have high temperature and hence in plasma state [4]. The energy of solar flares comes from the magnetic field in the solar atmosphere [1]. Major flares usually originate from solar active regions (ARs). In solar ARs, strong and concentrated bipolar magnetic field exists and manifests as dark sunspots in photosphere [2, 5].

The photospheric magnetic field acts as the bottom boundary of corona system and confines the magnetic structure of the corona [4, 5]. Because in the corona the plasmas are very tenuous, the magnetic force plays the dominant role, and other forces can be neglected. Thus, the plasmas in the corona are distributed along field lines and satisfy force-free condition (i.e., the Lorentz force being zero) for steady corona [6]. For complex ARs, the coronal magnetic field generally contains electric current around the magnetic polarity inversion lines (PILs), which corresponds to the nonpotential magnetic field and manifests as twisted field lines [7, 8].
The coronal magnetic field structure evolves as the response to the variations

of the photospheric magnetic field. This coronal evolution can be quasi-steady and approximated by the force-free condition [9, 10].

In some situations, the variations of photospheric magnetic field may cause sudden changes of topological structure of coronal magnetic field at certain sites in the corona. The plasmas at these sites lost equilibrium and are ejected from their original positions. This process is accompanied with magnetic reconnection and leads to the release of magnetic energy in the corona [1, 3, 11]. Part of the released magnetic energy is converted to the electromagnetic emission which manifests as sudden brightening across a broad range of electromagnetic wave spectrum, and hence the flare phenomenon is initiated [2]. The typical electromagnetic emissions include white-light flare in photosphere, optical flare in chromosphere, and soft X-ray flare in the corona [2, 4]. Other released magnetic energy is converted to the mechanical energy of the erupted plasmoid and is also carried off by the high-energy particle radiation [2–4], which might lead to the coronal mass ejections and the solar energetic particles associated with solar flares [12–14].

An overview on the relations between the variation of coronal magnetic field and the solar flare eruption is given in this chapter. In Section 2, the main observational properties of the solar flares are presented. In Section 3, the nonpotentiality of the coronal magnetic field associated with the solar flares is discussed. The process of the flare initiation caused by the variation of coronal magnetic field is described in Section 4. Section 5 provides the summary and conclusion.

2. Solar flare observations

The energy of solar flares comes from the magnetic field in the solar atmosphere. It is natural that almost all major flares were found to be located in solar ARs which possess strong and concentrated bipolar magnetic field and manifests as dark sunspots in photosphere [1–5].

The electromagnetic emissions of solar flares can be observed in the corona, chromosphere, and photosphere of the Sun [2, 4]. In the standard model of solar flare eruption [3, 4], the magnetic reconnection associated with a flare takes place in the corona, just above the AR PIL and beneath the erupting plasmoid. The heat produced in the space of magnetic reconnection is transmitted to the chromosphere along the field lines. At the foot points of the field lines, the matters in chromosphere are heated up to extremely high temperature (about 10^7 K) and manifest as two bright ribbons located at the two sides of the PIL in the filtergrams observed through the chromospheric spectral lines (such as Hα, Ca II H and K, etc.). The flare brightening in chromosphere is traditionally called optical flare since it can be observed via an optical device equipped with a band filter of the selected chromospheric spectral line (see **Figure 1** for an example of flare image in chromosphere). The heated chromospheric materials subsequently fill up the arcade system of field lines over the PIL, which manifests as bright flare loop arcade in the corona in the images observed in extreme ultraviolet (EUV) or soft X-ray band [2, 4] (see **Figure 2**).

The energetic electrons produced by the magnetic reconnections also transmit downward along the field lines and can reach as low as upper photosphere level and cause flare brightening in white-light band [2, 4]. In fact, the first observed major flare event by Carrington in 1859 is white-light flare [15].

Because the photosphere of the Sun is very bright, the white-light flares cannot be observed very easily and frequently [16], whereas the soft X-ray flares originate from the thermal radiation in the corona and have much low background radiation, thus, the solar soft X-ray flux is widely adopted as the basis for standard flare

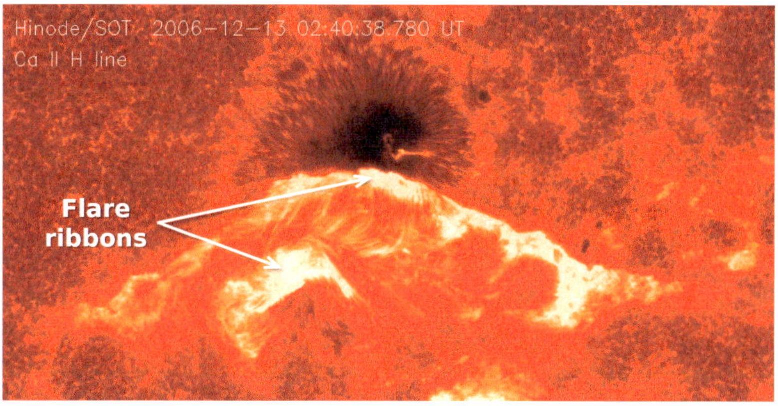

Figure 1.
Chromospheric image of a major flare event on 13 December 2006. The image was observed through the Ca II H spectral line by the Hinode satellite.

Figure 2.
Coronal image of the flare on 13 December 2006. The image was observed through the 195 Å EUV band by the SOHO spacecraft.

magnitude classification (i.e., A, B, C, M, X-class series of flare classification; see https://www.swpc.noaa.gov/products/goes-x-ray-flux) [14]. Other bands of electromagnetic spectrum, such as radio, ultraviolet (UV), hard X-ray, and γ-ray, are also commonly used for solar flare observations and studies [2, 4].

3. Nonpotentiality of coronal magnetic field

3.1 Force-free magnetic field of steady corona

Since the magnetic reconnection associated with solar flare eruptions takes place in the corona, the coronal magnetic field distribution is crucial for understanding the physical process of flares [1, 3, 4]. In fact, the photospheric magnetic field acts as the bottom boundary condition of corona system and confines the magnetic structure of the corona [4, 5].

In the corona, the temperature is very high (about 10^6 K), and the density of plasmas is very low [4]. Thus, the magnetic force (Lorentz force) dominates the coronal system, and other forces (such as gravity force, pressure, etc.) can be neglected. Then, in steady state of the corona, the plasmas are distributed along the field lines, and the Lorentz force is zero. This condition is called force-free condition, and the corresponding magnetic field is called force-free magnetic field [6]. Note that the essence of the force-free field concept is the dominant role of the magnetic field in a plasma system.

The force-free magnetic field $\boldsymbol{B}$ can be described by the following equation:

$$\nabla \times \boldsymbol{B} = \alpha \boldsymbol{B}. \tag{1}$$

The left part of Eq. (1) represents the electric current density vector:

$$\boldsymbol{j} = \frac{1}{4\pi} \nabla \times \boldsymbol{B} \tag{2}$$

(in electromagnetic CGS units). Then Eq. (1) indicates that the current density vector $\boldsymbol{j}$ and the magnetic field vector $\boldsymbol{B}$ are parallel to each other, and hence the Lorentz force is zero.

The proportional coefficient α in Eq. (1) is called force-free factor. The value of α can be positive, zero, or negative. If $\alpha > 0$, the electric current density $\boldsymbol{j}$ and the magnetic field $\boldsymbol{B}$ have the same directions, and if $\alpha < 0$, they have opposite directions. $\alpha = 0$ means that $\nabla \times \boldsymbol{B} = 0$, i.e., there is no electric current in the magnetic field (see Eq. (2)), and this special case is called potential magnetic field.

By taking the divergence of Eq. (1) and considering the divergence-free property of magnetic field, we have

$$\nabla \alpha \cdot \boldsymbol{B} = 0. \tag{3}$$

Eq. (3) means that along each field line, α is a constant. (Note that for different field lines, the values of α can be different.) This is an important character of the force-free magnetic field in corona.

Eqs. (1) and (3) together give the mathematical expression of the coronal force-free magnetic field model (two variable quantities α and $\boldsymbol{B}$ constrained by two equations) [6]. Given the observed photospheric vector magnetic field as the bottom boundary condition, the 3-D coronal magnetic field in steady corona can be calculated numerically based on Eqs. (1) and (3) [9, 10, 17–20].

3.2 Electric current and nonpotentiality of coronal magnetic field

The dominant force in the corona is magnetic force which acts on the electric current. In potential magnetic field, there is no electric current and hence no magnetic force. Thus, it is not possible for the potential magnetic field to produce eruptive phenomenon since there is no force to accelerate the plasmas.

To yield solar eruptions, the existence of electric current or nonpotential magnetic field in the corona is a necessary condition. In solar ARs with complex photospheric magnetic field, the coronal magnetic field generally contains electric current around the PILs (note that the magnetic field structure in the corona is confined by the photospheric magnetic field). This property (deviation from potential magnetic field) is called nonpotentiality of coronal magnetic field and is commonly employed to reflect the activity level (degree of possibility to produce eruptive events) of solar ARs [7, 8, 10, 11, 21]. The electric current density $\boldsymbol{j}$ (see Eq. (2)) is a physical measure to quantitatively describe the nonpotentiality of coronal magnetic field [10, 11] (see **Figure 3** for an example diagram of the electric current density spatial distribution in a solar AR).

3.3 Various manifestations of nonpotentiality

For a solar AR with single polarity magnetic field in photosphere, the coronal magnetic field is usually simple and tends to be potential field. The nonpotentiality

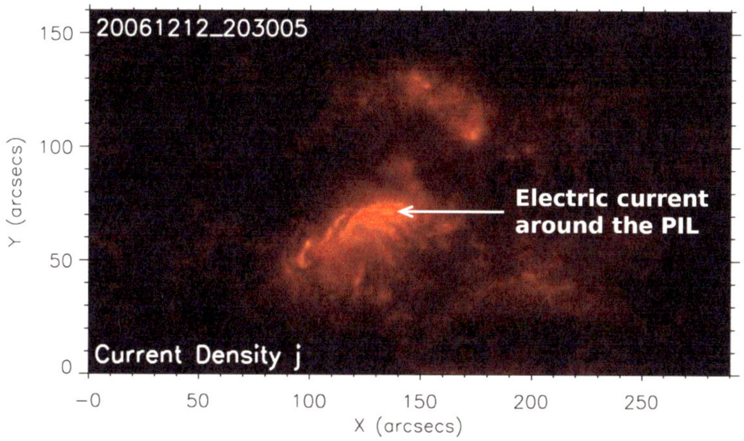

Figure 3.
Spatial distribution of electric current density in the source AR of the flare event on 13 December 2006 [11]. The electric current density values were derived from the 3-D coronal magnetic field data by using Eq. (2), and the coronal magnetic field data were calculated based on the force-free field model (see Eqs. (1) and (3)).

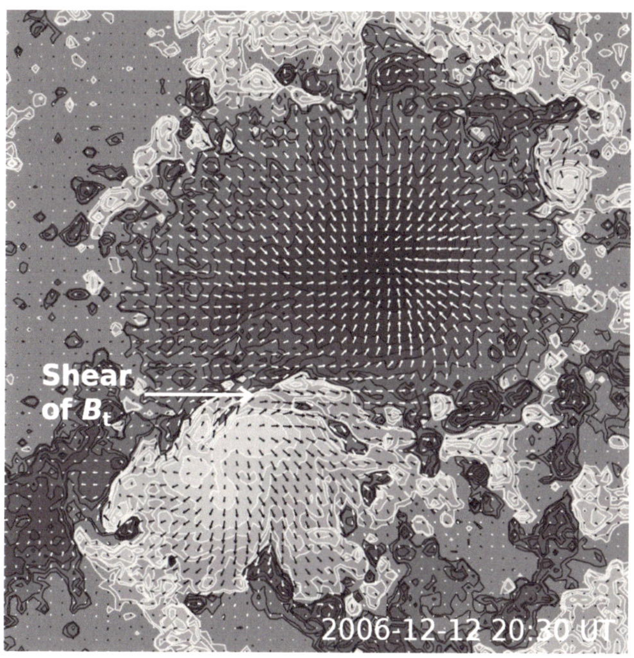

Figure 4.
Shear of the transverse component $\mathbf{B}_t$ of the photospheric magnetic field around the PIL in the source AR of the flare event on 13 December 2006 [11]. The small arrows indicate the directions of $\mathbf{B}_t$. The contours show the vertical component (and the two polarities) of the photospheric magnetic field. White contours represent the positive polarity, and black contours represent the negative polarity.

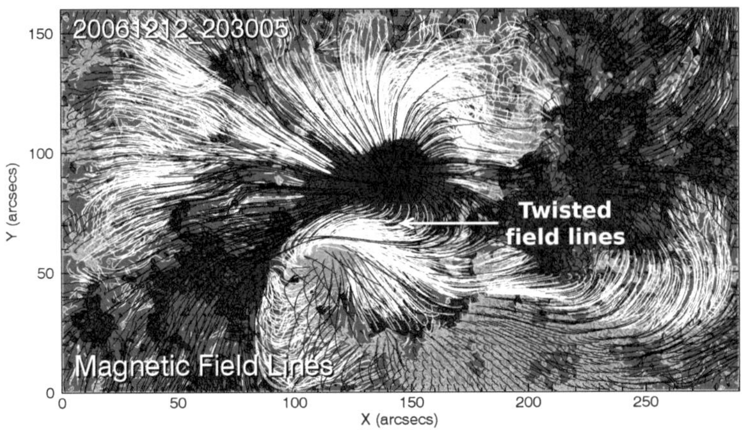

Figure 5.
Twisted field lines of the coronal magnetic field in the source AR of the flare event on 13 December 2006 [11]. Closed field lines are in white and open field lines are in black. The white and black background contours show the two polarities of the photospheric magnetic field. The coronal magnetic field was calculated based on the force-free field model.

is more prominent around the PILs of ARs with complex bipolar photospheric magnetic field. The typical characteristic of the photospheric vector magnetic field associated with the nonpotential magnetic field in the corona is the shear of the transverse component (denoted by $\boldsymbol{B}_t$) of the photospheric magnetic field. For potential magnetic field, the direction of $\boldsymbol{B}_t$ is generally vertical to the PILs, whereas for nonpotential magnetic field, the direction of $\boldsymbol{B}_t$ tends to be parallel to the PILs. The shear property reflects the direction deviation of the real $\boldsymbol{B}_t$ from the potential $\boldsymbol{B}_t$ [7] (see **Figure 4** for a diagram illustration).

Owning to the shear of $\boldsymbol{B}_t$ in photosphere, the field lines in the corona also show shear behavior around the PILs, i.e., the field lines tend to be parallel to the PILs. For a closed field line which starts at positive polarity, shears around the PIL, and ends at negative polarity, the three segments (start, shear, and end segments) of the field line compose a twist (S-shaped or inverse S-shaped) morphology (see **Figure 5** for an example diagram of the twisted field lines in corona). The twisted field lines indicate the existence of electric current in the corona and are another manifestation of the nonpotential magnetic field [8].

4. Variation of coronal magnetic field and solar flare eruption

4.1 Quasi-steady evolution of the corona

Because the photospheric magnetic field confines the magnetic structure in corona, along with the evolution of the photospheric magnetic field, the coronal magnetic field also evolves accordingly as the response to the variations in photosphere. The evolution in the photosphere is relatively slow owing to its relatively dense plasma, and the evolution in the corona is fast for its very tenuous condition. Thus, the corona can catch up the variations in photosphere promptly. If there are no eruptions on the Sun (quiet state), the coronal evolution (along with the evolution of the photosphere) can be quasi-steady and approximated by the force-free condition [4].

Provided a time series of observed photospheric vector magnetograms of a solar AR, by numerical modeling of the coronal magnetic fields based on the force-free field model for each magnetogram, we can obtain a time series of 3-D coronal magnetic field data associated with the photospheric magnetograms. These time series coronal magnetic field data quantitatively describe the quasi-steady evolution of the corona [9, 10]. By deriving the electric current density distributions from these coronal data, the time evolution of the nonpotentiality in the corona can also be revealed [10, 11].

4.2 Sudden change of coronal magnetic field structure and flare initiation

During the period of quasi-steady evolution of the corona, the plasmas in the corona are in quasi-equilibrium state, and the topological structure of coronal magnetic field evolves continuously. In some situations, the variations of photospheric magnetic field may cause sudden changes of topological structure of coronal magnetic field at certain sites in the corona [1, 3, 11]. The plasmas at these sites lost equilibrium and are ejected from their original positions. The magnetic reconnections occur beneath the erupted plasmoid, and then the flares are initiated. After the flare eruptions, the corona returns to equilibrium state and continues its quasi-equilibrium evolution [3, 11].

During the period of flare eruptions, the plasmas are in dynamic state, and the force-free condition is not satisfied. Out of the flare eruption period, the force-free field model is a well approximation to the coronal magnetic field [11].

Because the plasmoid ejection associated with the flare initiation needs magnetic force and the magnetic force (Lorentz force) only acts on the electric current, the sites of flare initiation are always located in the areas with strong electric current which in turn concentrates around the PILs of ARs. That is why the flare phenomenon is connected with the nonpotentiality (indicated by the electric current) of solar ARs, and flares always occur above the PILs.

4.3 Release of magnetic energy by flare eruption

Not all magnetic energy in the solar atmosphere can be depleted by flare eruptions. For example, the magnetic energy of a potential magnetic field cannot be consumed since no flare eruptions can occur in the potential magnetic field. Only the magnetic energy associated with the electric current in a nonpotential magnetic field can be accessed by flare eruptions. This part of available magnetic energy (energy bundled with the electric current) is called the free magnetic energy [4, 22].

During a flare eruption, a fraction of the total electric current around the PIL is ejected out together with the plasmoid eruption, and the corresponding part of the free magnetic energy is released. After the flare eruption, the total free magnetic energy decreases, and the twisted field lines around the PIL relax to a certain extent owing to the loss of electric current and the depletion of free magnetic energy [11].

A proportion of the released magnetic energy is converted to the electromagnetic emission which manifests as sudden brightening across a broad range of electromagnetic wave spectrum, and that is why the flare phenomenon is named [2]. Other parts of the released magnetic energy are converted to the mechanical energy of the erupted plasmoid and are also carried off by the high-energy particle radiation [2, 4]. The erupted plasmoid might lead to the coronal mass ejections (CMEs) accompanied with solar flares, and the high-energy particles might lead to the solar energetic particle (SEP) events associated with solar flares [12–14].

5. Conclusion

As a prominent eruptive phenomenon happening in solar atmosphere, solar flares usually come from solar ARs which possess strong and concentrated bipolar magnetic field in the photosphere. Between the two opposite polarities is the PIL of the magnetic field. Around the PIL is the sheared vector magnetic field in photosphere and twisted field lines in the corona, which are the manifestations of electric current distribution around the PIL and the existence of free magnetic energy bundled with the electric current. When a flare happens, the variation of the photospheric magnetic field causes sudden change of topological structure of coronal magnetic field at a site in the strong electric current area. The electric current and associated plasmas at this site lost equilibrium and are ejected from their original position. The magnetic reconnection occurs beneath the erupted plasmoid, and the flare is initiated. Part of the released free magnetic energy is converted to the electromagnetic emission of flares, which manifests as sudden brightening across a broad range of electromagnetic wave spectrum, such as white-light flare in photosphere, optical flare in chromosphere, and soft X-ray flare in the corona. Other released magnetic energy is transferred to CMEs and SEPs associated with flares. Big solar flares and the associated CMEs and SEPs can cause severe disturbances to the space weather condition in the solar-terrestrial space as well as in the whole heliosphere [14, 23–25].

Besides the Sun, flare phenomenon was also observed on the solar-type stars that possess magnetic activity [26–29]. In recent years, owing to the continuous light-curve observations for a large volume of stellar objects by the space missions, such as the Kepler space telescope [30], much more stellar flare samples were obtained and available for analysis [27, 29]. The understandings about the solar flares can provide a good physical framework basis for investigating the eruption mechanism of stellar flares [31, 32].

Acknowledgements

The author thanks the support of the Astronomical Big Data Joint Research Center, co-founded by the National Astronomical Observatories, Chinese Academy of Sciences and the Alibaba Cloud, the National Natural Science Foundation of China (NSFC; grants 11761141002, 11403044, 11303051, 40890160, 40890161, 10803011, and 10673017), and the Strategic Priority Research Program on Space Science, Chinese Academy of Sciences (grant XDA04060801).

Author details

Han He
National Astronomical Observatories, Chinese Academy of Sciences, Beijing, China

*Address all correspondence to: hehan@nao.cas.cn

References

[1] Priest ER, Forbes TG. The magnetic nature of solar flares. Astronomy and Astrophysics Review. 2002;**10**:313-377. DOI: 10.1007/s001590100013

[2] Benz AO. Flare observations. Living Reviews in Solar Physics. 2017;**14**:2. DOI: 10.1007/s41116-016-0004-3

[3] Shibata K, Magara T. Solar flares: Magnetohydrodynamic processes. Living Reviews in Solar Physics. 2011;**8**:6. DOI: 10.12942/lrsp-2011-6

[4] Aschwanden MJ. Physics of the Solar Corona: An Introduction with Problems and Solutions. 2nd ed. Chichester: Praxis; 2005. 892 p. DOI: 10.1007/3-540-30766-4

[5] Solanki SK, Inhester B, Schüssler M. The solar magnetic field. Reports on Progress in Physics. 2006;**69**:563-668. DOI: 10.1088/0034-4885/69/3/R02

[6] Wiegelmann T, Sakurai T. Solar force-free magnetic fields. Living Reviews in Solar Physics. 2012;**9**:5. DOI: 10.12942/lrsp-2012-5

[7] Wang J, Shi Z, Wang H, Lü Y. Flares and the magnetic nonpotentiality. The Astrophysical Journal. 1996;**456**: 861-878. DOI: 10.1086/176703

[8] Schrijver CJ. The nonpotentiality of coronae of solar active regions, the dynamics of the surface magnetic field, and the potential for large flares. The Astrophysical Journal. 2016;**820**:103. DOI: 10.3847/0004-637X/820/2/103

[9] Sakurai T. Computational modeling of magnetic fields in solar active regions. Space Science Reviews. 1989;**51**:11-48. DOI: 10.1007/BF00226267

[10] He H, Wang H, Yan Y. Nonlinear force-free field extrapolation of the coronal magnetic field using the data obtained by the Hinode satellite. Journal of Geophysical Research: Space Physics. 2011;**116**:A01101. DOI: 10.1029/2010JA015610

[11] He H, Wang H, Yan Y, Chen PF, Fang C. Variations of the 3-D coronal magnetic field associated with the X3.4-class solar flare event of AR 10930. Journal of Geophysical Research: Space Physics. 2014;**119**:3286-3315. DOI: 10.1002/2013JA019157

[12] Zhang M, Low BC. The hydromagnetic nature of solar coronal mass ejections. Annual Review of Astronomy and Astrophysics. 2005;**43**:103-137. DOI: 10.1146/annurev.astro.43.072103.150602

[13] Chen PF. Coronal mass ejections: Models and their observational basis. Living Reviews in Solar Physics. 2011;**8**:1. DOI: 10.12942/lrsp-2011-1

[14] Schwenn R. Space weather: The solar perspective. Living Reviews in Solar Physics. 2006;**3**:2. DOI: 10.12942/lrsp-2006-2

[15] Carrington RC. Description of a singular appearance seen in the Sun on September 1, 1859. Monthly Notices of the Royal Astronomical Society. 1859;**20**:13-15. DOI: 10.1093/mnras/20.1.13

[16] Kopp G, Lawrence G, Rottman G. The total irradiance monitor (TIM): Science results. Solar Physics. 2005;**230**:129-139. DOI: 10.1007/s11207-005-7433-9

[17] Schrijver CJ, De Rosa ML, Metcalf TR, Liu Y, McTiernan J, Régnier S, et al. Nonlinear force-free modeling of coronal magnetic fields. Part I: A quantitative comparison of methods. Solar Physics. 2006;**235**:161-190. DOI: 10.1007/s11207-006-0068-7

[18] Yan Y, Li Z. Direct boundary integral formulation for solar non-constant-α

force-free magnetic fields. The Astrophysical Journal. 2006;**638**:1162-1168. DOI: 10.1086/499064

[19] He H, Wang H. The validity of the boundary integral equation for magnetic field extrapolation in open space above a spherical surface. Monthly Notices of the Royal Astronomical Society. 2006;**369**:207-215. DOI: 10.1111/j.1365-2966.2006.10288.x

[20] He H, Wang H. Nonlinear force-free coronal magnetic field extrapolation scheme based on the direct boundary integral formulation. Journal of Geophysical Research: Space Physics. 2008;**113**:A05S90. DOI: 10.1029/2007JA012441

[21] Leka KD, Barnes G. Photospheric magnetic field properties of flaring versus flare-quiet active regions. II. Discriminant analysis. The Astrophysical Journal. 2003;**595**: 1296-1306. DOI: 10.1086/377512

[22] Régnier S, Priest ER. Free magnetic energy in solar active regions above the minimum-energy relaxed state. The Astrophysical Journal. 2007;**669**: L53-L56. DOI: 10.1086/523269

[23] Pulkkinen T. Space weather: Terrestrial perspective. Living Reviews in Solar Physics. 2007;**4**:1. DOI: 10.12942/lrsp-2007-1

[24] He H, Shen C, Wang H, Zhang X, Chen B, Yan J, et al. Response of plasmaspheric configuration to substorms revealed by Chang'e 3. Scientific Reports. 2016;**6**:32362. DOI: 10.1038/srep32362

[25] Lanzerotti LJ. Space weather: Historical and contemporary perspectives. Space Science Reviews. 2017;**212**:1253-1270. DOI: 10.1007/s11214-017-0408-y

[26] Schaefer BE, King JR, Deliyannis CP. Superflares on ordinary solar-type stars. The Astrophysical Journal. 2000;**529**:1026-1030. DOI: 10.1086/308325

[27] Maehara H, Shibayama T, Notsu S, Notsu Y, Nagao T, Kusaba S, et al. Superflares on solar-type stars. Nature. 2012;**485**:478-481. DOI: 10.1038/nature11063

[28] He H, Wang H, Yun D. Activity analyses for solar-type stars observed with Kepler. I. Proxies of magnetic activity. The Astrophysical Journal Supplement Series. 2015;**221**:18. DOI: 10.1088/0067-0049/221/1/18

[29] He H, Wang H, Zhang M, Mehrabi A, Yan Y, Yun D. Activity analyses for solar-type stars observed with Kepler. II. Magnetic feature versus flare activity. The Astrophysical Journal Supplement Series. 2018;**236**:7. DOI: 10.3847/1538-4365/aab779

[30] Borucki WJ. KEPLER Mission: Development and overview. Reports on Progress in Physics. 2016;**79**:036901. DOI: 10.1088/0034-4885/79/3/036901

[31] Benz AO, Güdel M. Physical processes in magnetically driven flares on the Sun, stars, and young stellar objects. Annual Review of Astronomy and Astrophysics. 2010;**48**:241-287. DOI: 10.1146/annurev-astro-082708-101757

[32] Choudhuri AR. Starspots, stellar cycles and stellar flares: Lessons from solar dynamo models. Science China Physics, Mechanics, and Astronomy. 2017;**60**:019601. DOI: 10.1007/s11433-016-0413-7

Chapter 13

Magnetohydrodynamics in Biomedical Applications

Hamid Farrokhi, David O. Otuya, Anna Khimchenko and Jing Dong

Abstract

This chapter discusses recent advances in biomedical applications of magnetohydrodynamics (MHD). The magnetohydrodynamic (MDH) effect is a physical phenomenon describing the motion of a conducting fluid flowing under influencing of an external magnetic field. The chapter covers four primary areas of research: (1) laser beam scanning, (2) nano-particle manipulation, (3) imaging contrast enhancement, and (4) targeted drug delivery. The state-of-the-art devices based on magnetohydrodynamic principles are also presented, providing a broad view of biomedical MHDs. As the field of biomedical MHDs continues to grow, advances towards micro-scale transitions will continue to be made, maintaining its clinically driven nature and motion towards real-world applications.

Keywords: magnetohydrodynamics, beam scanning, nanoparticles, imaging contrast, targeted drug delivery, magnetic constructs

1. Introduction

Magnetohydrodynamic (MDH) effect is a physical phenomenon that describes the motion of charge conducting fluid flowing which his influenced by an external magnetic field. Its applications have been studied extensively across multiple disciplines ranging from the study of solar winds [1, 2] to MHD-driven biomedical sensors [3] and actuators [4–7]. This chapter narrows the focus of MHD applications to biomedical sciences. The chapter introduces four primary MHD biomedical applications: (1) magnetohydrodynamic-based laser beam scanning, (2) nanoparticle manipulations for biomedical applications, (3) biomedical imaging contrast enhancement, and (4) targeted drug delivery.

This chapter aims to present the current state of the art in the field with regards to biomedical and clinical applications of the MHD effect.

2. Magnetohydrodynamic-based laser beam scanning

2.1 Introduction

Adaptive optics (AO) enables correcting of complex aberrations for a broad range of applications [1, 2]. Conventional AO systems use spatial light modulators [3, 4] or solid deformable mirrors (DM) [5, 6] to compensate the phase changes

 IntechOpen

resulting from non-uniformity in the properties of the medium through which light travels or from faults in the geometry of the component. There is two types of spatial light modulators operating in reflective and transmission modes. Due to using small liquid crystals, this type of wavefront shaping able to provide very high spatial resolution. However, there is a limitation in the correction magnitude which is usually in the range of a few micrometers. To resolve this issue, solid deformable mirrors have been developed and adopted widely to correct wavefront in optical systems. It consists of a solid reflecting surface connected to an actuator structure. By controlling the actuators, the shape of the reflective surface can be changed to compensate the wavefront distortions. The high cost of the actuators and intricate fabrication process are the main problems of the solid deformable mirrors. The current solid deformable mirrors can only provide small inter-actuator strokes, with the maximum deflection in the range of tens of micrometers.

AO systems have been used in laser beam shaping [7–9] and eye imaging systems [10–12] to effectively address both the low-intensity high-order aberrations and the high-intensity low-order aberrations. For example, high-resolution retinal imaging using AO plays a vital role in vision science and will help the early clinical diagnosis of eye diseases. For dealing with the ocular aberrations for a large population, e.g., myopic eyes, adaptive optics using two deformable mirrors have been designed [10–12]. The large-stroke DM with a limited number of actuators is used to correct large-intensity low-order aberrations. The second DM with a low stroke and a high spatial correction resolution is used to address the small-intensity high-order aberrations. However, its application in ophthalmology is limited by the complexity and the high cost. A new type of liquid deformable mirror [13–15] has been designed based on the actuation of the magnetic fluid. Although the liquid mirrors are constrained to the horizontal setup, the magnetic fluid deformable mirrors (MFDM) offer large strokes, low cost, and easy fabrication. Using the single inter-actuator, the strokes more than 100 μm can obtain with less power consumption. However, to produce a large mirror surface deformation, the size of the electromagnetic coils needs to be large. This makes the density of actuators low, and it is not suitable for high-order aberrations. A full-order correction with the high spatial resolution is achieved using a two-layer design layout with small electromagnetic coils has been designed. New wavefront corrector, MFDM, has major advantages such as large stroke, low cost, easy fabrication process, which can be easily customized for different applications. In the following sections, we will explain the technology and possible applications in *in-vivo* imaging and probing.

2.2 Magnetic fluid deformable mirror (MFDM)

The main elements of the MFDM are a layer of magnetic fluid, a thin film of a reflective material coated on the free surface of the fluid, a two-layer layout of the miniature electromagnetic coils placed beneath the fluid layer, and a Maxwell coil(s) (see **Figure 1**). The properties of the magnetic fluid used in the study are given in **Table 1**. Based on the definition, the magnetic fluid is a stable colloidal suspension of nano-sized, single-domain ferri/ferromagnetic particles. The fluid can be coated with a silver liquid-like thin film to improve the reflectance.

To achieve the correction of full-order aberrations with a high spatial resolution, the new design of MFDM is presented (see **Figure 1** for the details). The upper layer actuators of small size and high density are used to compensate for small-amplitude high-order aberrations, and the lower layer actuators with big size and low density are used to correct for large amplitude low order aberrations. The electromagnetic coils are circular coils wound on a cylindrical core. **Table 2** shows the physical parameters are of the coil. Each layer of the coil is arranged in a hexagonal array.

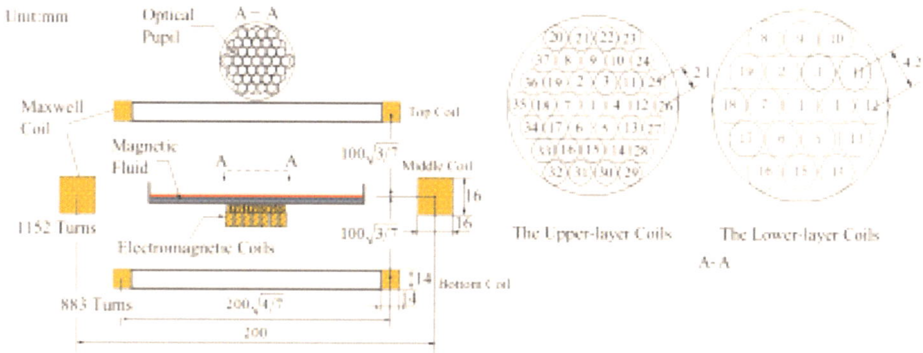

Figure 1.
Schematic design of the magnetic fluid deformable mirror (MFDM) (adapted from [20]).

Magnetic fluid	Parameters
Saturation magnetization	22 mT
Relative permeability	2.89
Density	1190 kg/m^3
Viscosity	3 cp
Thickness	1 mm

Table 1.
Parameters of the magnetic fluid.

Magnetic fluid	Parameters	
Position	Upper	Lower
Core-type	Air-cored	Air-cored
Material	Copper	Copper
Wire gauge	AWG37	AWG36
Internal diameter	1 mm	2 mm
External diameter	2 mm	4 mm
Length	1 mm	8 mm

AWG, American wire gauge.

Table 2.
Parameters of the miniature electromagnetic coil.

The upper layer coils are radially spaced at 2.1 mm from the center to center, and the lower layer coils are radially spaced at 4.2 mm, respectively.

To make the response of the actuators linear, the Maxwell coil was used to apply an external uniform magnetic field. The Maxwell coil consists of three individual coils, where both lateral/outer coils should have a radius of $\sqrt{4/7}$ R, at a distance of $\sqrt{3/7}$ R from the middle coil with a radius of R = 100 mm [16] (see **Figure 1**). **Table 3** shows the parameters of the coil. The three coils are wired using American wire gauge (AWG) 25 magnet wire. The turn ratio of 64:49 is used for the top and bottom coils relative to the middle coil [16]. Also, magnetic fluids typically show low reflectance to light and can be coated with silver liquid-like thin films to improve the reflectance [17, 18]. The self-assembly method has been usually used to

Magnetic fluid	Parameters
Nominal diameter of the middle coil	200 mm
No. of turns in the middle coil	1152
No. of turn in the top and bottom coil	883
Average resistance of the middle coil	71.2 Ω
Average resistance of the coils	42.3 Ω
Wire gauge	AWG 25
Wire material	Copper
Bobbin material	Aluminum

Table 3.
Parameters of the Maxwell coil(s).

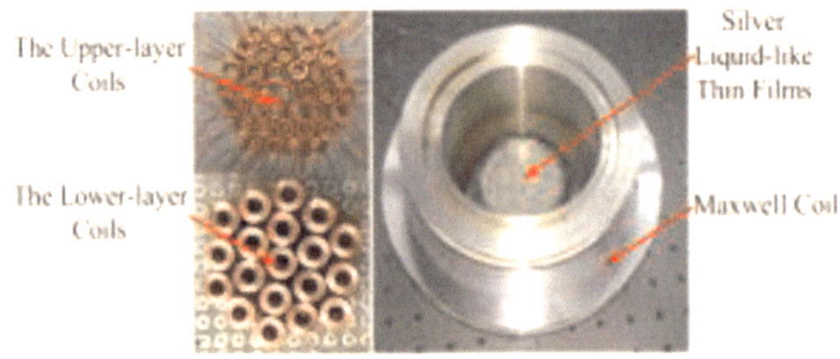

Figure 2.
Assembly of the prototype MFDM (adapted from [20]).

prepare the silver liquid-like thin film for the MFDM. Firstly, the solution of silver nanoparticles was dissociated by centrifugation to remove the supernatant, and ethanol was then infused to purify the silver nanoparticles. The obtained silver nanoparticles were mixed with the ethanol/dodecanethiol solution, kept at room temperature for 24 h, and then centrifuged. Then the ethyl acetate was added into the silver nanoparticles obtained from the previous step. This solution was then applied to the surface of the magnetic fluid. When the ethyl acetate evaporated, the hydrophobic dodecanethiol encapsulated silver nano-particles automatically stacked and spread on the surface of the magnetic fluid to form a large scale area of silver liquid-like film.

Figure 2 shows the fabricated mirror in which two-layer layout of the coils are installed within the Maxwell coil. Ferrofluid with layer thickness of about 1 mm is placed on top of the miniature coils, which is coated with the thin silver liquid-like film.

2.3 Working principle of MFDM

The MFDM is demonstrated by a cylindrical layer of a magnetic fluid as shown in **Figure 3**. The top free surface of the fluid layer is coated with a reflective film to be the deformable surface of the mirror. The surface deflection at the point (r_k, θ_k) is indicated by $\zeta(r_k, \theta_k, t)$, where $k = 1, 2, 3, \ldots$, k is a discrete number of surface points. The magnetic field produced by any specific coil, centered at the horizontal location (r_{ij}, θ_{ij}), is idealized as that of a point source of magnetic potential $\psi_{ij}(t)$, where $i = 1, 2$ is the ith layer of actuators, and j, $j = 1, 2, 3, \ldots, J_i$ is the jth coil of each layer.

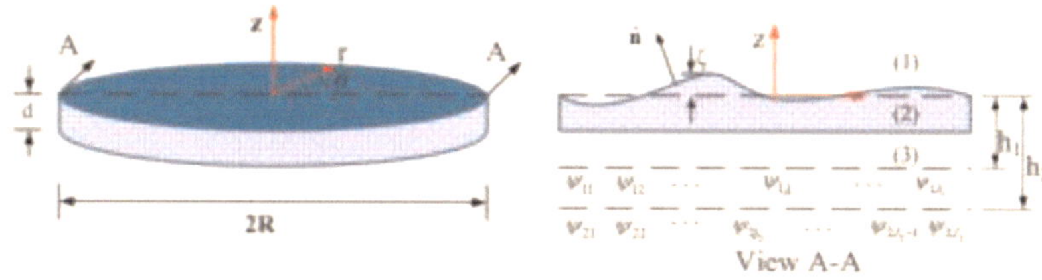

Figure 3.
Geometric representation of a circular MFDM (adapted from [20]).

Maxwell's equations govern the magnetic field. Since the magnetic field of the miniature coils is taken as point sources of magnetic potential located at the fluid boundary, electromagnetic field can be considered as a current-free one. Using this assumption, the displacement currents in the fluid are negligible, Maxwell's equations can be written as:

$$\nabla \times H = 0, \nabla \cdot B = 0 \tag{1}$$

where B is the magnetic flux density, which is related to the magnetic field H and the magnetization M by the following equation:

$$B = \mu H = \mu_0 \left(H + M\right) \tag{2}$$

μ and μ_0 are the magnetic permeability of the magnetic fluid and free space, respectively. Assuming the applied field linearly magnetizes the magnetic fluid, the magnetization vector M can be written as

$$M = \chi H \tag{3}$$

where $\chi = \mu/\mu_0 - 1$ is considered to be constant. Considering that the magnetic field extends into space above and below the fluid layer, Maxwell's equations are applied to all three sub-domains (1)–(3) as shown in **Figure 3**. The scalar potentials $\psi^{(1)}$, l = 1, 2, 3 describe the magnetic field vectors $H^{(1)}$ in these sub-domains as follows:

$$H^{(l)} = -\nabla\psi^{(l)}, l = 1, 2, 3 \tag{4}$$

Using Eqs. (2)–(4), the magnetic flux density $B^{(l)}$ in these sub-domains can be written in terms of the scalar potentials $\psi^{(l)}$as:

$$B^{(l)} = -\mu_0(1+\chi)\nabla\psi^{(l)}, l = 1, 2, 3 \tag{5}$$

The magnetic flux intensity (B) meets the principle of superposition. Assume the fluid is irrotational, then based on the principles of conservation of mass and momentum and the theories on magnetic fields, the perturbation part of the surface dynamic governing equations can be written as [19].

$$\nabla^2\psi = 0, \ -d \le z \le \varsigma \tag{6}$$

$$\nabla^2\psi^{(l)} = 0, l = 1, 2, 3 \tag{7}$$

$$-\rho\frac{\partial\phi}{\partial t} + \rho g\varsigma + \chi B_0\frac{\partial\psi^{(2)}}{\partial z} - \sigma\left(\frac{\partial^{(2)}\varsigma}{\partial r^{(2)}} + \frac{1}{r}\frac{\partial\varsigma}{\partial r} + \frac{1}{r^{(2)}}\frac{\partial^{(2)}\varsigma}{\partial\theta^{(2)}}\right) = 0, z = \varsigma \tag{8}$$

where ρ is the density of the fluid, σ is the surface tension, φ, and $\psi^{(l)}$, $l = 1, 2, 3$ are the perturbation components of the fluid velocity potential and the magnetic potential, respectively. Using the following two boundary conditions:

$$-\frac{\partial \Phi}{\partial z} = \frac{\partial \varsigma}{\partial t}, z = \varsigma \quad (9)$$

$$-\frac{\partial \Phi}{\partial z} = 0, z = -d \quad (10)$$

The solutions concerning the input $\psi_{ij}(t)$ thus are obtained as follows:

$$\zeta\,(r_k, \theta_k, t) = \widetilde{\zeta_{ij}}(t) J_m(\lambda r_k)\Theta(\theta_k) \quad (11)$$

$$\phi(r_k, \theta_k, z, t) = -\frac{1}{\lambda}\frac{cosh[\lambda(z+d)]}{sinh(\lambda d)}\frac{d\widetilde{\zeta_{ij}}}{dt} J_m(\lambda r_k)\Theta(\theta_k) \quad (12)$$

$$\phi^{(2)}(r_k, \theta_k, z, t) = -\left[A_{ij}(t)\left(\frac{\mu}{\mu_0} coshcosh(\lambda z) - sinhsinh(\lambda z)\right) + \frac{\chi}{\mu} B_0 \widetilde{\zeta_{ij}}(t) cosh(\lambda z)\right] J_m(\lambda r_k)\Theta(\theta_k) \quad (13)$$

where $J_m(\cdot)$ is the Bessel function of the first kind, λ is the separation constant, and

$$\Theta(\theta) = \{sinm\theta, m = 1, 2, 3 ... cosm\theta, m = 0, 1, 2, 3 ...\} \quad (14)$$

$$A_{ij}(t) = \frac{1}{Y(-\lambda h_i)} \times \left\{Z(-\lambda h_i) B_0 \widetilde{\zeta_{ij}}(t) + \frac{\kappa}{\pi R^2 [J_{m+1}(\varepsilon)]^2} \psi_{ij}(t) R(r_{ij}) \Theta(\theta_{ij})\right\} \quad (15)$$

$$Y(-\lambda h_i) = -\frac{1}{tanh(\lambda d) - coth(\lambda d)} \times \left\{\frac{\mu}{\mu_0}(\alpha + \chi) cosh\ cosh(-\lambda h_i) - \left(\frac{\mu}{\mu_0}(\alpha - \chi) - \frac{\chi^2}{\alpha}\right) sinh(-\lambda h_i)\right\} \quad (16)$$

$$Z(-\lambda h_i) = -\frac{1}{tanh(\lambda d) - coth(\lambda d)} \times \{\alpha cosh(-\lambda h_i) - \chi sinh(-\lambda h_i)\}\frac{\chi}{\mu}$$

$$\alpha = \frac{\mu}{\mu_0} tanh(\lambda d) - coth(\lambda d), \kappa = \{1\, m = 0 2\, m \neq 0\} \quad (17)$$

Considering that the miniature coils are located far from the walls of the fluid container, so at $r = R$ yields $J_m(\lambda R) = 0$, which can be solved numerically and yields an infinite number of solutions $\varepsilon_{mn} = \lambda R$, $m = 0, 1, 2, ..., n = 1, 2, 3, ...$, providing the eigenvalue λ_{mn} for each mode as $\lambda_{mn} = \varepsilon_{mn}/R$. Combining $J_m(\lambda_r)$ and $\Theta(\theta)$, we define the following mode shapes as $H_{mnc} = J_m(\lambda_{mn} r)\cdot \cos(m\theta)$ and $H_{mns} = J_m(\lambda_{mn} r)\cdot \sin(m\theta)$.

For any coil $\psi_{ij}(t)$ on each layer, based on Eq. (8) and the damping effect associated with the fluid viscosity η, the following surface dynamic equation with respect to the mode shape H_{mnc} can then be obtained as:

$$\frac{d^2 \widetilde{\zeta}_{ijmnc}(t)}{dt^2} + 4\frac{\eta}{\rho}\lambda_{mn}^2 \frac{d\widetilde{\zeta}_{ijmnc}(t)}{dt} \omega_{imn}^2 \widetilde{\zeta}_{ijmnc}(t) = -\frac{\chi}{\rho} B_0 \frac{tanh(\lambda_{mn} d)}{Y(-\lambda_{mn} h_i)} \lambda_{mn}^2 \frac{\kappa}{\pi R^2 [J_{m+1}(\varepsilon_{mn})]^2} \psi_{ij}(t) H_{mnc}(r_{ij}, \theta_{ij}) \quad (18)$$

where

$$\omega_{imn}^2 = gtanh(\lambda_{mn}d)\lambda_{mn} + \frac{\sigma}{\rho}tanh(\lambda_{mn}d)\lambda_{mn}^3 + \frac{\chi}{\rho}B_0^2 tanh(\lambda_{mn}d)\lambda_{mn}^2 \frac{Z(-\lambda_{mn}h_i)}{Y(-\lambda_{mn}h_i)} \quad (19)$$

$m = 0, 1, 2, \ldots$ and $n = 1, 2, 3, \ldots$

The main idea of the derivation of Eq. (18) is similar to the result of MFDM with a single-layer layout of actuators and more details can be found in [19]. A similar set of equations can be obtained concerning the mode shape H_{mns} as:

$$\begin{aligned} &\frac{d^2\tilde{\zeta}_{ijmns}(t)}{dt^2} + 4\frac{\eta}{\rho}\lambda_{mn}^2 \frac{d\tilde{\zeta}_{ijmns}(t)}{dt} \omega_{imn}^2 \tilde{\zeta}_{ijmns}(t) \\ &= -\frac{\chi}{\rho}B_0 \frac{tanh(\lambda_{mn}d)}{Y(-\lambda_{mn}h_i)}\lambda_{mn}^2 \frac{\kappa}{\pi R^2[J_{m+1}(\varepsilon_{mn})]^2}\psi_{ij}(t)H_{mns}(r_{ij}, \theta_{ij}) \end{aligned} \quad (20)$$

where m and $n = 1, 2, 3, \ldots$

The generalized displacements $\tilde{\zeta}_{ijmnc}(t)$ and $\tilde{\zeta}_{ijmns}(t)$, obtained from the solution of the second-order differential Eqs. (18) and (19) respectively, and the corresponding mode shapes H_{mnc} and H_{mns} evaluated at any desired location (r_k, θ_k), give the total surface displacement at the location as:

$$\begin{aligned} \zeta(r_k, \theta_k, t) &= \sum_{i=1}^{2}\sum_{j=1}^{J_i}\sum_{m=0}^{\infty}\sum_{n=1}^{\infty}\tilde{\zeta}_{ijmnc}(t)H_{mnc}(r_k, \theta_k) \\ &+ \sum_{i=1}^{2}\sum_{j=1}^{J_i}\sum_{m=0}^{\infty}\sum_{n=1}^{\infty}\tilde{\zeta}_{ijmns}(t)H_{mns}(r_k, \theta_k) \end{aligned} \quad (21)$$

Eqs. (18)–(19) show that the surface response $\zeta(r_k, \theta_k, t)$ has linear relation with the input $\psi_{ij}(t)$ applied by each coil. Note that the static surface response model of the mirror can be attained by the perturbed magnetic field of each actuator. For more details refer to Ref. [20].

This section introduced the current method to improve the correction performance of the MFDM for full-order aberrations, a new MFDM with a two-layer of actuators.

3. Nanoparticle manipulations for biomedical applications

The application of magnetic nanoparticles (NPs) has been extensively studied in biomedicine such as in imaging contrast enhancement [21] (see more detail in Section 4), magnetic separation [22], hyperthermia treatment [23], and targeted drug delivery [24] (see more detail in Section 5).

Due to their size, NPs can interact with biological systems at the molecular level and pass through biological barriers [24]. Using an external magnetic field to manipulate and to guide magnetic NPs to the wanted location has obtained special attention.

Several studies describe different magnetic nanoparticles as well as different delivery techniques [25]. Magnetite NPs (e.g., Fe_3O_4) are the most widely used due to several appealing properties [26]. This type of NPs exhibit superparamagnetism which means that they do not have magnetic moment when there is no external field but become magnetized when an external magnetic field is applied.

This property is of great value for a variety of biomedical applications as magnetite NPs do not form into a mass together and can move easily through blood vessels [26]. The typical shape of these NPs is a sphere with the size of around 5–100 nm. Most importantly, Fe_3O_4 NPs are biocompatible which makes it suitable for many biomedical applications [27].

Currently, one of the leading research interests is associated with the delivery of magnetic NPs to a target location. Several approaches have been proposed to address the issue. For example, the surface of magnetic NPs has been modified with biomolecules to identify and attach to target cells. Also, injection of magnetic NPs to some localized regions has been suggested [28]. Furthermore, using an external force to control the NPs in the desired direction remotely has been proposed [29]. When an electric current passes through the micro-coils, it generates a non-uniform magnetic field which attracts the NPs at its minimal. The magnitude of the magnetic field B relates to the size of the wire as $B \propto I/d$, gradients of the field as $\nabla B \propto I/d$ [22], and curvatures as $^2B \propto I/d$ [23], where I is the electric current through the wire [30]. The magnetic gradient makes it possible to trap magnetic NPs within a micrometer and sub-micrometer regions using small coils. The focusing depth of the micro-coils is in the range of tens of microns to several millimeters. The fluid containing magnetic NPs is usually placed on the top of the device allowing for magnetic manipulation of NPs. Such a method has been shown in several works for trapping and positioning of magnetic NPs by different types of micro-electromagnets [29, 31]. This provides the possibility of controlling the strength of the field using external parameters such the current passing through micro-coils.

As an example, this method [32] is used for focusing and selective destruction of red blood cells (RBCs) via magnetic NPs. The experiment was accomplished above the micro-electromagnet serving concurrently as a source of the magnetic field and as a local heater. The motion and focusing of the blood cells near wires which carry the electric current were observed in the previous study. It was also found that the increase of the electric current through the micro-electromagnet leads to the local cell hemolysis. All the previous results have been obtained for the case when the guiding magnetic field was produced by just one conducting contour. In this work, experimental data was obtained for the modified micro-electromagnet with two contours, where the current in each path can be individually controlled. This modification gives an extra degree of freedom to tune the magnetic field profile and to manipulate magnetic nanoparticles.

The schematics of the micro-magnetic prototype with biological substance and the external electric circuit are shown in **Figure 4(A)**. The device is fabricated on the top of a fused silica substrate. It consists of two Cu wires fabricated by standard optical lithography techniques. The width of the wires is 10 μm, and the height of the wires is 2 μm. The minimum spacing between the wires is 20 μm. Each of the two wires has individual contacts (1 mm × 1 mm) on the sides of the structure for connection with the outer electric circuit. The circuit includes a power source and two electrical relay switches allowing for the individual control of electric current in each wire. The wires are covered by a 300 nm thick layer of silicon dioxide deposited via PECVD which allows for both electrical insulations as well as corrosion protection for the Cu wires. The top part of the structure is the working area where the manipulation of biological cells takes place. As a test biological substance, human red blood cells (RBC) are obtained from fresh blood samples. RBC was selected both due to their availability as well as their robustness [33]. Samples were prepared by triple washing using centrifugation at 2000 rpm on an Eppendorf 5424 centrifuge in phosphate buffered saline solution (PBS) at room temperature. PBS was prepared from 10× PBS (Fisher Scientific, USA) concentrate solution with a final concentration of 2000–4000 cells/μl. From this RBC solution, 30 μl of the

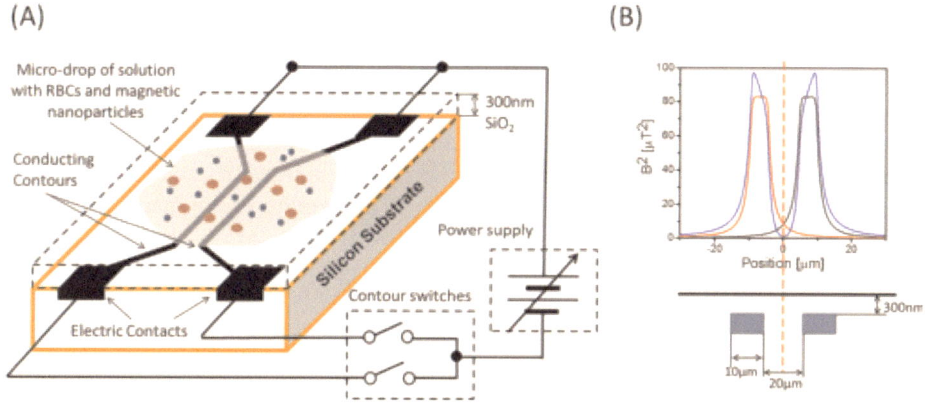

Figure 4.
(A) Schematics of the micro-electromagnet and the external electric circuit. The device is fabricated on the top of a fused silica substrate. It consists of two Cu wires fabricated by optical lithography. The width of the wires is 10 μm, and the height of the wires is 2 μm. The wires are covered by the 300 nm of silicon dioxide. The top part of the structure is the working area where solution with biological cells and magnetic nanoparticles is placed. (B) Results of numerical modeling showing the surface profile of the scaled magnetic energy B^2 produced by the current-carrying wires. The red and the black curves show the scaled magnetic energy when 100 mA electric current is flowing through one of the contours. The blue curve shows the scaled magnetic energy when an electric current is flowing through both contours (adapted from [34]).

solution are then mixed with 1 μl of commercially available ferrofluid from Ferrotec containing 1.1% magnetite (Fe_3O_4) particles in aqueous solution.

The electric current passing through the wire generates a non-uniform magnetic field around it. The gradient of this non-uniform magnetic field causes the magnetic NPs to be attracted towards the wires. Therefore, the movement of magnetic NPs pulls biological cells in the same direction [32]. **Figure 4(B)** shows the numerical results of the magnetic energy above the two straight wires. The red and black curves show the magnetic energy when the current is 100 mA in one of the wires. The blue curve shows the magnetic energy when the current passes through both wires. There is an overlap between the magnetic fields in the region and the wires (as depicted by the red dashed line in **Figure 4(B)**). This overlap is important for sequential trapping of magnetic NPs by applying a current through one or the other wires. The maximum of the field gradient is about 1.4 mT/μm in the vicinity of the current-carrying wire.

The individual control of electric current (hence magnetic field) in the two contours allows for a variety of experiments to be conducted on biological cell manipulation. This application shows that the biological cell manipulation is possible due to the presence of magnetic nanoparticles. To support this statement, some studies performed control experiments with RBC samples without adding magnetic nanoparticles [34]. It is shown that applying a magnetic field without nanoparticles initiates a slight motion of RBCs away from the field source which is attributed to local heating and the expanding solution. However, when the study is conducted by adding 1.1% magnetic nanoparticles, applying the magnetic field results in the cells focusing near the field source. The focusing is attributed to the drag effect, where a flow of magnetic nanoparticles drags the cells in the same direction. The potential application of this is that with time-varying magnetic field one can move RBCs towards or away from the specific region. This technique can be further evolved for precise controlling of cellular motion and drug-carrying cells. There are many questions related to the mechanism of magnetic nanoparticles interaction with living cells which deserve separate studies and it is out of the scope of this chapter. Nanoparticle manipulation using magnetic field potentially may be useful in various areas including molecular biology, medicine, gene engineering, and drug delivery technology.

4. Biomedical imaging contrast enhancement

MHD plays an important role in biomedical imaging, ranging from magnetic resonance imaging (MRI) that employs the endogenous magnetization contrast of water in tissue to magnetofluids acting as contrast agents in a number of imaging modalities. These contrast agents enhance the ability to detect tumors, infection, inflammation, infarction or lesions in the body. This section highlights the role of magnetic fluids as contrast enhancing agents in MRI, X-ray computed tomography (CT) and optical coherence tomography (OCT) imaging.

4.1 Contrast agents in MR imaging

MRI is one of the major biomedical applications of MHD. MRI is a non-invasive tomographic medical imaging technique based on nuclear magnetic resonance (NMR) that provides high-resolution images of soft anatomical structures such as brain, heart, ligaments, and eyes [35–41]. This information is vital in delineating healthy from diseased tissues or organs. MRI takes advantage of the inherent magnetic dipole moments of the atomic nuclei in our bodies, specifically hydrogen nuclei contained in water—that makes about 70% of our body mass. Three magnetic fields comprising of a static magnetic field (SMF), a time-varying gradient magnetic fields (GMF) and a pulsed radiofrequency field (RF) are used to probe the magnetization/demagnetization map of the body. The water molecules that occupies most of the tissue in the body consist of hydrogen atoms with nuclei that possesses a quantum-mechanical spin. The hydrogen nucleus spin is associated to a magnetic dipole moment (**Figure 5(A)**) that gets aligned to a strong magnetic field B_0 with a bulk magnetization M_{z0} along B_0 when the body is placed in it as shown in **Figure 5(B)**. Since these hydrogen nuclei dipoles have an angular momentum from the rotation about their own axes, they precess around the B_0 magnetic field axis with an angular frequency that follows Larmor equation $\omega = \gamma B_0$, where γ is the gyromagnetic ratio constant of the hydrogen nucleus. The precession gives rise to a transverse magnetic field that can be detected by a coil conveniently placed in the system to pick up this signal. The equation of the signal detected is proportional to $S(t) = \gamma B_0 \rho sin(\theta) \sin(\omega t)$, here γ is the gyromagnetic ratio constant of the hydrogen nucleus, B_0 the magnetic field intensity of the static magnetic field, ρ the density of hydrogen nuclei of the tissue, θ the flip angle between the axis of the SMF and the

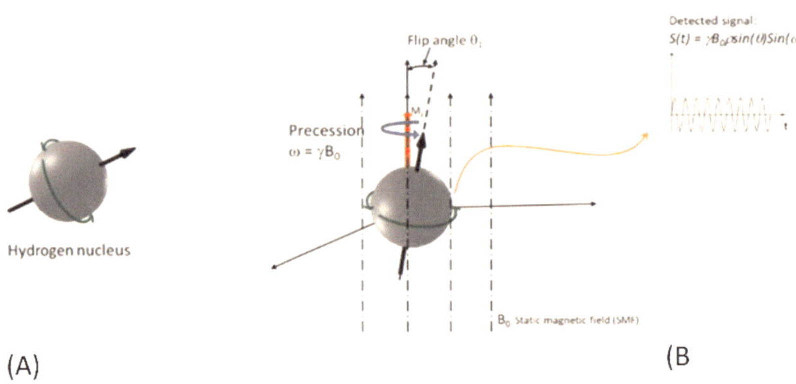

Figure 5.
(A) The hydrogen nucleus possesses a quantum-mechanical spin (angular momentum) that is associated with a magnetic dipole moment. (B) When the body is placed in a strong magnetic field B_o, the hydrogen nuclei dipoles get aligned with the field B_o with a net magnetization vector M_{Zo}. The dipoles precess around the axis of the magnetic field with an angular frequency that follows Larmor equation $\omega = \gamma B_0$, where γ is the intrinsic gyromagnetic ratio constant of the hydrogen nucleus.

rotation axis of the hydrogen nucleus and ω the Larmor angular frequency. When a resonant pulsed RF magnetic field is applied perpendicular to the SMF, the hydrogen nuclei process around the SMF axis in phase. After removal of RF pulse, the spins begin to diphase and so the detected transverse magnetic field M_T signal starts to decrease. It does so according to the equation $M_T(t) = M_T(0)\exp\left(\frac{-t}{T_2}\right)$, where T_2 is called the spin-relaxation time. The nuclei return to their initial equilibrium state before the RF pulse by emitting an MR signal which also occurs by stimulation from surrounding nuclei. The process is assumed to occur in a simple exponential manner according to the equation $M_Z(t) - M_0 = (M_{Z0} - M_0)\left(1 - \exp\left(\frac{-t}{T_1}\right)\right)$, where T_1, the time required for the nuclei system to return to 63% the equilibrium state, is called the spin-lattice relaxation time [42–44] (**Figure 6**).

Tissue/organ contrasts in MRI arises from differences in mainly these two basic physical parameters: the difference in the spin-relaxation time T_1 in the different organs required for the realignment to the SMF by the hydrogen nuclei of after the RF pulse is removed; the time constant T_2 with which the spins' signals arising from a given tissue will diphase after the pulsed RF signal is removed. **Figure 7** shows the variation in the longitudinal magnetization M_Z characterized by T_1 and the decay in

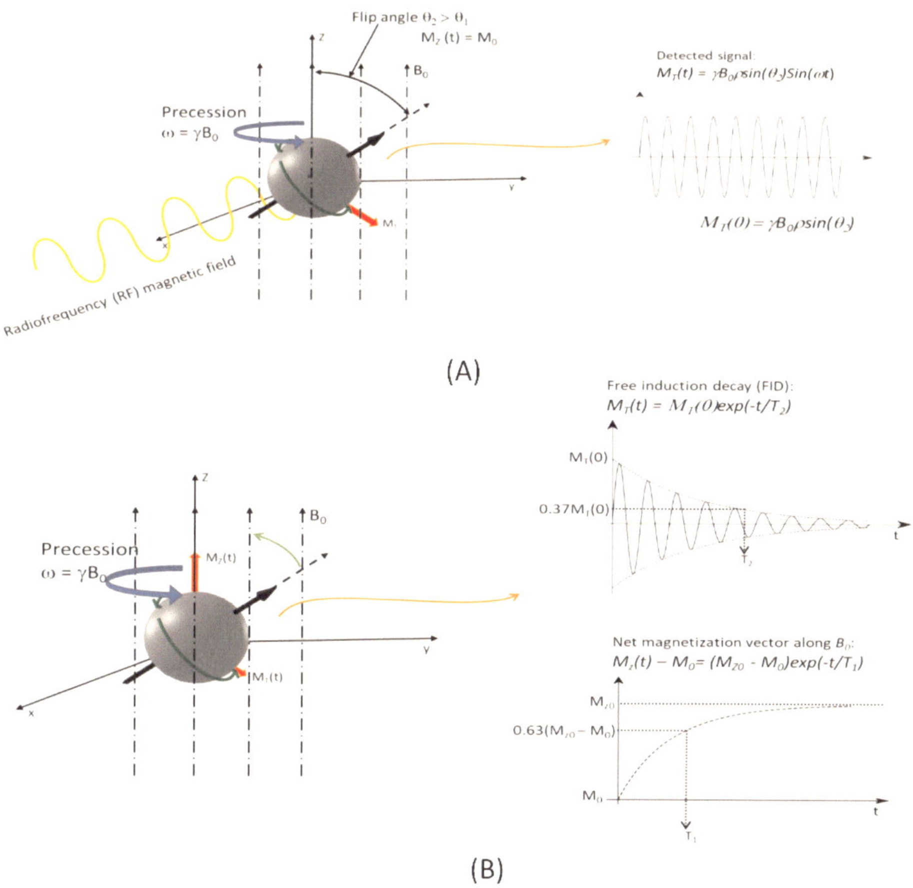

Figure 6.
(A) When a 90° RF pulse is applied to the aligned magnetic dipoles, the net magnetization gets tipped to the transverse xy-plane. Dephasing of the spins results in a quick decrease of the net magnetization in the xy-plane M_T. The dephasing occurs exponentially and characterized by T_2. (B) After the pulsed magnetic field is removed, the longitudinal magnetic field M_Z begins to grow exponentially to a maximum of M_{Zo}; this growth is characterized by a parameter called T_1, which the time it takes for the net magnetization along B_o to grow to 63% of the maximum value M_{Zo}.

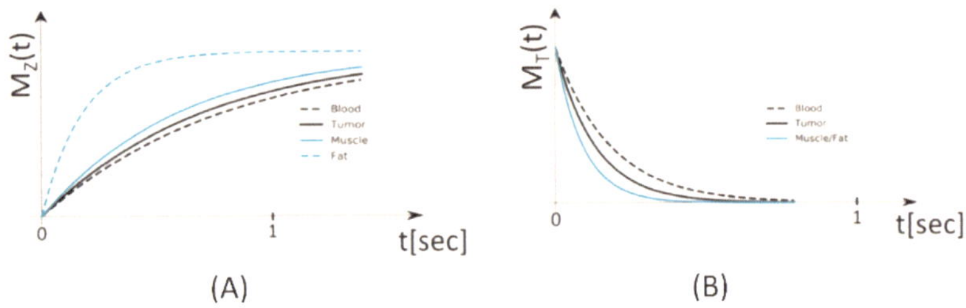

Figure 7.
(A) The T_1 relaxation time, known as the spin-lattice relaxation time, is the measure of how fast the net magnetization vector (NMV) occurs along the SMF. Different tissue in the body possesses different T_1 relaxation times (a) which act as a contrast for the different tissue and even of diseased tissues such as tumors. (B) T_2 relaxation refers to the progressive dephasing of spinning dipoles following the 90° pulse as seen in a spin-echo sequence due to tissue-particular characteristics (b).

the transverse magnetization M_T, characterized by T_2 in the different tissues in the body [45].

Magnetic fluids play an important role as contrast enhancing agents in MRI. These contrast agents serve to shorten the relaxation time of the water molecules surrounding the tissue or organs with the contrast agent molecules, which then increases the signal intensity detected and thereby providing a positive contrast. The contrast agents used in MRI can broadly be divided into two major categories: the T_1-weighted contrast agents and T_2-weighted contrast agents. T_1-weighted contrast agents shorten the T_1 relaxation time to improve the T_1 relaxation contrast, brightening the T_1-weighted image. T_1 contrast agents usually consist of gadolinium (Gd) compounds. On the other hand, T_2 contrast agents serve to brighten the T_2-weighted images by shortening the T_2 relaxation time. T_2 contrast agents consist mainly of superparamagnetic iron oxide and iron platinum. **Table 4** summarizes the major commercially available magnetic fluids used as contrast agents in MRI imaging today.

Compounds	Trade name	Target organs and tissue	Reference
T_1-weighted contrast agents			
Gadopentetate dimeglumine (Gd-DTPA)	Magnevist®	Glioma	[46, 47]
Gadoterate meglumine (Gd-DOTA)	Dotarem®	Brain and spine	[48]
Polyamidoamine	Dendrimer®	Angiography and tumor differentiation	[49]
Gadoxetate disodium (Gd-EOB-DTPA)	Primovist®	Liver	[50]
Gadodiamide (Gd-DTPA-BMA)	Omniscan®	Blood vessels	[51]
Gadobenate dimeglumine (Gd-BOPTA)	MultiHance®	Liver	[52]
Gadoteridol (GD-HP-DO3A)	ProHance®	Brain and spine	[53]
Gadoversetamide (C20H34GdN5O10)	OptiMARK®	Brain, spine, and liver	[54]
Gadobutrol (Gd-BT-DO3A)	Gadovist®/ Gadavist®	Angiography	[55]
Gadocoletic acid trisodium	Gadocoletic acid	Angiography	[56]

Compounds	Trade name	Target organs and tissue	Reference
Gadomelitol	Vistarem® (Gadomelitol)	Angiography	[57]
Gadoteric acid (Gd-DOTA)	ClariscanTM	Brain and spine	[58]
T_2-weighted contrast agents			
Ferumoxide (AMI-25) ($Fe_3O_4{\cdot}\gamma Fe_2O_3$)	Feridex®	Liver, spleen, Bone marrow	[59, 60]
Ferumoxsil (AMI-121) ($Fe_3O_4{\cdot}\gamma Fe_2O_3$)	Lumirem®	Liver, spleen, gastrointestinal tract	[61]
Ferumoxtran ($Fe_3O_4{\cdot}\gamma Fe_2O_3$)	Sinerem®	Lymph nodes, blood	[62]
Ferrixan (Fe_3O_4)	Resovist®	Liver	[63]

Table 4.
Commonly used MRI contrast agents for medical diagnosis.

4.2 Contrast agents in X-ray and CT

Computed tomography (CT) is a powerful non-invasive diagnostic imaging technique [64]. CT can be employed for imaging hard organs or tissues (e.g., bones) or soft ones such as the gastrointestinal (GI) tract, the cardiovascular system, renal tract, liver, lungs, cartilage, and tumorous tissue with the aid of contrast agents. A CT image is obtained by rotating an X-ray source(s) (or detector/detector array) around an object or vise verse, with a detector(s) positioned directly opposite the radiation source(s). Generally, X-ray scans are taken at small angular increments during rotation around the object over 360° or 180° [65]. An X-ray attenuation (or phase or scattering) map or projections are thus obtained. The projections are then processed mathematically to create a 3D rendering of the scanned object.

Another diagnostic imaging method related to CT is X-ray fluoroscopy—form of a projection imaging with contrast agent. Fluoroscopy allows for the acquisition of real-time, continuous images of the internal organs. Like in MRI, imaging contrast agents are often used in X-ray imaging for better contrast resolution. Usually, small iodinated agents are injected into blood vessels for use in fluoroscopic angiography, allowing for the evaluation of blood flow and visualization of the vasculature system, while barium contrast media are introduced orally or with an enema to investigate the anatomy (and pathology) of the gastrointestinal tract. The X-ray absorption coefficient μ can be expressed as:

$$\mu \approx \frac{\rho Z^4}{AE^3} \tag{22}$$

where ρ is the density of the material, Z the atomic number, A the atomic mass and E the X-ray energy. Therefore, materials of higher density and atomic number, higher density tissues, absorb X-ray better [66].

While lanthanide-based contrast agents are a common stay in MRI, their application in CT as contrast agents is being explored based on their high atomic numbers [64]. The two major reasons motivating the investigation of gadolinium-based compounds as CT contrast agents include use in patients who are contraindicated for iodinated agents based on allergic reactions or renal insufficiency and the fact that *Gd* has higher K edges than iodine providing better contrast enhancement at higher X-ray photon energies, potentially reducing radiation exposure to patients [67]. A summary of some of the gadolinium compounds used in MRI as contrast agents are listed in **Table 4**. Clinically approved gadolinium-based contrast agents

Compound	Trade name	Target organ	Reference
1. Gadopentate dimeglumine (Gd-DTPA)	Magnevist®	Urinary tract, aorta, blood vessels in cranium	[68, 69]
2. Gadodiamide (Gd-DTPA-BMA)	Omniscan®	Arterial angiography,	[70]
3. Gadoteridol (Gd-HP-DO3A)	ProHance®	Aorta, brain tumors,	[71]
4. Gadobutrol (Gd-BT-DO3A)	Gadovist®	Alternative to iodinated agents	[72]
5. Gadoxetate disodium (Gd-EOB-DTPA)	Primovist®	Liver, spleen, urinary tract	[73]

Table 5.
Lanthanide-based contrast agents for CT and X-ray.

have been tested in CT for imaging the cardiovascular system and for pulmonary and aortic angiography. **Table 5** summarizes some of the *Gd* compounds that have been tested as contrast agents in CT.

4.3 Magnetic contrast agents in OCT imaging

OCT is non-invasive cross-sectional imaging modality that uses light to interrogate tissue providing a 3D rendering of the tissue under investigation [74]. **Figure 8** depicts the principle of OCT. A low-coherence light from a laser source is split into two halves, with one half guided to a mirror (reference arm) and the other half used to illuminate the tissue under investigation (sample arm). The reflections from the reference arm and from the tissue are recombined by beam combiner and detected. The electrical signal from the photodetector is bandpass digitized and demodulated before being stored on a computer. The low-coherence in the light being used allows for interference between the two arms to occur only when the two optical path lengths are equal. Therefore, fine scanning of the reference arm mirror allows for interference to occur with light originating from the different depths of the tissue. Transverse scanning of the optical beam then allows for a 3D imaging of the tissue [75].

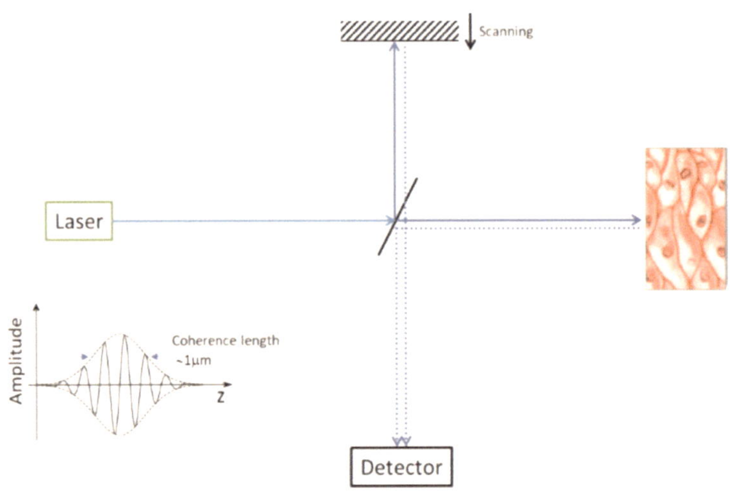

Figure 8.
Schematic illustrating the concept of low coherence interferometry. Using a short coherence length light source and a Michelson-type interferometer, interference fringes are observed only when the path lengths of the two interferometer arms are matched to within the coherence length (l_c) of the light source (adapted from [41]).

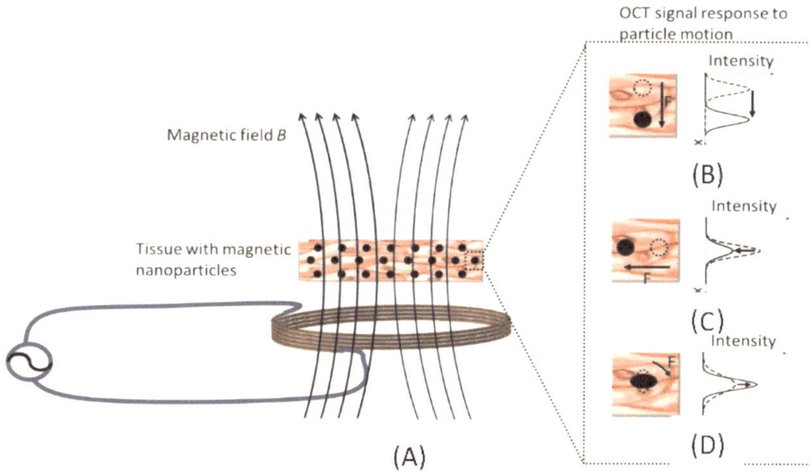

Figure 9.
Magnetomotive OCT tissue configuration. (A) A sample is doped with magnetic microparticles, placed in a magnetic field and shone with an optical imaging beam. The contrast mechanisms illustrated by two states of a magnetic microparticle (filled and dash-outlined shapes, respectively) within the imaging focal volume and the associated (A–D).

The ability of a tissue to be magnetized is quantified as magnetic susceptibility χ and can be used as means of contrast in OCT imaging. Biological tissue is comparatively non-magnetic $\left(|\chi| \sim 10^{-5}\right)$ when compared to ferromagnetic iron oxides, whose χ is 10^5 times greater than that of tissue. This means that the χ difference between tissue and magnetic particles can be exploited to provide contrast in OCT imaging of tissue doped with magnetic nanoparticles whose localized exogenously induced motion can be resolved by OCT and used as a contrast feature [76]. The motion of the nanoparticles in the tissue is caused by a magnetic gradient force:

$$\vec{F} = \frac{V(\chi - \chi_{med})\nabla\left|\vec{B}\right|^2}{2\mu_0} \quad (23)$$

where F is the force acting on the magnetic nanoparticles with susceptibility of χ within its surrounding tissue medium with susceptibility of χ_{med}, B being the magnetic flux density, V is the volume of the tissue nanoparticles and magnetic permeability in vacuum [77].

As shown in **Figure 9**, when the magnetic nanoparticles get laterally displaced by the magnetic force, the OCT the increase in the light scattered back also gets shifted laterally. **Figure 9(B–D)** shows the variation in the OCT signal variation that follow changes in the nanoparticles position arising from the magnetic force.

5. Targeted drug delivery

5.1 Introduction

As it was initially assumed by Widder et al. in 1978 [78], magnetic constructs can target specific locations, such as tumor sites, holding enormous potential for site-specific drug delivery, see **Figure 10**. This subchapter discusses a clinically driven application of MHD focusing on magnetic drug targeting. MHD drug targeting refers to the magnetically targeted and/or triggered therapeutic agent delivery

Figure 10.
Magnetic constructs (A) or constructs with evenly distributed magnetic material (B) applied parenterally (intravenously or intra-arterially), can be used as intravascular vehicles for targeted drug delivery if drug-loaded (C) or -coated (D).

method that augments the concentration of the agent in the target area. In the first part, drug-loaded magnetic constructs are discussed as individual intra-vascular vehicles for site-specific drug delivery. In the second part, magnetohydrodynamic (micro) pumps for controlling magnetic nanoparticles dispersed in a base fluid are discussed. These are reviewed and summarized separately based on the knowledge already available in literature.

5.2 Magnetic constructs

Due to their superior properties such as biocompatibility, biodegradability, large loading capacity, and controlled release ability, magnetic constructs have attracted a lot of attention. They offer mean to remotely direct therapeutic agents to well-localized sites of interest, improving control on dosing, reducing the concentration requirements up to 20% [79], associated toxicity, and fluctuation in circulating drug levels [80–82]. While drug release usually occurs by passive diffusion, triggering via enzymatic activity, physiological conditions [83] or magnetically [84–86], see **Figure 11**, can be performed. On-site controlled drug release is another important factor for effective therapeutics. For example, when the magnetic construct is exposed to an alternating magnetic field, drug release can occur. Via facilitated drug release, collateral tissue damage and toxic side effects may be further decreased or fully eliminated [87, 88]. This is important for applications balancing between efficacy and toxicity, e.g., chemotherapies [89].

The use of magnetic constructs as therapeutic agents has increased exponentially since the earliest studies by Senyei et al. [90] and magnetic carriers of increased sophistication have been developed [91]. While nanoparticles are the simplest magnetic constructs, currently, there are many different types of carriers for magnetic targeting. Generally, any biocompatible magnetic materials, magnetic materials coated by a biocompatible polymer or inorganic material, or magnetic materials precipitated inside the pores of a biocompatible polymer or inorganic material can be used. While the use of iron oxide particles in the form of magnetite (Fe_3O_4) or maghemite (γ-Fe_2O_3) predominates, any metal, e.g., cobalt or nickel, or

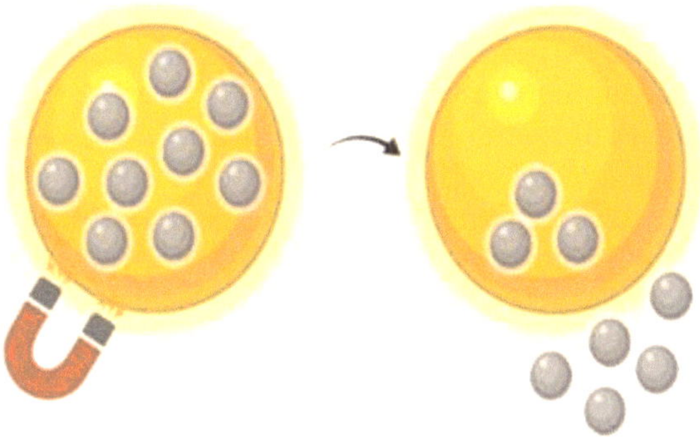

Figure 11.
Targeted drug release.

metal derivatives, e.g., μ-oxo N,N′-bis(salicylidene)ethylenediamine iron-Fe(salen) [92–94], are used. Magnetic constructs can be porous or hollow single magnetic nano/microparticles encapsulating the drug in pores and cavities [95], non-magnetic polymeric or inorganic nano/microparticles encapsulating both magnetic material and drug [96], magnetosomes [97], micelles [98] or liposomes, micro/nanoswimmers or micro/nanomotors prepared by layer-by-layer deposition [99] or 3D-printing [100], nanoparticle clusters [96, 101].

The basic principle of magnetically targeted drug delivery is based on the physical phenomenon of the translational motion of a magnetic construct when a magnetic field gradient is applied. Magnetic construct with a therapeutic agent encapsulated into the construct and/or conjugated on its surface is injected systemically (or delivered locally via catheter) and transported by the blood circulation to the target location. While intravenous injection is associated with higher clearance from the bloodstream, it is being used more frequently when compared to the intra-arterial routes. A strong, high-gradient magnetic field is then applied to allow translating magnetic constructs to the target location and capturing them. Literature suggests that flux density at the target location must be of the order of a few hundred millitesla with a few teslas per meter field gradient for slow blood flow arteries and up to a few hundred for fast blood flow arteries.

Magnetic constructs are driven by the magnetic force F_m dependent on the magnetic induction B:

$$B = \mu_0(H + M) = \mu_0 H \tag{24}$$

where μ_0 is a permeability of free space, H is a magnetic field strength and M is a total magnetization of the magnetic construct which could be neglected in case of diluted suspension.

In the case of supermagnetic constructs in a diamagnetic base solution, the magnetic moment m on a magnetic construct:

$$m = V_m M = V_m \Delta\chi H = V_m(\chi_m - \chi_w)H \tag{25}$$

where V_m is a volume magnetic construct, $\Delta\chi$ is an effective susceptibility given via a difference between susceptibility of a magnetic construct χ_m and susceptibility of a base solution χ_w.

Under the assumption that magnetic constructs are very small and could be assumed to be point-like particles:

$$F_m = (m \cdot \nabla)B = \frac{V_m\,\Delta\chi}{\mu_0}(B \cdot \nabla)B = V_{\mathrm{m}}\Delta\chi\nabla\left(\frac{B^2}{2\mu_0}\right) = V_m\Delta\chi\nabla\left(\frac{1}{2}\,B \cdot H\right) \tag{26}$$

On the one side, the magnetic force F_m dependents on the magnetic constructs size (amount of magnetic material). While magnetic constructs are flexible in their dimensions from a few nanometers up to a few micrometers, there is a trade-off between larger magnetic force and the appropriate physiological response [82, 102, 103]. On the other side, the magnetic force is proportional to the magnetic gradient. The magnetization of the magnetic constructs, however, increases with the external magnetic field only while the magnetic field is below the saturation field.

In order to increase magnetic force up to several orders of magnitude further, magnetizable implant (biocompatible wire, needle, stent, filament or seed) creating a high-gradient magnetic field at the target location under the influence of an external magnetic field could be used, see **Figure 12** [104]. While implant can solve problem of magnetic force strength and distance decay, as well as being successfully used at MRI facilities in applications related to cardiovascular, digestive and urinary systems under clinically feasible conditions, it makes the procedure of magnetic drug targeting minimally invasive. The concept has been successfully demonstrated *in vivo* [105] and is promising for magnetically targeted thrombolytic therapy.

Although magnetic drug targeting is appealing, most studies demonstrate only *in vitro* results for superficial organs due to difficulties in manipulating magnetic particles *in vivo* and rapid decrease of magnetic force with distance [79]. While some technical challenges remain, several drug delivery systems have been developed for the treatment of pulmonary disorders [106, 107], cancer [108–110], and cardiovascular diseases [111].

5.3 MHD (micro)pumps

The concept of MHD (micro)pump is relatively new and was developed by Jang and Lee [112] only in 1999 with an initial goal of applying in drug delivery applications. The working fluid of MHD (micro)pump is magnetic fluids. The term nanofluid was previously introduced by Choi and Eastman [113] in 1995 and describes colloidal suspensions of magnetic nanoparticles or nanotubes ($d_\varnothing$ < 100 nm) in a based solution e.g., water, oil, ethylene glycol mixture etc. Magnetic fluids, also called ferro- or nano-fluids, simultaneously exhibit liquid and magnetic properties, leading to the possibility to control their flows with magnetic fields.

In a typical setup, see **Figure 13**, a uniform magnetic field of strength B creates a controllable force (Lorentz force) F as a driving source in the flow for control and manipulation [114–119]. MHD (micro)pump can be fully described by a combination of Navier–Stokes equations of fluid dynamics and Maxwell's equations of electromagnetism. Considering the absence of moving parts, MHD (micro)pump

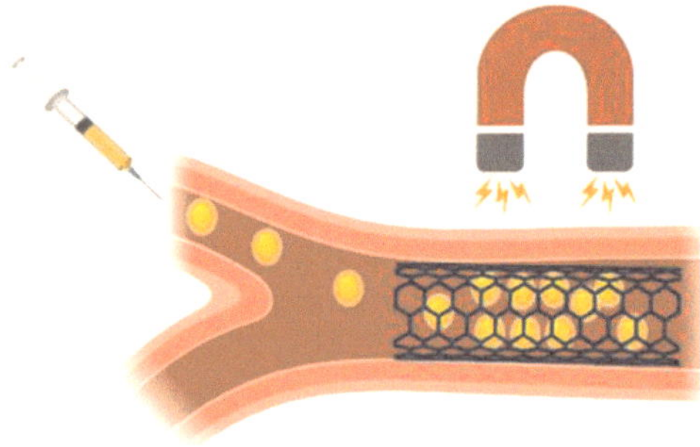

Figure 12.
Implant-assisted magnetic drug targeting with the magnetic stent implanted in the vasculature.

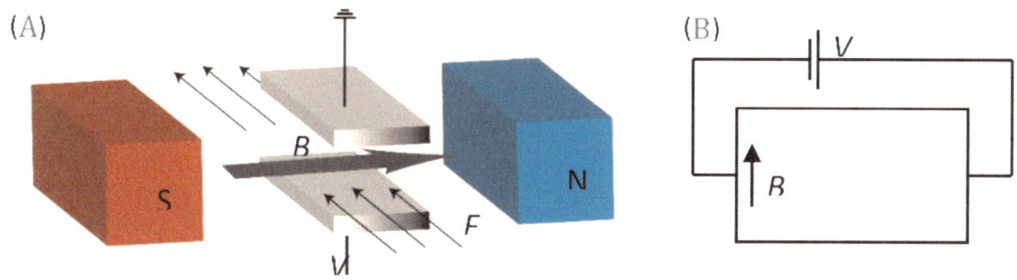

Figure 13.
Schematic of MHD (micro)pump (A) and its simplification (B). When an electric potential difference V is applied between the electrodes, an electric current circulates through the electrically conducting magnetic fluid perpendicularly to the uniform magnetic field of strength B so that driving Lorentz force F is produced.

possess a simpler fabrication process when compared to mechanical (micro)pumps [120]. MHD (micro)pumps have attracted the attention of many researchers as they could achieve high flow rates and produce bi-directional flow. To use MHD (micro) pumps for drug delivery, however, the stability of the flow rate is critical. The flow rate of MHD (micro)pumps depends on the current and the magnetic flux density [112]. Nevertheless the flux density from a permanent magnet is higher than that of an electromagnet, some authors hypothesized that electromagnets are more useful as their polarity could be reversed, leading to the change of the flow direction [82]. The use of direct current (DC) and alternate current (AC) MHD (micro)pumps have been demonstrated [120]. While the application of direct current (DC) (micro)pumps is often associated with bubbles leading to electrode corrosion, they can achieve higher flow rates (**Figures 14** and **15**).

5.4 Magnetic separation via particle labeling

Magnetic separation is being used in many biomedical applications, particularly cellular separation [121]. It is achieved via labeling the desired biological entity, e.g., red blood cells, with biocompatible magnetic nanoparticles. Labeled objects are separated from the base solution by passing the mixture through a high magnetic field gradient immobilizing the labeled entities via the magnetic force:

$$F_m = 6\pi\eta R_m \Delta v \tag{27}$$

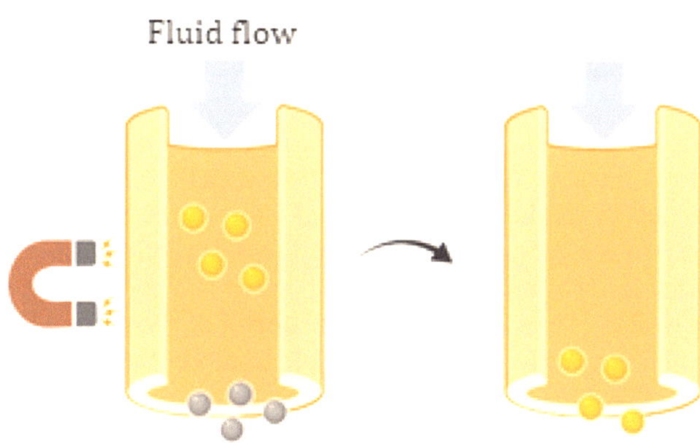

Figure 14.
Magnetic separation. Left: while magnetically labeled objects (yellow) are attracted by the magnet, non-magnetic objects (gray) and base fluid can be filtered out. Right: after removing the magnetic force captured objects are recovered.

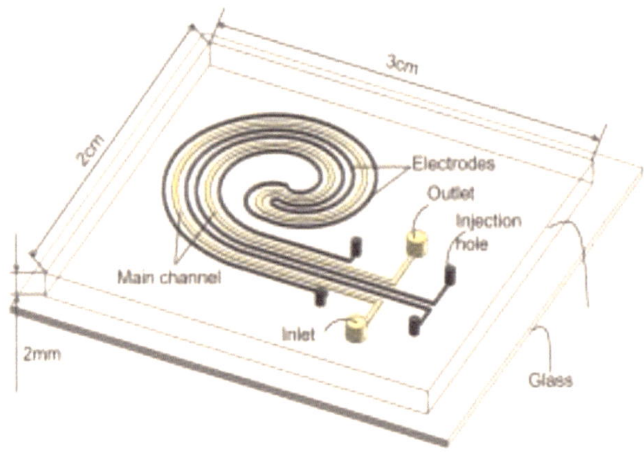

Figure 15.
Schematic of the polydimethylsiloxane (PDMS)-glass microfluidic chip for the liquid-metal-based magnetohydrodynamic (MHD) micropump (adapted from [122]).

where η is a base solution viscosity, R_m is a radius of magnetic nanoparticle, $\Delta \upsilon$ is a relative velocity of nanoparticle with respect to the base solution.

6. Current biomedical devices based on magnetohydrodynamics

In this section, we describe some of the applications of MHD such as pumps, integrated fluidic networks, stirrer, and micro-coolers.

6.1 MHD-based micro-pumps

One the best-known application of MHD is in the field of fluid pumping. The device consists of a conduit with two electrodes on either side. When a potential difference is applied across the electrodes, current flux J flows through the solution. In the presence of a magnetic field B, the electric field J interacts with the magnetic field B to generate a Lorenz force $J \times B$ which drives fluid motion.

Below is an example of MHD-based micro-pumped on PDMS platform.

6.2 MHD-based microfluidic networks

In lab-on-chip applications, it is often necessary to transport fluids and reagents across networks of conduits. Controlling the flow usually requires the use of pumps and valves. It is hard to implement mechanical pumps and valve in a lab-on-chip setting. MHD provides a proper solution that does not require a mechanical component. The basic idea is to equip many of the network's conduits, if not all, with individually controlled electrodes. By careful control of the electrode's currents and in the presence of a magnetic field, it is possible to direct the fluid flow along any desired direction [123, 124].

Figure 16 shows a simple example of an MDH microfluidic network fabricated using low temperature co-fired ceramic tapes (LTCT). By programming, electrodes can circulate the fluid around the tours at any desired direction and even maintain their temperature allowing for various biological processes like thermal cycling and potentially polymerase chain reaction (PCR) for DNA amplification.

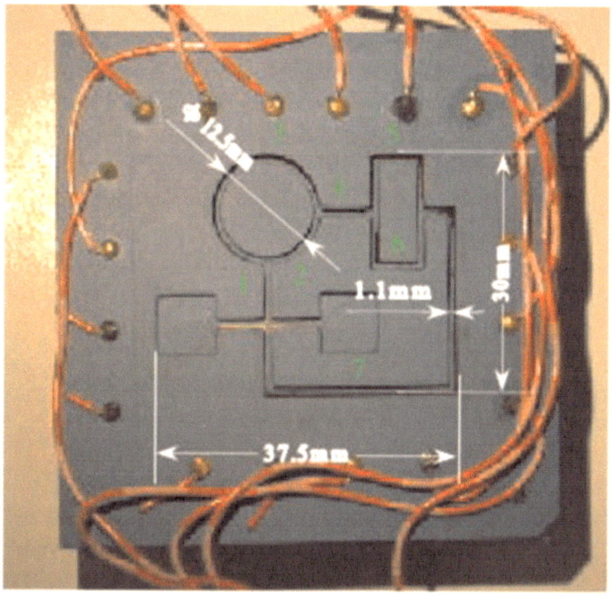

Figure 16.
A prototype of an MHD microfluidic network. The conduits are labeled with numbers (adapted from [125]).

6.3 MHD-based stirrer

Although the characteristic lengths associated with the microfluidic devices are small (e.g., in the order of 100 μm), diffusion alone does not allow sufficiently fast mixing. For example, at room temperature, myosin's diffusion coefficient in water is about 10^{-11} m^2/s, and the diffusion time along with a length of 100 μm is very large, about 10^3 s. Since Reynolds number (*Re*) of flows in micro-devices are usually very small ($Re \ll 1$), the flows are laminar, well-organized which cause poor mixing.

MHD provides us with rather easy means for mixing and stirring. Two different types of MHD stirrers have been reported in the literature. One relies on altering the flow direction to enhance dispersion [126, 127], while another type induces secondary flows to improve the mixing. For more details, refer to [126, 127].

In the current studies, it is shown that MHD is very suitable for providing better stirring in the microfluidic setting.

6.4 MHD-based micro-coolers

Since MHD can facilitate fluid circulation, it can be used to facilitate cooling. Liquid metals are particularly suitable for this purpose due to their high thermal conductivity, high boiling point temperature, ad large electric conductivity. Since MHD propulsion is easy to implement, miniaturize and does not require mechanical components, it is ideal for micro-cooling applications, such as those required in microelectronics. Although various patents address MHD micro-coolers, it is not known whether any products are in actual use.

7. Conclusions

In summary, this chapter covers important aspects of the MHD applied in the biomedical field. A new MFDM with the two-layer layout is proposed to improve the correction performance of the DM for full-order aberrations. The results showed the effectiveness of the method to correct full-order aberrations for adaptive optics systems. Furthermore, RBC manipulation by the magnetic field is demonstrated via an external magnetic field produced by a system of two current-carrying wires. It is shown that cell motion towards and away from the wires, as well as periodic motion in the region between the wires. This approach will likely have application in various fields including molecular biology, medicine, gene engineering, and drug delivery technology.

MHD plays an integral part in biomedical imaging, ranging from the endogenous magnetization properties of tissue that play an important role in MRI imaging to the ferromagnetic/superparamagnetic fluids that act as contrast enhancing agent in several imaging techniques such as MTI, CT/X-ray and OCT imaging. Several commercial contrast agents are in clinical use today. Research continues to be undertaken on new contrast agents and on the utility of MRI contrast agents in areas such as CT/X-ray and OCT imaging.

Moreover, the same magnetic constructs allow for a combination of enhanced diagnostic imaging (MRI, CT, OCT) and therapeutics (targeted drug delivery). While some technical challenges remain, several drug delivery systems have been successfully developed for treatment of pulmonary disorders, cancer, and cardiovascular diseases. At the end, we briefed current MHD-based devices with potential biomedical applications. MHD-based microfluidics operates at low voltages, can direct the liquid to flow along any desired path without a need for valves and pumps, and continuously circulate the sample in a closed loop, and furthermore can chaotically stir the sample without moving part.

Conflict of interest

Authors do not have a conflict of interest.

Author details

Hamid Farrokhi*, David O. Otuya, Anna Khimchenko and Jing Dong
Wellman Center for Photomedicine, Massachusetts General Hospital and Harvard Medical School, Boston, Massachusetts, USA

*Address all correspondence to: hfarrokhi@mgh.harvard.edu

References

[1] Priest ER. Solar Magnetohydrodynamics. Cambridge, UK: Cambridge University Press; 2003

[2] Freidberg JP. Ideal Magnetohydrodynamics. New York, NY: Plenum Press; 1987

[3] Laughlin DR. A magnetohydrodynamic angular motion sensor for anthropomorphic test device instrumentation. SAE Technical Paper. 1989:892428

[4] Jang J, Lee SS. Theoretical and experimental study of MHD (magnetohydrodynamic) micropump. Sensors and Actuators A: Physical. 2000;**80**:84-89

[5] Lemoff AV, Lee AP. An AC magnetohydrodynamic micropump. Sensors and Actuators B: Chemical. 2000;**63**:178-185

[6] Louis J, Lothrop J, Brogan T. Fluid dynamic studies with a magnetohydrodynamic generator. Physics of Fluids (1958–1988). 1964;7: 362-374

[7] Rosa RJ. Physical principles of magnetohydrodynamic power generation. Physics of Fluids (1958-1988). 1961;**4**:182-194

[8] Lei X, Wang S, Yan H, Liu WJ, Dong LZ, Yang P, et al. Double-deformable-mirror adaptive optics system for laser beam cleanup using blind optimization. Optics Express. 2012;**20**:22143-22157

[9] Sun B, Salter PS, Booth MJ. Pulse front adaptive optics: A new method for control of ultrashort laser pulses. Optics Express. 2015;**23**:19348-19357

[10] Carroll J, Kay DB, Scoles D. Adaptive optics retinal imaging—Clinical opportunities and challenges. Current Eye Research. 2013;**38**:709-721

[11] Sheehy CK, Tiruveedhula P, Sabesan R, Roorda A. Active eye-tracking for an adaptive optics scanning laser ophthalmoscope. Biomedical Optics Express. 2015;**6**:2412-2423

[12] Merino D, Loza-Alvarez P. Adaptive optics scanning laser ophthalmoscope imaging: Technology update. Journal of Clinical Ophthalmology. 2016;**10**:743-755

[13] Iqbal A, Amara FB. Modeling and experimental evaluation of a circular magnetic-fluid deformable mirror. International Journal of Optomechatronics. 2008;**2**:126-143

[14] Parent J, Borra EF, Brousseau D, Ritcey AM, Déry JP, Thibault S. Dynamic response of ferrofluidic deformable mirrors. Applied Optics. 2009;**48**:1-6

[15] Wu ZZ, Yuan S, Xie SR, Min LK, Huang MS. Response of magnetic fluid deformable mirror with large stroke surface deflection. Magnetohy-drodynamics. 2015;**51**:695-708

[16] Caprari RS. Optimal current loop systems for producing uniform magnetic fields. Measurement Science and Technology. 1995;**6**:593-597

[17] Ritcey AM, Borra E. Magnetically deformable liquid mirrors from surface films of silver nanoparticles. Chemphyschem. 2010;**11**:981-986

[18] Yen YT, Lu TY, Lee YC, Yu CC, Tsai YC, Tseng YC, et al. Highly reflective liquid mirrors: Exploring the effects of localized surface plasmon resonance and the arrangement of nanoparticles on metal liquid-like films. ACS Applied Materials & Interfaces. 2014;**6**:4292-4300

[19] Wu ZZ, Iqbal A, Amara FB. Modeling and Control of Magnetic Fluid

Deformable Mirrors for Adaptive Optics Systems. Berlin, Germany: Springer; 2013

[20] Wu Z, Kong X, Zhang Z, Wu J, Wang T, Liu M. Magnetic fluid deformable mirrors with two-layer layout of actuators. Micromachines. 2017;**8**(3):72

[21] Josephson L. Magnetic nanoparticles for MR imaging. BioMEMS and Biomedical Nanotechnology. 2006: 227-237

[22] Morisada S, Miyata N, Iwahori K. Immunomagnetic separation of scum-forming bacteria using a polyclonal antibody that recognizes mycolic acids. Journal of Microbiological Methods. 2002;**51**:141

[23] Hergt R, Dutz S, Muller R, Zeisberger M. Magnetic particle hyperthermia: Nanoparticle magnetism and materials development for cancer therapy. Journal of Physics: Condensed Matter. 2006;**18**:S2919

[24] Kim T, Hyeon T. Applications of inorganic nanoparticles as therapeutic agents. Nanotechnology. 2014;**25**(1): 012001

[25] Rand RW, Snow HD, Elliott DG, Snyder M. Thermomagnetic surgery for cancer. Applied Biochemistry and Biotechnology. 1981;**6**:265

[26] Wahajuddin, Arora S. Superparamagnetic iron oxide nanoparticles: Magnetic nanoplatforms as drug carriers. International Journal of Nanomedicine. 2012;**7**:3445

[27] Moroz P, Jones SK, Gray BN. Magnetically mediated hyperthermia: Current status and future directions. International Journal of Hyperthermia. 2002;**18**(4):267

[28] Nagesha D, Devalapally H, Sridhar S, Amiji MM. Multifunctional magnetic nanosystems for tumor imaging, targeted delivery, and thermal medicine. Fundamental Biomedical Technologies. 2008;**4**:381

[29] Lee CS, Lee H, Westervelt RM. Multifunctional magnetic nanosystems for tumor imaging, targeted delivery, and thermal medicine. Applied Physics Letters. 2001;**79**(20):3308

[30] Drndic M, Johnson KS, Thywissen JH, Prentiss M, Westervelt RM. Micro-electromagnets for atom manipulation. Applied Physics Letters. 1998;**72**:2906

[31] Lee H, Purdon AM, Chu V, Westervelt RM. Controlled assembly of magnetic nanoparticles from magnetotactic bacteria using microelectromagnets arrays. Nano Letters. 2004;**5**:995

[32] Gertz F, Azimov R, Khitun A. Biological cell positioning and spatially selective destruction via magnetic nanoparticles. Applied Physics Letters. 2012;**101**(1):013701

[33] Lovelock JE. The haemolysis of human red blood-cells by freezing and thawing. Biochimica Et Biophysica Acta. 1953;**10**:414

[34] Gertz F, Khitun A. Biological cell manipulation by magnetic nanoparticles. AIP Advances. 2016;**6**: 025308

[35] Hawkes RC, Holland GN, Moore WS, Worthington BS. Nuclear magnetic resonance (NMR) tomography of the brain: A preliminary clinical assessment with demonstration of pathology. Journal of Computer Assisted Tomography. 1980;**4**(5): 577-586

[36] Smith FW, Hutchison JM, Mallard JR, et al. Oesophageal carcinoma demonstrated by whole-body nuclear magnetic resonance imaging.

British Medical Journal (Clinical Research Ed.). 1981;**282**(6263):510-512

[37] Ogawa S, Lee T-M, Kay AR, Tank DW. Brain magnetic resonance imaging with contrast dependent on blood oxygenation. Proceedings of the National Academy of Sciences. 1990;**87**: 9868-9872

[38] Selskog P, Heiberg E, Ebbers T, Wigstrom L, Karlsson M. Kinematics of the heart: Strain-rate imaging from time-resolved three-dimensional phase contrast MRI. IEEE Transactions on Medical Imaging. 2002;**21**: 1105-1109

[39] Kupersmith MJ, Alban T, Zeiffer B, Lefton D. Contrast-enhanced MRI in acute optic neuritis: Relationship to visual performance. Brain. 2002;**125**: 812-822

[40] Moser T, Dosch J-C, Moussaoui A, Dietemann J-L. Wrist ligament tears: Evaluation of MRI and combined MDCT and MR arthrography. American Journal of Roentgenology. 2007;**188**:1278-1286

[41] Bashir A, Gray M, Hartke J, Burstein D. Nondestructive imaging of human cartilage glycosaminoglycan concentration by MRI. Magnetic Resonance in Medicine. 1999;**41**:857-865

[42] Westbrook C, Roth CK, Talbot J. MRI in Practice. 4th ed. London: John Wiley & Sons, Inc.; 2011

[43] Brown RW, Haacke EM, Cheng Y-CN, Thompson MR, Venkatesan R. Resonance Imaging: Physical Principles and Sequence Design. 2nd ed. Wiley Blackwell; 2014

[44] Gore JC, Kennan RP. Physical and physiological basis of magnetic relaxation. In: Stark DD, Bradley WG, editors. Magnetic Resonance Imaging. 3rd ed. Vol. 1. New York: C.V. Mosby Publishing Co; 1999. pp. 33-42

[45] Jeong Y, Sook Hwang H, Na K. Theranostics and contrast agents for magnetic resonance imaging. Biomaterials Research. 2018;**22**:20

[46] Nelson KL, Gifford LM, Lauber-Huber C, Gross CA, Lasser TA. Clinical safety of gadopentetate dimeglumine. Radiology. 1995;**196**:439-443

[47] Geraldes CF, Laurent S. Classification and basic properties of contrast agents for magnetic resonance imaging. Contrast Media & Molecular Imaging. 2009;**4**:1-23

[48] Herborn CU, Honold E, Wolf M, Kemper J, Kinner S, Adam G, et al. Clinical safety and diagnostic value of the gadolinium chelate gadoterate meglumine (Gd-DOTA). Investigative Radiology. 2007;**42**:58-62

[49] Strijkers GJ, Mulder WJ, Van Tilborg GA, Nicolay K. MRI contrast agents: Current status and future perspectives. Anti-Cancer Agents in Medicinal Chemistry. 2007;**7**(3):291-305

[50] Van Montfoort JE, Stieger B, Meijer DK, Weinmann H-J, Meier PJ, Fattinger KE. Hepatic uptake of the magnetic resonance imaging contrast agent gadoxetate by the organic anion transporting polypeptide Oatp1. The Journal of Pharmacology and Experimental Therapeutics. 1999;**290**: 153-157

[51] Tanaka H, Tanigawa T, Suzuki M, Otsuka K, Inafuku S. Effects of MRI contrast agents (Omniscan™) on vestibular end organs. Acta Oto-Laryngologica. 2010;**130**:17-24

[52] Kirchin MA, Pirovano GP, Spinazzi A. Gadobenate dimeglumine (Gd-BOPTA): An overview. Investigative Radiology. 1998;**33**: 798-809

[53] Runge VM, Kirsch JE, Burke VJ, Price AC, Nelson KL, Thomas GS, et al.

High-dose gadoteridol in MR imaging of intracranial neoplasms. Journal of Magnetic Resonance Imaging. 1992;**2**: 9-18

[54] Kim RJ, Albert TS, Wible JH, Elliott MD, Allen JC, Lee JC, et al. Performance of delayed-enhancement magnetic resonance imaging with gadoversetamide contrast for the detection and assessment of myocardial infarction: An international, multicenter, double-blinded, randomized trial. Circulation. 2008;**117**: 629-637

[55] Staks T, Schuhmann-Giampieri G, Frenzel T, Weinmann H-J, Lange L, Platzek J. Pharmacokinetics, dose proportionality, and tolerability of gadobutrol after single intravenous injection in healthy volunteers. Investigative Radiology. 1994;**29**: 709-715

[56] de Haën C, Anelli PL, Lorusso V, Morisetti A, Maggioni F, Zheng J, et al. Gadocoletic acid trisodium salt (b22956/1): A new blood pool magnetic resonance contrast agent with application in coronary angiography. Investigative Radiology. 2006;**41**: 279-291

[57] Hompland T, Ellingsen C, Rofstad EK. Preclinical evaluation of Gd-DTPA and gadomelitol as contrast agents in DCE-MRI of cervical carcinoma interstitial fluid pressure. BMC Cancer. 2012;**12**:544

[58] Bjørnerud A, Johansson LO, Ahlström H. Pre-clinical results with Clariscan™ (NC100150 injection); experience from different disease models. Magma. 2001;**12**:99-103

[59] Lee N, Hyeon T. Designed synthesis of uniformly sized iron oxide nanoparticles for efficient magnetic resonance imaging contrast agents. Chemical Society. 2012;**41**:2575-2589

[60] Clement O, Siauve N, Cuénod C-A, Frija G. Liver imaging with ferumoxides (Feridex): Fundamentals, controversies, and practical aspects. Topics in Magnetic Resonance Imaging. 1998;**9**: 167-182

[61] Bonnemain B. Superparamagnetic agents in magnetic resonance imaging: Physicochemical characteristics and clinical applications a review. Journal of Drug Targeting. 1998;**6**:167-174

[62] Sigal R, Vogl T, Casselman J, Moulin G, Veillon F, Hermans R, et al. Lymph node metastases from head and neck squamous cell carcinoma: MR imaging with ultrasmall superparamagnetic iron oxide particles (Sinerem MR)–results of a phase-III multicenter clinical trial. European Radiology. 2002;**12**:1104-1113

[63] Reimer P, Balzer T. Ferucarbotran (Resovist): A new clinically approved RES specific contrast agent for contrast-enhanced MRI of the liver: Properties, clinical development, and applications. European Radiology. 2003;**13**:1266-1276

[64] Lusic H, Grinstaff MW. X-ray-computed tomography contrast agents. Chemical Reviews. 2013;**113**(3): 1641-1666

[65] Kalender WA. Computed Tomography: Fundamentals, System Technology, Image Quality, Applications. 3rd ed. Munich, Germany: Publicis; 2011

[66] Lindsten J, editor. Nobel Lectures: Physiology or Medicine 1971–1980. Vol. 1. Singapore: World Scientific Publishing Co.; 1992

[67] Yu S, Watson AD. Metal-based X-ray contrast media. Chemical Reviews. 1999;**99**:2353-2377

[68] Janon EA. Gadolinium-DPTA: A radiographic contrast agent. American Journal of Roentgenology. 1989;**152**:1348

[69] Engelbrecht V, Koch JA, Rassek M, Modder U. Magnetic resonance tomography and localized proton spectroscopy in 2 siblings with Canavan's disease. RöFo: Fortschritte auf dem Gebiete der Röntgenstrahlen und der Nuklearmedizin. 1996;**165**:24

[70] Fobbe F, Wacher F, Wagner S. Arterial angiography in high-kilovoltage technique with gadolinium as the contrast agent: First clinical experience. European Radiology. 1996;**6**:224

[71] Staks T, Schuhmann-Giampieri G, Frenzel T, Weinmann HJ, Lange L, Platzek J. Pharmacokinetics, dose proportionality, and tolerability of gadobutrol after single intravenous injection in healthy volunteers. Investigative Radiology. 1994;**29**:709

[72] Bonvento MJ, Moore WH, Button TM, Weinmann HJ, Yakupov R, Dilmanian FACT. Angiography with gadolinium-based contrast media. Academic Radiology. 2006;**13**(8): 979-985

[73] Schumann-Giampieri G, Mahler M, Roll G, Maibauer R, Schmitz S. Pharmacokinetics of the liver-specific contrast ageng Gd-EOB-DTPA in relation to contrast-enhanced liver imaging in humans. Journal of Clinical Pharmacology. 1997;**37**:587

[74] Huang D, Swanson EA, Lin CP, Schuman JS, Stinson WG, Chang W, et al. Optical coherence tomography. Science. 1991;**254**(5035):1178-1181

[75] Fujimoto JG, Pitris C, Boppart SA, Brezinski ME. Optical coherence tomography: An emerging technology for biomedical imaging and optical biopsy. Neoplasia. 2000;**2**:9-25

[76] Oldenburg A, Toublan F, Suslick K, Wei A, Boppart S. Magnetomotive contrast for in vivo optical coherence tomography. Optics Express. 2005; **13**(17):6597-6614

[77] Oldenburg AL, Gunther JR, Boppart SA. Imaging magnetically labeled cells with magnetomotive optical coherence tomography. Optics Letters. 2005;**30**(7):747

[78] Widder KJ, Senyei AE, Scarpelli DG. Magnetic microspheres: A model system for site-specific drug delivery in vivo. Experimental Biology and Medicine. 1978;**158**(2):141-146

[79] Alexiou C, Jurgons R, Schmid R, Hilpert A, Bergemann C, Parak F, et al. In vitro and in vivo investigations of targeted chemotherapy with magnetic nanoparticles. Journal of Magnetism and Magnetic Materials. 2005;**293**(1): 389-393

[80] Zhang X, Le T-A, Yoon J. Development of a real-time imaging-based guidance system of magnetic nanoparticles for targeted drug delivery. Journal of Magnetism and Magnetic Materials. 2017;**427**:345-351

[81] Al-Jamal KT, Bai J, Wang JT-W, Protti A, Southern P, Bogart L, et al. Magnetic drug targeting: Preclinical in vivo studies, mathematical modeling, and extrapolation to humans. Nano Letters. 2016;**16**(9):5652

[82] Cherry EM, Maxim PG, Eaton JK. Particle size, magnetic field, and blood velocity effects on particle retention in magnetic drug targeting. Medical Physics. 2010;**37**(1):175-182

[83] Alexiou C, Arnold W, Klein R, Parak F, Hulin P, Bergemann C, et al. Locoregional cancer treatment with magnetic drug targeting. Cancer Research. 2000;**60**(23):6641-6648

[84] Derfus AM, Von Maltzahn G, Harris TJ, Duza T, Vecchio KS, Ruoslahti E, et al. Remotely triggered release from magnetic nanoparticles. Advanced Materials. 2007;**19**(22): 3932-3936

[85] Hu S-H, Liu T-Y, Liu D-M, Chen S-Y. Controlled pulsatile drug release from a ferrogel by a high-frequency magnetic field. Macromolecules. 2007;**40**: 6786-6788

[86] Kim D-H, Nikles DE, Johnson DT, Brazel CS. Heat generation of aqueously dispersed CoFe2O4 nanoparticles as heating agents for magnetically activated drug delivery and hyperthermia. Journal of Magnetism and Magnetic Materials. 2008;**320**: 2390-2396

[87] Braddock M. Nanomedicines: Design, Delivery, and Detection. Cambridge: RSC; 2016

[88] Alexiou C, Schmidt A, Klein R, Hulin P, Bergemann C, Arnold W. Magnetic drug targeting: Biodistribution and dependency on magnetic field strength. Journal of Magnetism and Magnetic Materials. 2002;**252**(1–3): 363-366

[89] Mahmoudi M, Simchi A, Imani M, Milani AS, Stroeve P. Optimal design and characterization of superparamagnetic iron oxide nanoparticles coated with polyvinyl alcohol for targeted delivery and imaging. The Journal of Physical Chemistry. B. 2008;**112**(46): 14470

[90] Senyei A, Widder K, Czerlinski G. Magnetic guidance of drug-carrying microspheres. Journal of Applied Physics. 1978;**49**(6):3578-3583

[91] Gregory TS, Wu KJ, Yu J, Box JB, Cheng R, Mao L, et al. Magnetohydrodynamic-driven design of microscopic endocapsules in MRI. IEEE/ASME Transactions on Mechatronics. 2015;**20**(6):2691-2698

[92] Eguchi H, Umemura M, Kurotani R, Fukumura H, Sato I, Kim J-H, et al. A magnetic anti-cancer compound for magnet-guided delivery and magnetic resonance imaging. Scientific Reports. 2015;**5**

[93] Sato I, Umemura M, Mitsudo K, Fukumura H, Kim J-H, Hoshino Y, et al. Simultaneous hyperthermia-chemotherapy with controlled drug delivery using single-drug nanoparticles. Scientific Reports. 2016;**6**(1)

[94] Ohtake M, Umemura M, Sato I, Akimoto T, Oda K, Nagasako A, et al. Hyperthermia and chemotherapy using Fe(Salen) nanoparticles might impact glioblastoma treatment. Scientific Reports. 2017;**7**

[95] Chen P, Cui B, Bu Y, Yang Z, Wang Y. Synthesis and characterization of mesoporous and hollow-mesoporous $M_xFe_{3-x}O_4$ (M = Mg, Mn, Fe, Co, Ni, Cu, Zn) microspheres for microwave-triggered controllable drug delivery. Journal of Nanoparticle Research. 2017; **19**(12):1-11

[96] Wang G, Zhao D, Li N, Wang X, Ma Y. Drug-loaded poly(ε-caprolactone)/Fe_3O_4 composite microspheres for magnetic resonance imaging and controlled drug delivery. Journal of Magnetism and Magnetic Materials. 2018;**456**:316-323

[97] Zhang F, Zhao L, Wang S, Yang J, Lu G, Luo N, et al. Construction of a biomimetic magnetosome and its application as a SiRNA carrier for high-performance anticancer therapy. Advanced Functional Materials. 2018;**28**(1)

[98] Zheng S, Han J, Jin Z, Kim C-S, Park S, Kim K-P, et al. Dual tumor-targeted multifunctional magnetic hyaluronic acid micelles for enhanced MR imaging and combined photothermal-chemotherapy. Colloids Surfaces B Biointerfaces. 2018;**164**: 424-435

[99] Park B-W, Zhuang J, Yasa O, Sitti M. Multifunctional bacteria-driven

microswimmers for targeted active drug delivery. ACS Nano. 2017;**11**(9):8910

[100] Xu H, Medina-Sánchez M, Magdanz V, Schwarz L, Hebenstreit F, Schmidt OG. Sperm-hybrid micromotor for targeted drug delivery. ACS Nano. 2018;**12**(1):327-337

[101] Kralj S, Potrc T, Kocbek P, Marchesan S, Makovec D. Design and fabrication of magnetically responsive nanocarriers for drug delivery. Current Medicinal Chemistry. 2017;**24**(5): 454-469

[102] Häfeli UO, Pauer GJ. In vitro and in vivo toxicity of magnetic microspheres. Journal of Magnetism and Magnetic Materials. 1999;**194**(1):76-82

[103] Moghimi SM, Hunter A, Murray J. Long-circulating and target-specific nanoparticles: Theory to practice. Pharmacological Reviews. 2001;**53**(2): 283-318

[104] Fernández-Pacheco R, Marquina C, Gabriel Valdivia J, Gutiérrez M, Soledad Romero M, Cornudella R, et al. Magnetic nanoparticles for local drug delivery using magnetic implants. Journal of Magnetism and Magnetic Materials. 2007;**311**(1):318-322

[105] Polyak B, Fishbein I, Chorny M, Alferiev I, Williams D, Yellen B, et al. High field gradient targeting of magnetic nanoparticle-loaded endothelial cells to the surfaces of steel stents. Proceedings of the National Academy of Sciences. 2008;**105**(2):698

[106] El-Sherbiny I, Elbaz N, Sedki M, Elgammal A, Yacoub M. Magnetic nanoparticles-based drug and gene delivery systems for the treatment of pulmonary diseases. Nanomedicine. 2017;**12**(4):387-402

[107] Price DN, Stromberg LR, Kunda NK, Muttil P. In vivo pulmonary delivery and magnetic-targeting of dry powder nano-in-microparticles. Molecular Pharmaceutics. 2017;**14**(12): 4741-4750

[108] Li D, Ren Y. High-gradient magnetic field for magnetic nanoparticles drug delivery system. IEEE Transactions on Applied Superconductivity. 2018;**28**(6):1-7

[109] Kohler N, Sun C, Wang J, Zhang M. Methotrexate-modified superparamagnetic nanoparticles and their intracellular uptake into human cancer cells. Langmuir. 2005;**21**(19): 8858

[110] Jain T, Morales MA, Sahoo S, Leslie-Pelecky D, Labhasetwar V. Iron oxide nanoparticles for sustained delivery of anticancer agents. Molecular Pharmaceutics. 2005;**2**(3):194-205

[111] Wickline SA, Neubauer AM, Winter PM, Caruthers SD, Lanza GM. Molecular imaging and therapy of atherosclerosis with targeted nanoparticles, Journal of Magnetic Resonance Imaging. 2007;**25**:667-680

[112] Jang J, Lee SS. Theoretical and experimental study of MHD (magnetohydrodynamic) micropump. Sensors and Actuators A: Physical. 2000;**80**(1):84-89

[113] Choi SUS, Eastman JA. Enhancing thermal conductivity of fluids with nanoparticles. In: Energy. USDO, editor. Int Mech Eng Congr Exhib San Fr. 1995

[114] Nisar A, Afzulpurkar N, Mahaisavariya B, Tuantranont A. MEMS-based micropumps in drug delivery and biomedical applications. Sensors and Actuators B: Chemical. 2008;**130**(2):917-942

[115] Huang L, Wang W, Murphy MC, Lian K, Ling Z. LIGA fabrication and test of a DC type magnetohydrodynamic

(MHD) micropump. Microsystem Technologies. 2000;**6**(6):235-240

[116] Heng K-H, Wang W, Murphy MC, Lian K. UV-LIGA microfabrication and test of an ac-type micropump based on the magnetohydrodynamic (MHD) principle. In: SPIE Proceedings. Vol. 4177. 2000. pp. 161-171

[117] Lemoff AV, Lee AP. An AC magnetohydrodynamic micropump. Sensors and Actuators B: Chemical. 2000;**63**(3):178-185

[118] Wang Y-N, Fu L-M. Micropumps and biomedical applications—A review. Microelectronic Engineering. 2018;**195**: 121-138

[119] Abhari F, Jaafar H, Yunus NAM. A comprehensive study of micropumps technologies. International Journal Of Electrochemical Science. 2012;7: 9765-9780

[120] Gregory TS, Cheng R, Tang G, Mao L, Tse ZTH. The magnetohydrodynamic effect and its associated material designs for biomedical applications: A state-of-the-art review. Advanced Functional Materials. 2016;**26**(22):3942-3952

[121] Hola K, Markova Z, Zoppellaro G, Tucek J, Zboril R. Tailored functionalization of iron oxide nanoparticles for MRI, drug delivery, magnetic separation and immobilization of substances. Biotechnology Advances. 2015;**33**(6):1162-1176

[122] Zhou X, Gao M, Gui L. A liquid-metal based spiral magnetohydrodynamics micropump. Micromachines. 2017;**8**:365

[123] Bau HH, Zhu J, Qian S, Xiang YA. Magneto-hydrodynamic microfluidic network. In: Proceedings ASME International Mechanical Engineering Congress & Exposition; New Orleans. 2002. p. 33559

[124] Bau HH, Zhu J, Qian S, Xiang Y. A magneto-hydrodynamically controlled fluidic network. Sensors and Actuators B. 2003;**88**:207-218

[125] Qian S, Bau HH. Magneto-hydrodynamic based microfluidics. Mechanics Research Communications. 2009;**36**(1):10-21

[126] Gleeson JP, West J. Magnetohydrodynamic micromixing. Technical Proceedings of the International Conference on Modeling and Simulation of Microsystems. 2002: 318-321

[127] Gleeson JP, Roche OM, West J, Gelb A. Modeling annular micromixers. SIAM Journal of Applied Mathematics. 2004;**64**:1294-1310